Springer Texts in Statistics

Advisors:
George Casella   Stephen Fienberg   Ingram Olkin

Springer Texts in Statistics

Robert E. Weiss

Modeling Longitudinal Data

With 72 Figures

 Springer

Robert E. Weiss
Department of Biostatistics
UCLA School of Public Health
Los Angeles, CA 90095-1772
USA
robweiss@ucla.edu

Cover illustration: The image appears as figure 7.11 in text.

ISBN 0-387-40271-3   Printed on acid-free paper.

Printed in the United States of America.   (MP)

9 8 7 6 5 4 3 2 1

Springer is a part of Springer Science+Business Media
springeronline.com

To

Maria Benedita Cecilio
Sofia Isabel Cecilio Weiss
Michael Diniz Cecilio Weiss

with love

Contents

Preface

Before the beginning of years
There came to the making of man
Time with a gift of tears,

          –Algernon Charles Swinburne

If we offend, it is with our good will.
That you should think, we come not to offend,
But with good will. To show our simple skill,
That is the true beginning of our end.

          – William Shakespeare

Longitudinal data occurs when we repeatedly take the same type of measurement across time on the subjects in a study. My purpose in writing this textbook is to teach you how to think about and analyze longitudinal data.

As a graduate student, I joined the American Statistical Association and began to subscribe to professional journals. I was aware that most people did not read their journals, and in the natural exuberance of early graduate-student-hood I vowed to be different. I opened my first journal with the express intent to read it cover to cover; and quickly discovered not every article was interesting. However, I did read one article thoroughly. Ware (1985) had published an article titled "Linear Models for the Analysis of Longitudinal Studies." I spent a lot of time trying to understand that article, and in a real sense I am still working on it today. This book is the outcome of my interest in longitudinal data, that began with that article.

Why This Book?

This is a textbook, not a monograph. Included material must be directly helpful when analyzing longitudinal data. Mathematical presentation is kept to a minimum although not eliminated, and statistical computing is not covered.

This book has several key features that other books on longitudinal data do not have. First of all, this book has chapter-length treatments of graphical methods, covariance modeling, and modeling the effects of covariates. These chapters are often only a small section in most other texts currently on the market. The effects of covariates requires at least one full chapter on top of what students have learned about covariates from their linear regression courses.

Many current texts are unbalanced in their coverage of this material. Many texts spend a lot of space on discrete data analysis–an entertaining and important topic. However, like courses on linear regression and generalized linear regression, students should cover linear regression in depth before moving on to logistic and Poisson regression. One book spends more than 25% of its space on missing data modeling. Understanding missing data and bias is an important part of statistical data analysis of longitudinal data. I do provide an introduction to missing data here, but first students need to know how to model regular longitudinal data before spending time learning about missing data.

Texts on longitudinal data from the 1980s and even 1990s are already out of date, usually concentrating on generalizations of analysis of variance rather than on generalizations of regression. The techniques they cover are often archaic. There are also several doctoral-level monographs on longitudinal data that cover multivariate analysis at a more advanced mathematical level, usually including substantial effort on computation and inference, but this is at the expense of not covering the nuts and bolts of data analysis, and those books cannot be read by master's-level students.

A number of texts treat longitudinal data as a special case of repeated measures or hierarchical or multi-level data. Those books emphasize the random effects approach to modeling to the detriment of other covariance models. Random effects models are powerful and flexible, and several sections of this text are devoted to random effects models. However, polynomial random effects models often do not provide the best fit to longitudinal data. Consequently, I treat random effects models as one of several covariance models to be considered when modeling the covariance matrix of longitudinal data.

Computation

I assume that computation will be handled by a software package. Statistical textbooks at the master's level typically do not cover statistical computation, and this book is no exception. My discussion of computation tries to aid the data analyst in understanding what the software does, why it may or may not work, and what implications this has for their own data analysis. I do not discuss code for particular packages because software changes too rapidly over time. It is altered, often improved, and eventually replaced. I am thankful to the vendors that supply programs for analyzing longitudinal data, and I wish them a long and successful run. Extensive software examples will be available on the book's Web site. A link to the book Web site will be located at `http://www.biostat.ucla.edu/books/mld`. You will find data sets, example code, example homework problem sets, computer labs, and useful longitudinal links.

Initially, example code for fitting these models in SAS® Proc Mixed® and Proc Nlmixed® will be available on the course Web site. Sets of computer labs will also be available for teaching longitudinal data analysis using SAS.

Mathematical Background

I have kept the mathematical level of the text as low as I could. Students really should be comfortable with the vector form of linear regression $Y = X\alpha + \delta$ where X is a matrix of known covariates with n rows and K columns, α is a K-vector of coefficients, and Y and δ are n-vectors of observations and residual errors, respectively. I use α rather than the more common β for the regression coefficients. Linear algebra beyond $X\alpha$ is rarely required, and those spots can be readily skipped. In chapters 5 and 6, I write down some likelihoods and the weighted least squares estimator for the regression coefficients in longitudinal data. This requires a few matrix inverses. This material is partly included to assuage my guilt had it been omitted and to provide hooks into future mathematical material should the reader cover more advanced material elsewhere. But this material is not central to the main theme. If the students do not swallow that material whole, it should not impede understanding elsewhere. I do review linear regression briefly, to remind the reader of what they learned before; one can't learn regression fresh from the review, but hopefully it will serve to exercise any neurons that need strengthening.

Multivariate Data and Multivariate Data Courses

Because longitudinal data is multivariate, you will learn something about multivariate data when you read this book. Longitudinal data is not the only type of multivariate data, although it is perhaps the most common type of multivariate data. One of the (dirty little?) secrets of statistical research in classical multivariate data methods is that many methods, while purporting to be multivariate, are actually illustrated on, and mainly useful for, longitudinal data.

Many statistics and biostatistics departments have courses in multivariate data analysis aimed at master's-level students and quantitative graduate students from other departments. These courses cover multivariate analysis of variance (MANOVA) and multivariate regression, among other things. I strongly recommend replacing such a course with a course in longitudinal data analysis using this book. The value of longitudinal data analysis to the student will be much greater than the value of MANOVA or multivariate regression. I often think of this course as a "money course." Take this course, earn a living. I hire many students to analyze data on different projects; it used to be that I required familiarity with regression analysis. Now familiarity with longitudinal data analysis is the most usual prerequisite.

Target Audience

Graduating master's students in statistics and biostatistics are very likely to be analyzing longitudinal data at least some of the time, particularly if they go into academia, the biotech/pharmaceutical industry, or other research environment. Doctoral students and researchers in many disciplines routinely collect longitudinal data. All of these people need to know about analyzing longitudinal data.

This book is aimed at master's and doctoral students in statistics and biostatistics and quantitative doctoral students from disciplines such as psychology, education, economics, sociology, business, epidemiology, sociology and engineering among many other disciplines. These are two different audiences. The common background must be a good course in linear regression. A course at the level of Kutner, Nachtsheim, and Neter (2004), Fox (1997), Weisberg (2004) or Cook and Weisberg (1999) is a necessary prerequisite to reading this book. The seasoning provided by an additional statistics or biostatistics course at this level will be exceedingly helpful. I have taught this material to students from other disciplines whose mathematical background was not up to this level. They found this course rewarding but challenging.

The statistics and biostatistics students bring a deeper knowledge of mathematics and statistics to the course, but often little knowledge of longitudinal data other than perhaps knowledge that longitudinal data is likely to be in their future or on their comprehensive exam. Students from outside stat/biostat tend to have much less mathematical and statistical background. Instead, they bring with them the motivation that comes from having data in hand and needing to analyze it, often for their dissertation. The two different backgrounds can both lead to success in learning this material.

Applied researchers with a good regression course under their belt and some added statistical sophistication should be able to read this book as well. For anyone reading this book, the single best supplemental activity when reading the text would be to have your own data set and to draw all the relevant plots and fit all the relevant models you read about to your own data.

An Overview

This overview is for anyone; but I'm writing it as if I were talking to another teacher.

Chapter 1, *Introduction*, introduces longitudinal data, gives examples, talks about time, discusses how longitudinal data is different from linear regression data, why analyzing longitudinal data is more difficult than analyzing linear regression data and defines notation.

Chapter 2, *Plots*, discusses the plotting of longitudinal data. Intertwined with the plots are ways of thinking about longitudinal data, issues that are naturally part of longitudinal data analysis. Even if you do not wish to cover every last piece of this material in a course, I recommend that the students read the whole chapter.

Chapter 3, *Simple Analyses*, discusses things like paired *t*-tests and two-sample *t*-tests and the two-sample *t*-test on paired differences, called the difference of differences, (DoD) design. These simple analyses are done on various subsets of the data or on summaries of the data. The ideas are re-used in the chapter on specifying covariates. Chapter 4, *Critiques of Simple Analyses*, complains about these analyses and explains some of the problems. Perhaps the real cost of simple analyses is the loss of the richness of multivariate data.

Chapter 5, the *Multivariate Normal Linear Model*, starts with the iid multivariate normal model for data, then introduces parameterized covariance matrices and covariates and the basic aspects of and techniques for drawing conclusions.

Chapter 6, *Tools and Concepts*, contains a grab-bag of useful tools (likelihood ratio tests, model selection, maximum likelihood and restricted maximum likelihood, back-transforming a transformed response, an introduction to design) and discussions about issues with longitudinal data analysis (assuming normality, computation). These tools may be skipped at first reading. However, my suspicion is that those readers who only read a section or two out of the entire book are most likely to dip into this chapter or into one of the topics chapters at the end. Many readers will come back to the various sections of chapter 6 when needed or interested. Most readers will continue on to chapters 7 and 8, coming back to pick up material on model selection, computation, inference as needed.

Chapters 7 and 8, *Specifying Covariates* and *Modeling the Covariance Matrix*, respectively, are the chapters that allow the flavor and beauty of longitudinal data analysis to come to full bloom. As best as possible, I have tried to write these chapters so they could be read in either order. I have tried both orders; my preference is to study covariates first. Covariate specification in longitudinal data analysis requires additional modeling skills beyond what is taught in linear regression and is where the science usually comes in when analyzing longitudinal data. I prefer to have that as early as possible so students can start thinking about their own longitudinal data problems and how to specify the scientific questions. Another reason is that otherwise we are well past the mid-quarter mark before having talked about covariates and that is too long in the quarter to put off talking about covariates. Because there are many short references to covariance matrix specification in chapters 5 and 7, it allows for a softer introduction to the material on covariance models. The downside of this order is that students tend to ask a lot of questions about covariance models before you are ready to discuss them.

Chapter 9, *Random Effects Models*, discusses the random effects model as a hierarchical model, with discussions of random effects estimation and shrinkage. Longitudinal data sets frequently have subjects nested inside larger groups, for example students in classrooms or children in families. We explain how to model this data as well.

Chapter 10, *Residuals and Case Diagnostics*, presents current knowledge about residuals and case diagnostics with emphasis on residuals in random effects models as more is known (by me at any rate) about residuals there than in the general multivariate linear regression model.

Chapter 11, *Discrete Longitudinal Data* introduces discrete longitudinal data models. I discuss the random intercept model for binary data and for count data.

Chapter 12, *Missing Data*, is an introduction to issues surrounding missing data in longitudinal data. We talk about intermittently observed data and dropout and missing at random and variants.

Finally, chapter 13, *Analyzing Two Longitudinal Variables*, introduces bivariate longitudinal data, when you measure two variables repeatedly over time on subjects and wish to understand the interrelationship of the two variables over time.

Teaching from This Book

I teach this book as a quarter course, covering essentially the entire text. Lectures are supplemented with a computer lab that covers the use of a computer program for analyzing longitudinal data.

I have also taught precursors of this material as a subset of a quarter course on multivariate analysis for biostatistics doctoral students. In this course, I cover material from chapters 1, 2, 7, 8, and 9 in three to four weeks, concentrating on the mathematical presentation. I replace chapter 5 with a substantially higher level of mathematical rigor. Chapters 1 and 2 are shortened and the material tightly compacted. Next time I teach that course, I plan to require that students read the entire book and may add parts from chapter 11 and 13 to lectures as well.

A number of homework problems are included. That is how you can tell this is a textbook and not a monograph. The most important homework problems should lead students through a complete analysis of a simple data set. I use the Dental data for this first set of homework problems, which is why it does not appear in the text. Students should first plot and summarize the data, then explore the fixed effects, model the covariance matrix, look at the residuals and finally put their results all together in a report. This can be over a set of three homework assignments. The next assignment(s) can either be a report on the complete analysis of a somewhat more complicated longitudinal data set or another three homework assignments analyzing a data set with unbalanced or random times and more covariates. The last project should be the analysis of a still more complex data set supplied by the teacher or a data set supplied by the student. I do not give exams when I teach this material as a stand-alone course. Report writing supplies a useful form of training that was often historically lacking in statistical training. Ironically, the initial motivation for chapter 7 came from observing the difficulty that many very smart biostatistics doctoral students had in setting up even simple covariate matrices for longitudinal data during comprehensive exams. The Web site has homework assignments that I have used.

Feedback

Comments are actively solicited; especially comments that will help me make the reading and learning experience more helpful for future readers.

Acknowledgments

Many people have provided assistance in the writing of this book, in ways large and small. A number of colleagues have helped indirectly by talking with me about longitudinal data and directly with information for and comments on various drafts of the book. My apologies for omitting way too many of them from these acknowledgments. My thanks to my colleagues at UCLA both inside the Department of Biostatistics and outside for putting up with, encouraging, ignoring, and abetting. Particular thanks to Robert Elashoff, Bill Cumberland, and Abdelmonem Afifi for early encouragement.

Students in the courses Biostat 236 and Biostat 251 at UCLA have sat through many presentations of this material over a number of years and have contributed much through their questions, enthusiasms, homework answers, report writing, yawns, laughter, and typo reports. I'd like to thank them for being willing participants in reading early versions as they were written and in particular for letting me read the notes to them and for helping me catch the typos on the fly. A number of students have written master's papers and doctoral dissertations with me on longitudinal data analysis; every one has helped me understand the subject better.

I have had tons of assistance in data management and in writing code using SAS, R©/Splus®, and ARC©/xlispstat©. Thanks to Charlie Zhang, Zhishen Ye, Yunda Huang, Susan Alber, Leanne Streja, Lijung Liang, Luohua Jiang, Jim Sayre, John Boscardin, Scott Comulada, Zhen Qian, Wenhua Hu, and others who I have unfortunately omitted.

I'd like to thank Sandy Weisberg for LaTeX® help and for always answering my questions about almost anything; John Boscardin for programming, LaTeX, and longitudinal help; Marc Suchard for detailed comments, and many many discussions, breakfasts, bagels, and pushes; Steve West; Eric Bradlow; Bill Rosenberger; Billy Crystal in Throw Momma from the Train: "A writer writes: always!"; Lynn Eberly for particularly helpful comments and for encouragement; several anonymous reviewers for comments both general and detailed; Susan Alber for comments on the writing, help with SAS, and teaching me about longitudinal data analysis. A big thanks to John Kimmel for his patience, encouragement, stewardship, and for finding the right answers when it mattered.

I have gotten data sets from a number of places. I'd like to thank Dr. Lonnie Zeltzer for the Pediatric Pain and Vagal Tone data; Dr. Mary Jane Rotheram for the BSI and other data sets; Dr. Charlotte Neumann for the Kenya data; Robert Elashoff for the Weight Loss data.

Finally, I would like to thank my multi-generational family for putting up with me while I worked on this. You have often asked when this book would be done. If you are reading this for the first time, check your watch and you will have your answer.

Robert Weiss
Los Angeles
2005

1
Introduction to Longitudinal Data

Fate, Time, Occasion, Chance, and Change? To these
All things are subject

— Percy Bysshe Shelley

What is past is prologue.

— William Shakespeare

Overview

In this chapter, we cover the following

- Longitudinal data
  - What is longitudinal data?
  - What are related types of data?
  - What types of inferences can we make with longitudinal data?
  - What are the benefits of collecting longitudinal data?
  - How does longitudinal data compare with cross-sectional data?
  - Complexities in the analysis of longitudinal data.

- Time
  - Time frames for collecting longitudinal data.
  - The language of time.

- Notation

1.1   What Is Longitudinal Data?

Longitudinal data is ubiquitous in a wide range of fields: medicine, public health, education, business, economics, psychology, biology, and more. Economists call it panel data. Other names are repeated measures and time series, which are related data types.

The design of the simplest longitudinal study uses a simple random sample of subjects. A single measurement is collected repeatedly over time on each subject in the study. The temporal ordering of the measurements is important because measurements closer in time within a subject are likely to be more similar than observations farther apart in time. As an example of longitudinal data, we may measure a subject's weight weekly for the duration of a weight loss study. We collect all the weekly weights on all our subjects into a single data set; this forms a longitudinal data set. In other longitudinal examples, we may measure blood pressure whenever patients happen to come into our clinic for other purposes; or we may assess math achievement in elementary school children at the end of every school year for six years.

Longitudinal data are *multivariate*. Multivariate has two meanings, one in statistics and another in fields where statistics is applied. In statistics it means that there is more than one response per subject. In areas where statistics is applied to produce scientific results, multivariate means that more than one variable per subject is used in the analysis. Multiple linear regression is *univariate* according to statisticians and multivariate in application fields. Longitudinal analyses are multivariate according to both definitions of multivariate. Longitudinal data contrasts with traditional statistical multivariate data where we collect a number of measurements on each subject and we analyze the multiple measurements as a single multivariate outcome. For example we could analyze height, weight, head circumference, lean body mass and upper arm circumference as a multivariate observation on subjects. In traditional multivariate data, subjects contribute the same number of measurements per subject. In longitudinal data, subjects need not contribute the same numbers of measurements each.

Longitudinal data has a natural *hierarchical* or *multi-level* structure, with observations at the bottom level *nested* or *clustered* within subjects at the top level. An *observation* is a single measurement on a subject. A subject refers to all of the observations from that subject. Other words for subject are *case* or *experimental unit* and we will often use the word *person*, because the greatest interest of human beings is human beings and most of our data has been collected over time on people. If subjects are rats or countries or wallabies, then of course we will call subjects rats or countries or wallabies.

The defining feature of longitudinal data is that the multiple observations within subject can be ordered across time. Longitudinal surveys of humans use calendar time, months or years, as the dimension separating observations on the same subject. Long-term medical studies may have years between consecutive observations on a subject, whereas short-term studies may have seconds, minutes, or hours separating observations. We may use linear metrics other than time for spacing measurements. For example we could take repeated measures along a plant root. Time is then supplanted by distance in millimeters from the tip of the root. The repeated measures may be consecutive trials in an experiment, as when a rat is repeatedly timed in running a maze or people repeatedly attempt a task. Some longitudinal data sets may become easier to analyze when time is replaced by the natural log of time. As long as the metric or *metameter* is linearly ordered, whether time, log(time), centimeters or trials, we have longitudinal data. We will call the dimension that separates our measurements *time*. The *feasible times* where we can take measures may be all real numbers, only positive numbers, all times in a range for example from the beginning of our study to the end, or a subset of the integers as when our data come from consecutive trials or when we measure time in integer days or months since an event.

1.2 Related Data Types

Longitudinal data are a particular form of *repeated measures data*. In repeated measures data, a single measurement is collected repeatedly on each subject or experimental unit. Observations may or may not be distributed over time. For example, repeated measures may be spatially distributed in two or three dimensions, as measurements of ground water contamination spread over a state or a three-dimensional magnetic resonance image of a rat brain. Repeated measures data may be *clustered* or nested inside of larger units. Household surveys may sample data from all persons living in a single household. Responses on persons sharing a single household are clustered inside the household, which is the experimental unit. In educational studies, children are clustered in classrooms. The set of children's test scores are repeated measures on the experimental unit, which may be the classroom, teacher, or school.

These other types of repeated measures data differ from longitudinal data in the relationships between the repeated measures. Longitudinal repeated measures can be ordered along a line. Spatial and three-dimensional repeated measures cannot always be linearly ordered, although any two measurements are at well-defined distances and directions from each other. Clustered data are not ordered, and in the simplest setup, each subunit is related to the others symmetrically. Each type of data is interesting and useful and worthy of a full book.

Time series is another data structure related to longitudinal data. Traditional time series are longitudinal data observed over a long time period on a single experimental unit. Well-known time series data sets include (a) the number of observed sunspots, measured daily for centuries and (b) the annual number of lynx trapped in Canada from 1821 to 1934. Longitudinal data consist of many time series measured on a sample of subjects. Longitudinal data time series usually have fewer repeated measurements than traditional time series data. Longitudinal data may have as few as two observations per subject, although the full flavor of longitudinal data analysis only begins to be apparent starting with three and four observations per subject. In data sets with variable numbers of observations per subject, some subjects may contribute but a single observation, although the average number of observations per subject will more commonly be anywhere from 3 to 10 or more.

Structure in longitudinal repeated measures data sets can get quite complicated. Suppose at each data collection visit, we measure a person's heart rate before exercise (resting rate) then their peak heart rate during exercise and one more measure after exercise (recovery rate). We proceed to do this Monday through Friday for two weeks. Is this $3 \times 10 = 30$ repeated measures? Or is it a set of 10 longitudinally repeated trivariate observations? In clinical trials with longitudinal measures, we repeatedly observe multiple variables, for example systolic and diastolic blood pressure, heart rate, and total cholesterol. This we call multivariate longitudinal data, meaning that the repeated measure is itself a multivariate observation. In this book we mainly discuss analysis of a single measurement taken repeatedly over time. This is univariate longitudinal data to distinguish it from situations where we want to study the multivariate measures in longitudinal fashion. The first heart rate example is perhaps best studied as a trivariate longitudinal measure. When we are willing to reduce the trivariate observation to a single measure, such as peak rate minus resting rate, or recovery defined as $(\text{after} - \text{resting})/(\text{peak} - \text{resting})$, then that single measure, repeatedly measured 10 times over time, is something that we can directly analyze with the tools of this book. Chapter 13 discusses bivariate longitudinal data, longitudinal data with two different measurements at each time point.

This text provides a thorough introduction to the analysis of continuous longitudinal repeated measures data. Repeated binary or count longitudinal data are common, and we provide an introduction to these data in chapter 11.

1.3   Inferences from Longitudinal Data

Suppose for a moment that we have univariate observations rather than longitudinal. That is, we take a single measurement y_i on each subject i in a sample of size n with i running from 1 up to n. The most general inference we can make is about the entire distribution of the y_i. This includes the mean, variance, skewness (third moment), kurtosis (fourth moment), various quantiles, and so on. If we also have covariates x_i, then the most general inference we can attempt is how the entire distribution of y_i varies as a function of the covariates x_i. This is in general rather difficult. We usually restrict interest to simpler interpretable inferences, for example how the population mean of y_i varies as a function of the covariates and, occasionally, how the population variance varies as a function of covariates.

There are two basic types of inferences for longitudinal data that we wish to make: (i) inference about the population mean and (ii) individual variation about the population mean. If we can accomplish that, we may wish to add in (iii) the effects of covariates on (a) the population mean and (b) individual variation, and (iv) predictions of new observations. We discuss these briefly in turn.

1.3.1 The Population Mean

With univariate observations y_i on a simple random sample of size n from a population under study, we may use the sample mean

$$\bar{y} = \frac{\sum_{i=1}^{n} y_i}{n}$$

of the observations to estimate the population mean μ, which is a scalar.

With longitudinal observations, we have a number of observations taken on subject i at various times. The population mean is no longer a single number. Rather, there is a population mean at all feasible times where we can collect data; the population mean is a function $\mu(t)$ of time. The function $\mu(t)$ may be constant over time or it may trend up or down; it may be cyclic or otherwise vary in complex fashion over time. If we have a set of observations Y_{ij} on a random sample of subjects $i = 1, \ldots, n$ at a particular time t_j, we may use the sample average

$$\bar{Y}_j = \frac{\sum_{i=1}^{n} Y_{ij}}{n}$$

of these observations to estimate $\mu_j = \mu(t_j)$, the population mean at time t_j.

1.3.2 Individual Variability

In the univariate simple random sample, the population variance, σ^2, is a scalar, that is, a single number. It is estimated by the sample variance

$$s^2 = \frac{\sum_{i=1}^{n} (y_i - \bar{y})^2}{n - 1}$$

of the observations. The population standard deviation σ tells us how far observations will vary from the population mean. In particular, with normal data, we expect 68% of our observations within plus or minus one standard deviation of the mean and 95% of the observations within plus or minus two standard deviations. Experience suggests that these numbers are surprisingly accurate as long as the data histogram is even approximately bell-curve shaped. The variance also provides information on how precisely we estimate the population mean. The standard deviation of the mean is $n^{-1/2}\sigma$.

For longitudinal data, the population variance is no longer a single number. At each time t_j we have a population variance σ_{jj} and standard deviation $\sigma_{jj}^{1/2}$. The population variance σ_{jj} describes the variability in the responses Y_{ij} at a particular time. Putting two subscripts on σ_{jj} and not having a superscript 2 is helpful when we introduce covariance parameters next.

1.3.3 *Covariance and Correlation*

Between each two times t_j and t_l, we have a population covariance $\sigma_{jl} = \sigma(t_j, t_l)$. The population covariance is a function not of a single time but of two times t_j and t_l. The absolute value of the covariance σ_{jl} is bounded above by the product of the two corresponding standard deviations

$$0 \le |\sigma_{jl}| \le \sigma_{jj}^{1/2}\sigma_{ll}^{1/2}.$$

Dividing covariance σ_{jl} by the product $\sigma_{jj}^{1/2}\sigma_{ll}^{1/2}$ restricts the resulting quantity to range between -1 and 1, and this quantity is the *correlation* ρ_{jl} between the observations at time t_j and t_l

$$\rho_{jl} = \frac{\sigma_{jl}}{\sigma_{jj}^{1/2}\sigma_{ll}^{1/2}}.$$

A correlation is usually easier to interpret than a covariance because of the restricted range.

A positive correlation ρ_{jl} tells us that if observation Y_{ij} is greater than its population mean μ_j, then it is more likely than not that observation Y_{il} will be greater than its population mean. The farther Y_{ij} is above the mean, then the farther we expect Y_{il} to be above its mean. If Y_{ij} is at the population mean μ_j, then Y_{il} has even chances of being above or below its population mean μ_l.

If the correlation between Y_{ij} and Y_{il} is zero, then knowledge of Y_{ij} does not tell us whether Y_{il} is above or below its mean and Y_{il} again has even chances of being above or below its mean. If the population correlation ρ_{jl} between observations at times t_j and t_l is very high and positive, we expect Y_{il} to be almost the same number of population standard deviations above μ_l as Y_{ij} is above μ_j. Let $\mathrm{E}[Y_{il}|Y_{ij}]$ be the expected value of observation Y_{il} at time t_l on subject i given that we know the value of the observation Y_{ij} at time t_j. Knowing Y_{ij} changes our best guess, or *expected value* of Y_{il} as long as the correlation is non-zero. In general, assuming multivariate normal data,

$$\frac{\mathrm{E}[Y_{il}|Y_{ij}] - \mu_l}{\sigma_{ll}^{1/2}} = \rho_{jl}\frac{Y_{ij} - \mu_j}{\sigma_{jj}^{1/2}}.$$

The factor on the right multiplying ρ_{jl} is the number of standard deviations that Y_{ij} is above its mean. The term on the left is the expected number of standard deviations that Y_{il} is above its mean. We do not quite expect Y_{il} to be exactly the same number of standard deviations above its mean as Y_{ij} was above its mean, and the shrinkage factor is exactly the correlation ρ_{jl}. If ρ_{jl} is negative, the formula still applies, but we now expect Y_{il} to be below its mean if Y_{ij} is above its mean.

The population *profile* is the plot of μ_j against time t_j. A subject profile is a plot of observations Y_{ij} plotted against times t_j. Under the assumption of multivariate normality, the set of all subject profiles varies about the population mean profile in a manner determined by the population variances and covariances. We will see many illustrations of this in chapter 8.

We calculate the sample *covariance* s_{jl} between observations Y_{ij} and Y_{il} taken on subjects $i = 1, \ldots, n$ at times t_j and t_l by

$$s_{jl} = (n-1)^{-1}\sum_{i=1}^{n}(Y_{ij} - \bar{Y}_j)(Y_{il} - \bar{Y}_l)$$

The sample variance at time j is

$$s_{jj} = \frac{\sum_{i=1}^{n}(Y_{ij} - \bar{Y}_j)^2}{n-1}$$

which is exactly the same formula as for s_{jl} but with $j = l$. We put a circumflex, or *hat*, ˆ, on $\hat{\rho}_{jl}$ to indicate the estimate of the unknown parameter ρ_{jl}.

1.3.4 Covariates

A data set with a univariate response y_i and a vector of covariates x_i on a random sample of subjects of size n leads to a regression model to determine how the distribution of y_i changes with x_i. Standard multiple linear regression assumes that the mean of y_i changes with x_i and the variance of y_i is constant for different x_i. When we measure covariates and a longitudinal response, the covariates may affect the population mean in ways simple or complex. Covariates may further affect the variance and correlation functions, however those models have not been worked out thoroughly and are the subject of current research.

1.3.5 Predictions

Longitudinal data models contain complete information about possible profiles of individual subjects. This can be used to predict the future for new subjects. If we take a few observations on a new subject, we can find which old subjects' profiles are most similar to the new subject's observations and make predictions of the new subject's future values. The population mean forms a base for the predictions. The variance/covariance of the measurements combined with the first few observations forms an adjustment to the population mean that is specific to the new subject.

1.4 Contrasting Longitudinal and Cross-Sectional Data

The main alternative to a longitudinal study design is a cross-sectional study. In a cross-sectional study, we collect a univariate response y_i and covariates x_i on subject i where i runs from 1 up to the sample size n. We might analyze this data using regression analysis. Let x_i be a vector with elements x_{ik}, where k runs from 1 up to K, and usually with a first element $x_{i1} = 1$ corresponding to the intercept in the model. The standard linear regression model is

$$y_i = x_i'\alpha + \delta_i \tag{1.1}$$

where $\alpha = (\alpha_1, \ldots, \alpha_K)'$ is a vector of unknown regression coefficients with kth element α_k, x_i' represents the transpose of the vector x_i so that $x_i'\alpha = x_{i1}\alpha_1 + \ldots + x_{iK}\alpha_K$, and δ_i is a random error with mean 0 and variance σ^2. Observations are assumed independent given the unknown parameters α and σ^2. Often we presume the residual δ_i is distributed normally

$$\delta_i \sim N(0, \sigma^2).$$

Under model (1.1), the expected value of a response y_i with covariates x_i is

$$E[y_i] = x_i'\alpha.$$

This is the average of all responses from all subjects in the population with covariate vector equal to x_i. The interpretation of coefficient α_l for $l > 1$ is the change in the population average when we switch from subjects with a covariate vector x_i to subjects with covariates x_i^*, where $x_{ik}^* = x_{ik}$ except for $x_{il}^* = x_{il} + 1$. The change in mean is the difference in average response between the old population subset with covariate vector x_i and the new population subset with covariate vector x_i^*. It does not talk about a change in a single person's response y_i if we were to manage to change that person's lth covariate value from x_{il} to $x_{il} + 1$.

One exception to this interpretation is in randomized trials, where subjects are randomly assigned to a control group $x_{ik} = 0$ or a treatment group $x_{ik} = 1$. In this special case, the treatment/control group assignment is not intrinsic to the subject but imposed by the investigator, and α_k is the expected change in response of an individual if we were to have assigned them to the treatment rather than the control.

Suppose we have people of various ages in our cross-sectional study. The coefficient of age in our regression talks about differences among differently aged subjects and does not directly address the effects of aging on a single individual's response. As a simple example, consider a dental study where the response measure is the subject's total number of cavities. In many populations, the coefficient of age in years will be positive $\alpha_{\text{age}} > 0$ as opposed to zero. But if we take all 27-year-olds and check on them 10 years later, we will not find that they have had an average increase of $10\alpha_{\text{age}}$ cavities. Or so one would hope. Rather α_{age} is 1/10th the average number of cavities that people aged 37 at the time of the original study had minus the number that the 27-year-olds had. Cavity counts are something characteristic of the age group, and changes (decreases hopefully!) in younger adults are due to improvements (hopefully!) in fluoridation, food consumption (well, maybe), dental hygiene, and education over the years.

Except for the randomized treatment assignment coefficient, regression coefficients do not tell us how individuals in the population might change under modification of the covariate value. This problem still holds for longitudinal data models with one incredibly important exception, namely time. If we collect repeated measurements over time on the patients in our study, then we are directly measuring the effects of time and aging on our individual patients. Assuming we have the statistical tools to properly summarize the effects of time, we will be able to describe patterns of change over time of a population of people and also the patterns of change for individuals. Suppose we assess cavities over time. We might well find that all adults have modest increases in number of cavities over time, but older adults have more cavities than younger adults both at the beginning of our study and at the end.

What other benefits accrue from using a longitudinal design? Suppose we have samples of two groups, say men and women, or Hispanic and non-Hispanic, or treatment and control. In a cross-sectional study, we can only describe the difference in level between the two groups at the current time. With a longitudinal study, we will additionally be able to describe the separate trends over time of each group. Questions we can answer include

- Is the average response equal in the two groups at all times?

- If not, is the average response pattern over time the same in the two groups, apart from a constant level shift?

- If not, are the differences increasing or decreasing over time? and

- If not monotone, just how do the differences change over time?

1.5   Benefits of Longitudinal Data Analysis

The nature of our study determines the value of collecting longitudinal data. When we are interested in the trend of a response over time, then longitudinal data are a mandatory component of the design. Sometimes longitudinal data can be valuable even when we are not interested in time trends or when there are no time trends in our data.

1.5.1   Efficiency and Cost

In section 6.7, we numerically illustrate why longitudinal data collection can be cheaper than a simple random sample. Without giving details now, let us note that collecting data costs money. It is usually much cheaper to collect additional data from a subject already in our study than it is to recruit a new person and get them to participate in the first place. And often there is useful information to be gained from multiple measurements of the same type on a single person. When this happens, longitudinal data collection can be helpful.

1.5.2   Prediction

Longitudinal data collection opens the possibility of predicting how a new subject will trend given previous measurements. This is not possible from a cross-sectional study as our cavity example was intended to illustrate. Suppose our longitudinal data shows that most people increase from their level on study entry. When the next person enters the study, we measure their level at study entry, and it is a reasonable generalization that their response will also increase at later measurements.

1.5.3   Time Trends

If we are interested in individual trends over time, then longitudinal data are required. How else to find out how fast individuals decrease or increase? How else to find out if those who start low increase at the same rate as those who start high?

1.6   Time Frames for Observing Longitudinal Data

In this section, we consider a number of familiar measurements and various time intervals between measurements. These are listed in table 1.1. We

Measure	Time interval	Changing?
Human height	Minutes through days	No
	Weeks	Perhaps newborns?
	Month(s)	Infants
	Year(s)	For children/adolescents
	Decades	Not sure: maybe for adults/seniors?
Human weight	Minutes through hour(s)	No
	Days	Maybe not, depending On accuracy of scale
	Week(s)	Yes, in weight-loss program
	Months	Certainly (kids)
Wage/salary	< 1 hour	Not for salaried workers
	Day	If on commission or piece-work
	Month/year	If on monthly pay plan
Housing cost (rent or mortgage)	Annual	Yes
income	Hourly	For businesses
Gender	Any	No, only need once
Race	Any	No, only need once
DNA	Any	No, only need once with exceptions
Blood pressure	Minutes	Potentially, depending on study goals
	Years	Same
3-month recall of illicit drug use	Every 3 months	Yes for drug users
Depression score on psychological exam	Monthly	Yes in study of treatment for depression

Table 1.1. Evaluation of some measurements and time intervals between measurements.

make a judgment whether longitudinal measurements on humans taken with the given intervals would show change or not. If there is no or almost no change over those intervals, then taking longitudinal measurements at those intervals surely has little value. If there is change, or change for some population, we indicate a population where it would be reasonable to take measures at that interval.

In looking at table 1.1, one should start thinking about different types of measurements and practice making judgments about how fast those measurements might change over time. This sort of intuition is very important for several aspects of statistical practice. Having designed and executed a study, we will often produce estimates of changes in the response per unit time. It is important to know whether the estimates we have produced are reasonable or not. Unreasonable estimates may be due to mistakes in the data file, errors in computation, or mistakes in understanding. A good way to reduce mistakes of understanding is through familiarity with the measurement and timescale and through reasoning about how fast the measure might change.

For example, consider human height. In designing a longitudinal study to assess growth in human height, we will want to wait long enough between measures to be able to observe changes in heights. Very roughly, a human infant may start out at around 20 inches, and, by age 18, will reach roughly 66 to 72 inches. This makes for approximately $50 = 70 - 20$ inches of growth in 18 years, or about two and a half inches per year. There are growth spurts as well as periods of slower growth so some age groups may provide interesting data either with more or fewer observations per year. We need to be cognizant of these variations for the particular ages in our population.

To see growth over time within a single subject, we would need to have enough growth between measurements to be noticeable using our ruler. If our ruler is accurate to the nearest half an inch, we would probably not wish to take measurements less often than $1/5$ of a year or roughly every three months, that giving us the time needed for a child to grow $1/2$ an inch on average. In practice, we might want to wait more like every 6 months, giving an average growth of somewhat over one inch between measurements. Measuring height more often than every 6 months would likely waste resources and not reveal much interesting information.

If we were comparing two groups that we expect to grow at different rates, the differences in growth rate might not be apparent in less than a couple of years. Suppose one group was growing at one quarter an inch per year faster than the other on average. It might well be two years before we could even begin to tell the difference between the two groups, depending on how much variability there is among individuals and how large the sample size is.

1.7 Complexities in Analyzing Longitudinal Data

Longitudinal data sets are usually larger and more complex than data sets from cross-sectional studies that can be analyzed using multiple linear regression. Longitudinal data analyses (as opposed to the data set) are more difficult than linear regression analyses. The models are more complex and more decisions need to be made in the analysis. We briefly discuss some of the issues in this section.

1.7.1   Correlation

The correlation of observations across time within a subject gives the study of longitudinal data its special flavor and separates longitudinal data analyses from cross-sectional regression analyses. We expect longitudinal heart rate measurements on the same person to be more similar to each other than would two separate measurements on different people. For example, if a person is higher than the population average on their first measurement, we anticipate that person to be higher on their second and future measurements, although it is not 100% certain. We usually expect positive correlation between longitudinal measurements. This correlation must be modeled in the analysis to properly assess the information in the data set about the population mean response and about the affect of covariates on the mean. If we erroneously treat the observations within a subject as independent, we will over- or under-estimate, possibly quite strongly, the strength of our conclusions. A number of models for the correlations of longitudinal data are covered in chapter 8.

1.7.2   Size

Suppose we have 100 subjects whom we interview every three months for two years for a total of nine longitudinal measurements, and we collect 10 variables measured at each interview. That will make for 9000 numbers for us to keep track of in our database. Now suppose not 100 but 1000 subjects, and, more realistically, that we collect not 10 but 1000 numbers at each visit. Suddenly we have 9 million numbers to keep track of. Some studies can have tens of thousands of subjects, measured over years and sometimes even decades. The data management problem is enormous. This text does not discuss data management, but it is a substantial and important problem in its own right.

Suppose that we are to analyze data from a study of large size, with thousands of variables to choose from in our analysis, a key problem is selection of which variables to study. What responses we analyze and which covariates we use need to be thought through thoroughly. We cannot analyze the entire data set, so a coherent and complete yet parsimonious set of responses and covariates must be selected. In typical regression textbook homework problems, this step has already been done. Constructing the working data set that we analyze may require substantial work, and we will not want to reconstruct the working data set multiple times and rerun all of the analyses too many times. Once we have chosen what longitudinal data to analyze, this book discusses how to analyze it.

1.7.3 Profiles Over Time

The population profile is the average response at a particular time plotted against time. In many data sets the population profile may be flat, as for example adult human height, or nearly flat, as adult weight, or it may be linear as for example a child's body size (height, arm circumference) over moderate periods of time. Some data sets have patterns that are quite complex. Mice weight increases steadily during the first few days after birth, accelerates, has a linear period, then slows again after about two weeks. The concentration of a drug in the bloodstream after inhaling a medicine does not change at first, then takes off, has a rapid rise, a peak, then drops off, first rapidly, then at a slower and slower pace. These patterns in time must be modeled, and they can be quite non-linear.

An interesting feature of many longitudinal data sets are the profiles of the responses within a subject over time. The subject profile is a plot of the individual subject's response plotted against time. On top of the population average, individuals may have trends that differ in pattern from the population average, and these patterns must also be modeled for proper inference in a longitudinal data set.

We will discuss ways to model population trends over time as a function of covariates in chapter 7.

1.7.4 Complex Data Structures

As mentioned earlier, the basic longitudinal data structure is a simple random sample of subjects and longitudinal repeated measures within subject. Many studies have a longitudinal component as part of a more complex data design. Multivariate longitudinal data is one example.

Another common complexity is *clustering* of subjects. Consider a study of grade school children. Each year from first through sixth grade, we measure their mathematical abilities. This study design supplies us with standard longitudinal data. Children from the same school can be expected to be more similar than children from different schools. Typically we might have an average of n_1 children from each of n_2 schools. The children within school are clustered, and we must account for this clustering to properly model the data. Furthermore, we may have schools clustered within school district, and school districts may be clustered within states. Each level of clustering induces some similarity or correlation amongst the subjects inside that cluster. We discuss clustering in section 9.2.

Other sorts of complexity can occur in data sets; these need to be accommodated by the analysis. Longitudinal data collection is a feature of many data sets; learning how to analyze the basic longitudinal data set properly provides us with an important tool. The models we learn about in this book then become important components of more complex models needed for analyzing more complex data structures.

1.7.5 Missing Data

Not all observations occur as planned. For example, the spacing between observations may be wrong. We may have planned for observations to be taken every three months say at 0, 3, and 6 months, but someone shows up one month late for an appointment, so that their observations are at 0 months, 4 months, and 6 months. Some people may not be available for data collection, so instead of three observations we only have two for some people. Our mouse may be available for weighing on day 7 as planned, but the research assistant does not show up, and the mouse does not get weighed that day.

Some people may drop out from the study. We wanted observations every three months for two years, but some people will move and not leave a forwarding address, some will stop answering our phone calls or emails, some will be too sick to respond, and some may die. We may be intending to follow people for two years, but if we recruit someone one year before the end of the study, they will have the first few measurements that we wanted, but after some point they will not have any more measurements. Someone may be inaccessible for a short period and so miss a scheduled visit but continue to fill out questionnaires at later visits. These events are all causes of missing data.

These examples illustrate missing values in the responses that we intended to collect. Sometimes missing data may occur by design. Due to the heavy burden on survey participants, many surveys only collect partial information at any one visit. If responses are correlated across visits, then longitudinal data analysis may be used to *fill in* or *impute* the missing values. The actual data, along with these filled in or imputed values may be supplied to other research groups for analysis. By filling in the values in proper fashion, the data structure is simplified and the tasks of future analysts are simplified. When we are analyzing the data directly ourselves, longitudinal data analysis can be used on the original data set without needing to fill in or impute responses.

All of these forms of missing data are common in longitudinal data sets, due to the complexity and size of the data collection tasks involved, and the fact that human beings are involved and are neither infallible nor perfectly compliant with research protocols. We need methods for data analysis that will accommodate these different forms of missing data, and we will discuss how well our various models can accommodate it. Chapter 12 discusses missing data concepts.

Another form of missing data is missing data in the covariates. Consider that a paper survey may be damaged so that we cannot read the age of a person, a person may not respond to certain questions such as ethnicity or income, or a person may respond in a fashion outside the parameters of the survey. If we wish to use that person's other data in the analysis, the analysis will need to know how to deal with missing data in the covariates.

Missing covariate data are a different form of missingness than missing response data. One important approach to missing covariate data is called *multiple imputation*. It is beyond the scope of this text to discuss missing covariates.

1.7.6 Non-constant Variance

Many longitudinal data sets have non-constant variance. The variance of the observations may change across time. Variances may be different in different groups. Many models for longitudinal data naturally incorporate non-constant variance, but whether the most common models correctly model the variability in our particular data set is an issue to be resolved separately for each data set.

1.7.7 Covariates and Regression

The population mean response must be modeled as a function of time and covariates. This mean response may be simple or complex. Because of the presence of time, in many data sets, we must model interactions between time and covariates, making the analysis more complicated than what we have in linear regression. Chapter 7 discusses modeling of the population mean as a function of time and covariates.

Covariates may well affect the variance of the response. A simple covariance model allows different groups to have different covariance models. More complex models that allow covariates to affect the correlation matrix are the subject of current research.

1.8   The Language of Longitudinal Data

In this section we introduce the language we use to describe longitudinal data.

1.8.1 Responses

The jth observation on subject i is Y_{ij}. The entire set of observations for subject i is Y_i, defined as

$$Y_i = \begin{pmatrix} Y_{i1} \\ \vdots \\ Y_{ij} \\ \vdots \\ Y_{in_i} \end{pmatrix},$$

which is a *vector* of length n_i.

The number n_i is the *within-subject sample size*; it is the number of observations on subject i. If the n_i are all equal to 1, then we have univariate data. We have a sample of n subjects with i indexing subjects and running from 1 up to n.

The total number of observations is

$$N = \sum_{i=1}^{n} n_i .$$

We have two *sample sizes* associated with longitudinal data. The total number of *observations* or *measurements* is N and the total number of subjects n. Both sample sizes are important and reflect on the size of the study and the difficulty of the analysis. Some studies can have n in the thousands or hundreds of thousands. Other studies may have $n < 10$, while the n_i can be in the thousands. And in a few other studies, both n and the n_i are enormous.

Although typically we have most n_i greater than or equal to 3, it is certainly possible to have some n_i equal to 1 or 2. There is information in these subjects. We do not delete them from our analysis merely because they have few observations.

There is no requirement that n_i be the same for different subjects i. When the n_i are all equal, we will let J be the common number of observations for each subject. Then $N = nJ$ is the number of observations and Y_{iJ} is the last observation on each subject.

1.8.2 It's About Time

Observation Y_{ij} is measured at time t_{ij}. The jth observation Y_{ij} is measured after the $(j-1)^{\text{st}}$ observation $Y_{i(j-1)}$. We arrange observations within a subject by increasing t_{ij} so that $t_{i(j-1)} < t_{ij}$. The vector of times for subject i is

$$t_i = \begin{pmatrix} t_{i1} \\ \vdots \\ t_{in_i} \end{pmatrix}.$$

1.8.2.1 Units of Time and Baseline

Time t_{ij} may be measured in seconds, minutes, days, months, or years. The units of t_{ij} are the same units as the natural units of clock or calendar time that separate measurements. The *zero time* may be a subject-specific time or the zero point may be common to all subjects. It is quite common for the first measurement to be taken at exactly time zero, in which case we find $t_{i1} = 0$ for all subjects. In an intervention study, $t_{ij} = 0$ may be set either at the first observation or, my slight preference, at the onset of the intervention. Often these are the same time. The choice of zero time, although unimportant in theory, may in practice make for easier plotting and modeling of the response.

Study baseline refers either to a key event such as study entry where $t_{ij} = 0$ or it may refer to all times up to or including the zero time, with any $t_{ij} \leq 0$ called a baseline observation. Any data collected at this time is called baseline data. In an intervention study, we may collect one or several responses Y_{ij} with $t_{ij} \leq 0$; all of these are baseline responses. In some intervention studies, the intervention may occur at a positive time; observations prior to this time are referred to as baseline responses.

The calendar date where $t_{ij} = 0$ is usually different for all subjects. An exception occurs when the zero time is defined by an important communal event. For example, a school lunch intervention commences on a single date for everyone, so the zero time will be the same calendar day for everyone. Longitudinal assessment of stress caused by a natural disaster would have a zero time that is the same for everyone, the date of the disaster. Assessment of the impact on consumers of a pricing or marketing change will have a single date of onset also. Time could be days since the study was initiated in which case $t_{i1} > 0$ for all subjects, and again the zero time is the same calendar date for everyone.

Another choice for time could be age. Time zero would be birth of each subject. Depending on the subjects in the trial, we might measure time in years for people over 5 years, months for children under 10, in days for infants in the first year of life, and in hours for newborns.

In dosage trials, experimenters give subjects a drug at various doses, with doses in a random order and then take a measurement for each dose. Some analysts then re-order the responses Y_{ij} by dosage level d_{ij} with $d_{i(j-1)} < d_{ij}$ rather than by trial number. Dosage level is clearly a key predictor, but we do not want to ignore time in favor of dosage. The proper analysis keeps track of both time and dosage in the analysis. The correlations among observations will be determined by the time order, while dosage level and order will contribute to the mean of the response.

1.8.2.2   Integer Times and Nominal Times

Time often takes on integer or other discrete values. An example we saw in the first section was when observations corresponded to consecutive trials in an experiment. If there are J trials, then feasible times are integers from 1 to J. Another situation where time is discrete is when data collection visits are scheduled on a periodic basis, perhaps every few months. Time of visit t_{ij} takes on integer values when measured in units of the period.

Study designs often call for regular data collection visits. We might collect data every 3 months for 2 years for 9 visits in total at times 0, 3, 6, 9, 12, 15, 18, 21, and 24 months; or every 5 months for 6 data collection visits ending 25 months after the first visit. Times like 0, 5, or 25 months are the *nominal* times of measurements. The first visit will be at time zero, $t_{i1} = 0$. Future visits between investigators and subjects are scheduled as close to the nominal times as possible. The nominal times contrast with the *actual* times the measurements are recorded. Either nominal time or actual time or both may be used in data analysis. Nominal time can be considered a rounded version of the actual times. Whether using nominal times in place of actual times has much effect on the final analysis depends on many things including the differences between nominal and actual times, correlations among observations, and the change in level over time of the response. I recommend using actual times when possible and not rounded or nominal times, particularly when the difference between the rounded integer units and actual times is large.

Vagaries of scheduling can cause nominal times to approximate actual times only roughly. If actual times are different in consistent fashion, for example always later, then nominal times may not have much value. If we calculate a slope based on nominal times, and actual times are 50% later than the nominal times, the correct slope estimate will be two thirds the reported slope. Current software implementations sometimes require using nominal or otherwise rounded times to get to a model that describes the data as correctly as possible.

1.8.2.3   Balance and Spacing

The times when observations are taken is part of the study design. We often intend that all times for all subjects be the same. *Balanced data* occurs when all subjects have the same number of observations $n_i = J$ and all subjects have observations taken at the same times: $t_{ij} \equiv t_j$ for all j and all subjects $i = 1, \ldots, n$.

Equal spacing occurs when the time between consecutive observations is the same for all observations and subjects. Mathematically, this is when $t_{ij} = t_{i1} + (j - 1)\Delta$ for all i and j. The time between observations is $\Delta = t_{i2} - t_{i1}$. To have balanced as well as equally spaced data, the starting times must be the same for all subjects, $t_{i1} = t_1$, and all subjects must have the same number of observations, $n_i = J$.

1.8.2.4   Balanced with Missing Data

Balanced with missing data is when all t_{ij}s are a subset of a small number of distinct possible times. The number of missing measurements should be small compared with N. Similarly, *balanced and equal spacing with missing data* occurs when all t_{ij}s are a subset of a discrete grid of times t_j with $t_j = t_1 + (j - 1)\Delta$, and again where the number of missing measurements should be small compared with N.

What is the definition of small? Rather than give a percentage of missing observations, we give another definition. Let $n(j, l)$ be the count of the number of subjects where observations are taken at both times t_{ij} and t_{il}.

We use the term balanced with missing data for data sets where $n(j, l) \geq 5$ for all j and l. The value 5 is chosen somewhat arbitrarily. Four or possibly even three may be substituted for 5. If any $n(j, l)$ is 1 or 2, it is difficult to estimate the covariance of observations at times t_{ij} and t_{il} when $n(j, l) = 2$ without making parametric assumptions.

Many longitudinal data sets are balanced with missing data; the balance occurs by experimenter design. The missing data occurs because people do not always appear for data collection visits. This balanced with missing data assumes that the nominal times are the times. When we have balanced with missing data, there are two ways to think about the data set. We can omit the missing data entirely. Then each subject's Y_i vector has J or fewer observations depending on the number of missed visits, and each Y_i vector has $n_i \leq J$ observations. This fits the way we described the Y_i data vector at the beginning of this section.

A second way of thinking about the data is to have every vector Y_i to have length J. Most observations are observed, but some are missing. The missing observations are not filled in, but we can still talk about a subject's second observation, whether missing or not.

1.8.2.5   Simultaneous Measurements

It is rare in practice to have two measurements taken at exactly the same time on a single subject. One exception is when multiple measurements are taken for quality control purposes. For example, in a 24-hour diet recall study, we may send two different interviewers to the same household at the same time and have them repeat the diet measurements. An analysis is then run to determine how close the two measurements are to each other and to provide an estimate of the sampling error of the measurements. For many covariance models, many software packages will assume that observations taken at exactly the same time must be identical; for those models and software packages, we will have to specify only a single observation from the two taken at the same time.

1.8.2.6   Geometric Spacings

Measurements are often separated by approximately geometrically increasing time intervals. For example, clinic visits for data collection in a clinical trial may be scheduled at 0, 1, 3, 6, 12, and 24 months. This particular type of design occurs when we expect more rapid change early in the study and less and less change as time moves on. Measurements of blood drug concentration in a pharmacokinetic (PK) trial may be taken at 0, 1, 3, 7, 15, and 31 minutes. In this case, we can get even spacing by transforming time. Let the new time equal the log of the old time after adding 1. Using log base 2 gives integer times. When measurements are taken at geometric spacings, we refer to this as *log-time spacing*.

Time interval	Measures
Milliseconds	Neuron firing
Seconds	REM sleep
Minutes	Heart rate
Hours	Pharmacokinetics: blood
	Concentration after drug dosage
Days	Blood pressure, stock prices
Weeks	Weight loss, store purchases
Months	Growth, 3-month recall
Fractions of 24 hours	Circadian rhythms
Fractions of 30 days	"
Fractions of 365 days	"
Years	Growth and aging
	Country gross domestic product
Quarters	Educational studies
Semesters	"
Grade in school	"

Table 1.2. Examples of time spacings and research study areas where those spacings might be used.

1.8.2.7 Unbalanced and Randomly Spaced Data

Unbalanced data or *randomly spaced* data occurs when the times of observations have a distribution that cannot be conveniently thought of as balanced with missing data. There is a natural continuum from balanced to balanced with a modest amount of missing measurements to balanced with lots of missing measurements to random spacing. The boundaries are not well defined. Unbalanced or randomly spaced data occur naturally in observational studies where we do not have control of subjects. One example occurs when patients visit the doctor for colds or other illnesses; heart rate, blood pressure, and weight measures are routinely recorded in the patient's chart at these visits. A retrospective study of subjects at a clinic might use these measurements. A grocery store might test a new marketing campaign in several cities. The impact of the campaign on shoppers could result in measurements on the total number of dollars spent at each visit. The store has no control over visits but can collect data on purchases when they do occur.

1.8.2.8 Lag

The *lag* between two observations Y_{ij} and Y_{il} is the time $|t_{ij} - t_{jl}|$ between them. When observations are equally spaced, we more commonly define lag to be the number $|j - l|$.

Study	Units of time	Nominal times
Cholesterol response to drug intervention	Weeks	0 1 4 12
Children of HIV+ Parents		Three monthly first 2 years, every six months next 3 years
Young adult HIV+ drug users	Months	0 3 6 9 15 21
Kenya school lunch Cognitive	Semesters	−1 1 2 4 6
Anthropometry	Monthly	Before intervention monthly in year 1, bimonthly year 2
Autism parenting intervention	−	0, 11 weeks, 1 year
	Weekly	weeks 0 − 10

Table 1.3. Specific studies with time intervals and nominal times of measurements.

1.8.2.9 Examples

Table 1.2 gives examples of general time spacings and some research areas that might use those times. The table is hardly exhaustive, rather it is intended to introduce the breadth of possible studies where longitudinal data may be collected.

Table 1.3 lists nominal times of measurements for various studies I have recently collaborated. In the Cholesterol study, a subset of only about 25% of the patients were sampled at weeks 1 and 4. In the Cognitive data set from the Kenya school lunch intervention study, cognitive measures were taken at terms −1, 1, 2, 4, and 6. These are the first three terms starting before the intervention onset and then every second term after term 2. Anthropometry measures from the same study were taken monthly the first year and bimonthly the second year. In the parenting intervention for caretakers of autistic children, one set of measures are taken before a 10-week training session, after the training session and 1 year later. Another set of measurements are taken weekly at each training session.

1.8.3 Covariates

We use covariates to predict our responses Y_{ij}.

Two distinctions that we make are between *time-fixed covariates* and *time-varying covariates*. Time-fixed covariates do not change over time, examples include gender, race, ethnicity, disease status at study entry, age at study entry. Time t_{ij} is the canonical time-varying covariate. Very often time and functions of time t_{ij}^2 are the only time-varying covariates in longitudinal analyses. *Baseline covariates* are covariates measured at baseline or study entry. Baseline covariates are by definition time-fixed. Age at study entry, race, and gender are typical baseline covariates. It is possible to have similar covariates one of which is time-varying and one of which is time-fixed, for example, age is time-varying while age at baseline is time-fixed.

A treatment/control indicator variable may be either time-fixed or time-varying. If treatment is assigned and delivered before the first measurement, then the treatment/control indicator variable will be time-fixed. If the baseline measure is taken before the onset of treatment, then at baseline no one is in the treatment group. Those who are randomized or otherwise assigned to the treatment group will presumably have a different trend over time after treatment onset from those in the control group.

1.8.4 Data Formats

Longitudinal data can come in two main formats, *long* format and *wide* format. Wide format only accommodates balanced data or balanced with missing data. In wide format, row i in the data set contains all the data for subject i. There will be J columns reserved for the J longitudinal observations. Covariate values are contained in another set of K columns. There may or may not be a subject id for each row. If there is missing data, a special missing data code is entered into the data array, this may be a dot ".", not available "NA", a question mark "?" or other special code, depending on the software program that created the data set and depending on the software that will be reading the data set in. With n subjects and $K + J$ pieces of data per subject, the data fits into an array that is n rows by $K + J$ columns. In balanced or balanced with missing data, the jth observation is taken at time t_j for all subjects. The time t_j is associated with the jth column, it is not encoded in the data set itself.

Long format has one row per observation rather than one row per subject. Long format can accommodate random times as well as balanced or balanced with missing data. In long format, there is a single column for the longitudinal outcome measure. Another column gives the subject id so that we can identify which observations came from the same subject. A third column gives the time at which the outcome was measured. Covariates again take up K columns in the table. One column is reserved for each covariate. Time-fixed covariates such as gender for subject i are repeated n_i times in the n_i rows for that subject. Time-varying covariates are poorly accommodated in wide format; when there are time-varying covariates, it is best to use long format.

With random times, there is no practical version of wide format. For balanced with missing data there are two different versions of the long format. In the more compact version, there are $N = \sum_i n_i$ rows, and if there is a missing response datum for subject i, then there is no row associated with subject i and that time point. In this compact long form, it is strictly necessary to have a column with time identified in it. In the less compact long form, there are nJ rows in the data set, and every subject has J rows of data, whether or not they provided all J intended observations. As with the wide format, a missing data symbol is entered in those rows where the longitudinal response is missing.

Most longitudinal software uses long format to enter data. Often the software will make assumptions about the time spacing among observations, and often this is that observations are balanced and equally spaced. If we have balanced with missing data, then it will be necessary to use the non-compact long format with this software. More complex models and software will allow us to enter the times for each observation; often the same software accommodates both formats. only

1.9 Problems

1. How many observations? How long is our study?
    (a) We plan to take l observations spaced apart by m units of time. How long is there between the first and last observation?
    (b) We take observations on people every three months for two years. How many observations do we take per person?
    (c) We take observations every m units of time. Our study lasts for T time units. How many observations do we get per person?
    (d) We enroll subjects continuously for two years. We wish to follow them for two years. How long must our study be?

2. Work out the same sort of calculation for human weight as we did for height at the end of section 1.6.

3. Can you work out the same sort of calculation for blood pressure as we did for height at the end of section 1.6?

4. Sometimes we cannot execute a paired design and we must therefore use a two-sample design. Why, even if we can take longitudinal data, might we not be able to use a paired design? Alternatively, in some circumstances, even if we can do a paired design, why might it not be a good design? Consider that the second measurement, even in the absence of a treatment, may have a different population mean than the first observation.

5. In table 1.1:
    (a) Which line in the table do you disagree with most, and why?
    (b) Add a section for (adult human) heart rate to the table.
    (c) For a measurement that you are familiar with, add your own section to the table.

(d) Typical human height measurements are to the nearest centimeter (roughly half an inch), and typical weight measures are to the nearest pound (roughly half a kilogram). Suppose that we can measure human height or human weight much more accurately than these usual measurements. How would this affect judgments in the table? In particular, what studies might be interesting that measure human weights on an hourly basis or human height on a daily or even half-daily basis?

6. For table 1.2, think of a completely new set of studies that might use the same time spacings given in the table.

7. Why might recruiting patients at a second clinic complicate the analysis of a study? What aspects of the data might be different at the second clinic?

8. (Continued) Why might recruiting patients at a second clinic involve substantial costs over only having one clinic? Consider that a new doctor or doctors must be involved, must be willing to adhere to recruitment protocols, must be willing to enroll patients and collect the necessary data. Costs are involved, possibly new personnel to be hired. Consider that running a study involving people requires human subjects committee approval at each institution and probably other paperwork as well. Requirements at each institution are typically different, and someone must learn about all paperwork that is required.

9. Inspect the articles in a journal in your discipline. Read each article and determine how many of them collected longitudinal data and what the responses were. For two articles that use longitudinal data, determine

   (a) The definition of time: units, zero point, number of measurements, times between measurements, number of measurements actually used in the article;
   (b) Whether they used nominal times or actual times in the analysis;
   (c) What covariates they considered in the analysis;
   (d) What software they used to fit the data; and
   (e) The name(s) of the model(s)/analyses used to fit the data.

10. Explore the Anthropometry data.

   (a) What is the earliest time of any measurement?
   (b) What is the latest time?
   (c) What are the nominal times of observations?
   (d) How many subjects have observations that are taken within two days after the zero point? Ten days? Twenty days?
   (e) How much impact could the treatment possibly have on height or on weight after twenty days?

(f) For the weight outcome measure, explore the actual times of observations. Is there any sort of pattern to the times? How often are observations taken? Is it the same frequency in the first year and the second year?

(g) For the height outcome, repeat the previous item.

(h) Compare the frequencies of the height and weight measurements. Why do you suppose the investigators did this?

11. Repeat problem 10 for the Raven's outcome in the Cognitive data; omit the last two parts.

12. For each of the following data sets, decide if the data set is balanced, is balanced with missing data, or has random times:

(a) Small Mice
(b) Big Mice
(c) Pediatric Pain
(d) Weight Loss
(e) BSI total
(f) Cognitive

2
Plots

"Do you see the big picture, [Meehan]?"
"Never have, Your Honor," Meehan told her. "I'm lucky if I make sense of the inset."
– From Donald E. Westlake's Put a Lid on It.

There I shall see mine own figure.
Which I take to be either a fool or a cipher.
– William Shakespeare

Overview

In this chapter, we cover the following

- Plotting longitudinal data
  - What we want from our graphics
  - Defining profile plots
  - Interpreting profile plots
  - Variations on profile plots

- Empirical residuals

- Correlation
  - Correlation matrix
  - Scatterplot matrices
  - Correlograms

- Empirical summary plots
- How much data do we have?

This chapter presents exploratory data graphics for longitudinal data. Most graphics in research reports are used to present inferences. Long before we draw conclusions, we must understand our data and determine the models we will use. In drawing exploratory graphics, we plot the data in ways that shed light on modeling decisions. Longitudinal data are more complicated than cross-sectional data, and our plots will be more complex as well.

Our discussions will assume balanced or balanced with missing data. Most ideas apply equally well to random time data. The problem with explicitly including random time data in the discussion is that there is a significant increase in notational complexity without much corresponding benefit.

2.1   Graphics and Longitudinal Data

General multivariate data are difficult to plot in a way that provides insight into the data. Much ingenuity has been expended on creating such plots. In contrast, a number of useful plots for longitudinal data exist. The reason for the difference is that the units of measurement for longitudinal observations Y_{i1}, Y_{i2}, ..., are all identical. Within-subject comparisons make no immediate sense for general multivariate data. It is hard to compare heart rate and blood pressure as part of a multivariate observation; the units are not directly comparable. Only between-subject comparisons of corresponding measurements such as $Y_{i1} - Y_{l1}$ or $Y_{i2} - Y_{l2}$ make sense. In contrast, with longitudinal data we can take differences between observations within subjects. The difference $Y_{i2} - Y_{i1}$ is the increase in the response from the first to the second observation, and we want plots to show that change. We may take differences of similarly timed measures between subjects. Suppose that $t_{ij} = t_{lj}$ for subjects i and l, then $Y_{ij} - Y_{lj}$ is the amount that subject i is higher than subject l at time j. Even if times t_{ij} and t_{lk} are different, the difference $Y_{ij} - Y_{lk}$ is still interpretable.

What are the quantities we want our graphics to show? Some basic quantities we are interested in are the value of a particular observation Y_{ij} and the average response from subject i

$$\bar{Y}_i = \frac{1}{n_i} \sum_{j=1}^{n_i} Y_{ij}.$$

We want to compare observations within a subject. For example, we want to evaluate the difference $Y_{ij} - Y_{i(j-1)}$, and we need to compare observations across subjects at a particular time as in $Y_{ij} - Y_{lj}$. We want to answer questions such as which observations Y_{ij} are highest or lowest and which subjects are highest or lowest on average. We would like to assess the average response across subjects at a single time j

$$\bar{Y}_{.j} = \frac{1}{n} \sum_{i=1}^{n} Y_{ij}$$

and the sample standard deviations

$$s_{jj} = \left[\frac{1}{n-1} \sum_{i=1}^{n} (Y_{ij} - \bar{Y}_{.j})^2 \right]^{1/2}$$

of the Y_{ij} at a specific time j. We want to know if these means and standard deviations are increasing, constant, or decreasing over time. The ratio

$$\gamma_{ij} = \frac{Y_{ij} - Y_{i(j-1)}}{t_{ij} - t_{i(j-1)}} \tag{2.1}$$

is the slope of the line segment between observation $j-1$ and j for subject i. We will want to compare these slopes for different j within subject and also across different subjects at similar or different times. Are the slopes increasing over time or decreasing? Is the typical subject's slope increasing or decreasing at time t? What is the average of γ_{ij} over subjects i? We would like to see which subjects have similar profiles on average $\bar{Y}_i = \bar{Y}_l$ or have profiles with similar patterns over time such as $Y_{ij} - \bar{Y}_i = Y_{lj} - \bar{Y}_l$ even though their averages may not be the same.

So far we have mentioned basic features of our data; observation level Y_{ij}, subject average level $\bar{Y}_i$, across subject within time average level $\bar{Y}_{.j}$; differences $Y_{ij} - Y_{ik}$, $Y_{ij} - Y_{lj}$, standard deviations s_{jj}, and slopes $(Y_{ij} - Y_{i(j-1)})/(t_{ij} - t_{i(j-1)})$. Thinking now not about features of the data, but features of the models we will be creating, what are the basic components of our models? We want our plots to help us with specification of these components. The basic features of our models will be

- the population mean response at a particular time,
- the population variance or standard deviation of the responses at a particular time,
- the correlations between observations within subjects, and
- the effects of covariates on these quantities.

We want our plots to show us information so we can specify our models appropriately. Mainly we want to make qualitative judgments from our plots; quantitative judgments are reserved for the output of our models. We do not need to learn that the mean of the observations at day 2 is 200 and that at day 20 it is 950. Rather, we need to learn from our plots if the mean response is increasing over time or not. If the mean response is increasing, is the increase linear or something more complicated?

2.2   Responses Over Time

Time permeates all longitudinal data analyses. The first graphic we make plots the longitudinal response against time. The obvious first plot we might consider plots all responses Y_{ij} against time t_{ij}. Figure 2.1 shows this plot for the *Big Mice* data. The response is the weight in milligrams for $n = 35$ mice with each mouse contributing observations from various days starting at birth, day 0, through day 20. Thirty-three of the mice were weighed every three days for a total of seven observations each. Eleven mice in group 1 were weighed beginning on day 0, ending on day 18; group 2 has 10 mice weighed beginning on day 1 ending on day 19; and group 3 has 12 mice weighed beginning on day 2 ending on day 20. The last two mice are in group 4 and were weighed daily from day 0 to day 20. A subset of the Big Mice data forms the *Small Mice* consisting of the group three mice plus the group four observations on the same days. The Small Mice form a balanced subset of the data, whereas the Big Mice data are balanced with lots of data missing. Each mouse comes from a separate litter, so it is reasonable to treat mice as independent. All weighings were performed by a single person using a single scale.

See the data set appendix for details about any given data set. We will discuss data sets in the text as we need the information. To make it easy to find this information at a later time, data set descriptions are kept in the data set appendix. The mice data set description is in section A.2.

2.2.1   Scatterplots

Figure 2.1 is a scatterplot of weight against time with all $33 \times 7 + 2 \times 21 = 273$ observations plotted. The weights start out low and grow rapidly. On days 0 and 1, the weights are all less than 200 milligrams (mg), by day 5 the average weight has more than doubled, and the weights more than double again by the end of the study at day 20. Somewhere around day 10 the daily increase in weight appears to slow down although the exact pattern of increase is unclear.

2.2.2   Box Plots

There is a fair amount of over-plotting of circles in figure 2.1, and with smaller page size, poorer quality graphics or larger data sets, over-plotting can be even worse. One solution that people have used is to plot repeated box plots over time. Rather than attempting to plot all of the observations, the box plot summarizes the observations at each time point and does a careful job of presenting the summary.

Figure 2.2 shows 21 repeated box plots of the mice data. Each box plot summarizes the observations taken on one particular day. The central divided rectangular box plots the lower quartile (lowest line), the median

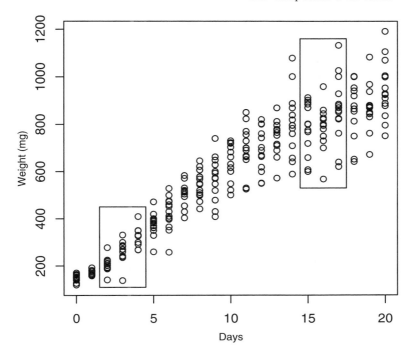

Figure 2.1. Scatterplot of Big Mice weights in milligrams against time in days. Thirty-three of the mice contribute 7 observations each and 2 mice contribute 21 observations each. The boxes are explained in section 2.3.1.

(middle line), and the upper quartile (upper line) of the observations measured on that day. The lower (upper) quartile is the observation with at least 25% (75%) of the observations at or below it and at least 75% (25%) of the observations at or above it. The *whiskers* are dashed lines extending from the lower and upper quartiles to the minimum and maximum values indicated by the short horizontal lines. The box plot shows the interquartile range, the upper quartile minus the lower quartile, and it shows the range, the maximum minus the minimum. The box plot is less crowded than 2.1. Figure 2.1 tries to show every data point, while the box plot displays five summary statistics of the data at each time point.

We again see the sharp rise in the weights over time. The increase accelerates around days 2–6. The medians, for example, increase rapidly each day until around day 11, when they grow less quickly. Around days 12–14 the rise in the medians continues, but perhaps at not such a sharp rate. Thereafter the increases are uneven. At days 14–16 the median weight is nearly constant and again for days 17–19.

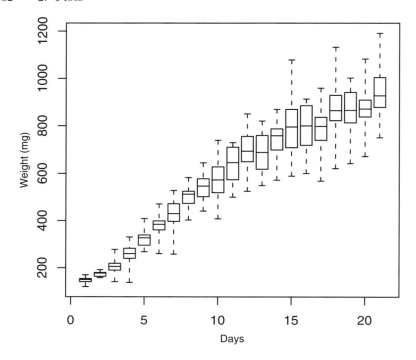

Figure 2.2. Repeated box plots of the Big Mice weights over time. Each box plot summarizes the distribution (minimum, lower quartile, median, upper quartile, maximum) of weights observed on that day.

The variability of observations on a given day increases as the mice get older. As the mice age, we see in figures 2.1 and 2.2 that the variability in weights increases up till perhaps around day 9, then at some point, possibly day 14, it increases again. In figure 2.2, the range and the interquartile range appear to increase over time as well. As the mice grow, the range increases and then appears to stabilize around day 9 or so.

2.2.3   Flaws in Scatterplots and Box Plots for Longitudinal Data

The data points and data summaries in figures 2.1 and 2.2 are not independent. One may have experience looking at a scatterplot of a response y versus a predictor x and deciding whether there is a significant or important differences in the response as functions of time. Because our observations come from the same mice at different time points, our intuition based on independent observations may not apply. For neighboring days, the observations are almost independent except for the two mice in group 4 who contribute data to all days. If we wish to compare the data from two days that are multiples of three days apart, we have correlated data. If we were to compare the means between days 4 and 7 for example, we would do it with a paired t-test, not a two-sample t-test. Figures 2.1 and 2.2 do not show the connection between observations from the same mouse.

It is a flaw of these figures that we cannot tell which observations come from the same mouse. Three of the four largest weights are the largest weights at days 14, 17 and 20. Because most mice are weighed every three days, we suspect, but cannot tell, that these observations belong to the same mouse. The largest observations on days 14, 17, and 20 weigh much more than the largest observations at days 15, 16, 18. Day 14's maximum is even slightly higher than the maximum at day 19. Similarly, days 2 and 5 have measurements distinctly lower than the other observations, and we suspect, but cannot tell, that we are looking at two measurements of one mouse. There is a similar low pair of observations at days 3 and 6.

Additional features of our data that we cannot identify include the differences $Y_{ij} - Y_{i(j-1)}$ or the slope between consecutive observations within a mouse, nor can we identify whether a particular mouse is high or low compared to the remaining mice. Longitudinal data have a natural hierarchical structure that should be reflected in our plots. Observations within subject are correlated and the nesting or clustering of observations within subject should be encoded in our plots.

2.2.4   Profile Plots

A *profile* is the set of points (t_{ij}, Y_{ij}), $j = 1, \ldots, n_i$. A *profile plot* improves on the basic scatterplot 2.1 by using line segments to connect consecutive observations (t_{ij}, Y_{ij}) and $(t_{i(j+1)}, Y_{i(j+1)})$ within a subject. No lines are drawn between observations from different subjects. Profile plots are useful because the clustering of observations within subject is directly visible. In a profile plot, the basic plotting unit is not the observation (t_{ij}, Y_{ij}), rather it is the entire profile (t_i, Y_i). The profile plot in figure 2.3 displays the mice data. We can see that a single mouse is heaviest at days 14, 17, and 20. It was not the heaviest mouse at time 11 or earlier. The second heaviest mouse at days 14 and 17 is outweighed by yet another mouse at day 20. We see that the mouse that was heaviest at days 3–6 was one of the two mice that were measured daily. It ends up among the heaviest mice but is not the heaviest.

Generally we see that the mice all grow in parallel; if mouse A is heavier than mouse B at an earlier time, it has a tendency to be heavier at a later time. This is particularly clear after day 9 or 10; the plot is cluttered before day 9 and it is not so easy to see if this is true for observations from before day 9. Mice that are close in weight may change rank from day to day, but if mouse A is more than 100 milligrams greater than mouse B after day 9, it is unlikely to ever be lighter than mouse B.

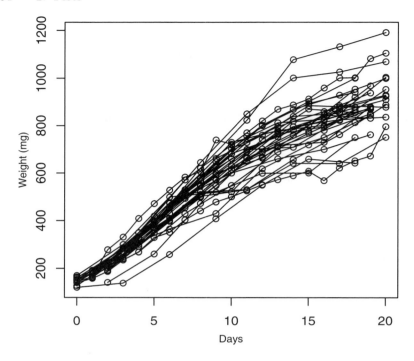

Figure 2.3. Profile plot of mice weights against time. Consecutive observations within a mouse are connected by line segments.

In figure 2.3, we also notice that on a few occasions, mice actually decrease in weight from one observation to the next, something we could only infer from the previous two plots and only for a few special circumstances. Examples include the heaviest mouse at day 9, which lost weight at day 10, and the second lightest mouse at day 15, which is the lightest mouse on days 16 and 17.

2.2.5   The Need to Connect-the-Dots in Profiles

The need for line segments in profile plots is illustrated in figure 2.4. Five fictional subjects contribute six observations each to figure 2.4(a). From this plot we do not know which observations belong to which subjects. We can only learn about the *marginal distribution* of the data at any given time point. Pick any point t on the time axis, and look at the collection of observations above t, and perhaps within a window slightly to either side, say observations with times in the range $t - \Delta, t + \Delta$ for some modest value of Δ. Average the responses in the window, and look at a number of windows centered at different times t. We learn that the average value appears to be fairly constant across time. To the right of the middle of the time axis, there appears to be possibly less variability in the response values or possibly there are merely fewer observations at that time. We do not know the reason for this lower variability. From this plot we do not learn about the *joint distribution* of observations within a subject. In 2.4(a) we do not know, for example, if the largest half a dozen observations across all time points belong to the same or different subjects.

Figure 2.4(b) presents a possible profile plot for the data in 2.4(a) with observations within subjects connected by consecutive line segments. Profiles are labeled by subject id from 1 to 5 at the left of each profile. Subject 1 has the highest response values at all times, and subject 5 has the lowest responses. Generally, observations within subject at different times have similar responses Y_{ij} across time. If subject A begins higher than subject B at the left side of the plot (early times), then A's observations are higher in the middle (middle times) and again at the right side at the latest times.

In contrast, figure 2.4(c) represents a different assignment of observations to subjects. At the earliest times, the subject profiles are the same as in figure 2.4(b). However, somewhere in the middle of time, subjects who start low tend to rise, while subjects who start high tend to fall, and at the late times, subject 5 who started lowest is highest, while subject 2 for example, who started second highest ends as second lowest. Subject 3 has a flat profile throughout: subject 3 had an average response in the beginning is still average at the end.

Figures 2.4(b) and 2.4(c) suggest different explanations for the reduced variance of the responses in the late middle time region. In figure 2.4(b), it appears that the reason for the gap is that there were few observations on subjects 4 and 5 around that time, whereas there were plenty of observations for subjects 1, 2, and 3. If we had observations on subjects 4 and 5 to the right of the middle time we would expect them to be low; we expect the variability of the responses across subjects to be roughly constant across time. Figure 2.4(c) is different; we see that each subject's responses appear to be following their own line as a function of time. Each subject has a different slope and intercept. Subjects 1 and 2 have negative slopes, 3 is flat, and subjects 4 and 5 have positive slopes. In figure 2.4(c), it appears that no matter how many observations we collected from these 5 subjects, we would see the same decrease in variance to the right of the middle; observations around the point where the lines cross will always be tightly clustered, and thus the variability of the responses will be lower to the right of the middle time.

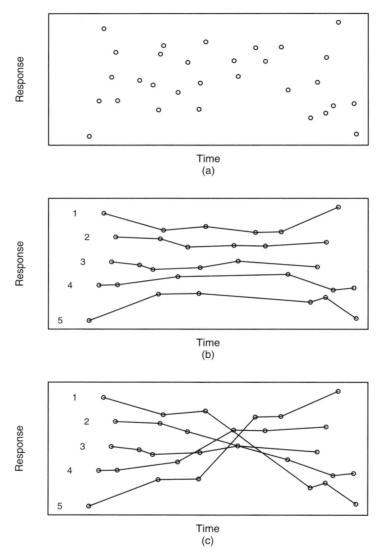

Figure 2.4. Plotting longitudinal data. (a) Scatterplot of responses Y_{ij} against t_{ij} for 5 fictional subjects of 6 observations each. (b) Possible profile plot based on the observations in plot (a). (c) Alternative profile plot based on the observations in plot (a). Subjects are labeled 1–5 to the left of their earliest observation in (b) and (c).

2.3 Interpreting Profile Plots

We can read basic information about subjects' response patterns from a profile plot. Figure 2.5 illustrates four different situations. Each subfigure displays data from eight hypothetical subjects. Subjects are measured at random times, usually with 5 observations per subject. Individual observations are plotted using a circle and, as before, line segments connect consecutive observations within a person. Later on we will drop the circles and only use the connected line segments. Figure 2.5(a) shows a very common situation not unlike that in 2.4(b). We see that each profile is roughly flat, with observations near a subject-specific mean. Individual profiles generally do not cross other profiles, that is, they are roughly parallel. There are but a few exceptions in the middle of the data where one profile crosses another.

If we extrapolate each subject's profile back in time to the time $t = 0$ axis, each profile would intersect the axis at a subject-specific intercept. If subjects are a random sample from the population of interest, then any subject-specific characteristic is also a sample from the population of possible values of that characteristic. In particular, the intercepts are a sample from the population of intercepts. We say that the data in 2.5(a) has a *random intercept*. The term random is used in the same way as when we said that the subjects in our study are a random sample from the population under study. Another way to say random intercept is to say that each subject has their own subject-specific mean response and that observations vary around the mean response.

In figure 2.5(b), the profiles are again parallel, but this time each has a linear time trend with a positive slope. If we extrapolate by eye back to the origin, the profiles all appear to have different intercepts, and again we conclude that the data has a random intercept. When we look at the slopes of the profiles, all of the slopes appear to be about the same. Here, we have a *fixed slope*, a slope that does not vary by subject. We conclude that the population also has a fixed slope; each subject's responses increase at the same rate over time.

The data in figure 2.5(c) illustrate a different pattern of responses. Most of the profiles start low at time $t = 1$, and grow larger as time progresses. There is one unusual profile that starts high and does not grow over time. We identify that subject as an outlier, and would strongly consider removing it from the data set before fitting models to this data. The remaining profiles are linear with similar initial values at the earliest measurement but they increase over time at different rates. We conclude that we have a random slope and a fixed intercept in the population. The unusual subject's earliest observation is a univariate outlier; we can identify univariate outliers on a profile plot when a single observation (t_{ij}, Y_{ij}) is the most extreme Y-value, either highest or lowest at time t_{ij}, or, for random times,

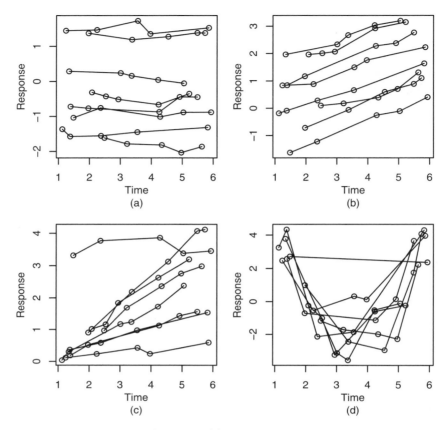

Figure 2.5. Example profile plots. (a) Random intercept, constant population mean; (b) random intercept, positive fixed population time trend; (c) random slope, fixed intercept, with one outlying profile; (d) fixed quadratic.

if Y_{ij} is the most extreme Y-value for all observations within a narrow window of time centered on t_{ij}.

Bivariate outliers (t_{ij}, Y_{ij}), (t_{ij+1}, Y_{ij+1}) can be identified if the line segment connecting them is unlike all the other line segments in the same region of time. The unusual subject in figure 2.5(c) also begins with a bivariate outlier, as no other subject has a high followed by a high first two observations. Bivariate outliers not necessarily need be univariate outliers. Imagine in figure 2.5(a) a subject with points at (t, y) equal to $(2, -1)$ followed by $(3, +1)$. Neither y-value of -1 or $+1$ is unusual, but the line segment connecting them would have the largest slope of any other line segment in the plot, by a substantial margin, indicating that this was an unusual pair of observations.

The final figure 2.5(d) shows subjects following a quadratic trend in time. The shape of the quadratic appears to be the same for each subject, and we conclude that the data follows a fixed quadratic path. At first glance there appears to be a single outlying subject. Closer inspection reveals that that subject has but two observations, one early at $t \approx 1.5$ and one late at $t \approx 6$. The $\approx$ is read *approximately equal to*. The impression of an outlier is given because we use a line segment to interpolate between the two observations, while the bulk of the observations follow a distinct non-linear trend between those times.

Many other possible patterns of profiles exist. We can imagine a plot where every profile is approximately linear, each with a different slope and intercept. The data would have random intercepts and slopes. We can imagine many forms of curvature over time too innumerable to even begin to discuss. Problem 9 presents a few possibilities.

2.3.1 Sample Means and Standard Deviations

Key features we can estimate informally from profile plots are population quantities such as the population mean or population standard deviation of the data as a function of time. Suppose we have a balanced data set with no missing data, and subjects are a random sample from our population. We might take a mean of all observations at each time point where we have data. These means estimate the *population mean* as a function of time. If we hypothetically had observed all subjects, then the mean of all observations at a given time is the population mean at that time! Depending on need, we may plot these sample means over time, or we might roughly eyeball them merely by viewing the profile plot. Inspection of the means will indicate to us whether the population mean is constant over time or if it is increasing or decreasing and whether the population trend is linear or not. If the linear trend is modest, we may need to resort to a formal statistical test to determine significance, and when presenting our conclusions to others we almost always supplement our informal judgments with statistical tests.

The population standard deviation at a given time is the standard deviation of a set of observations, one per subject, if we had observed all subjects at a single time. Given a sample of subjects, we have an estimate of the population standard deviation. The population standard deviation measures the within-time across-subject variability.

When we have random times or balanced with many missing data, or just sparse data, we may not have enough data to calculate a mean or standard deviation at a given time or there may be too few observations to get a reliable estimate. Instead, we may pool responses taken from observations with similar times to calculate our mean or standard deviation (sd). In particular, we might take all observations $Y(t)$ within a *window* along the time axis. The window has a midpoint at time t_M, a width w, a left endpoint $t_L = t_M - w/2$, and a right endpoint $t_R = t_M + w/2$. We collect all the

Y_{ij} values from observations whose t_{ij} are in the window, $t_L \leq t_{ij} \leq t_R$; and we perform some statistical operation, for example, mean, sd, min, max, or median on those observations. We plot the resultant mean, sd, or other quantity against the midpoint t_M. Next we move the window along the t axis, moving t_M from one end of the data set to the other. For each window, we calculate the same statistical operation, and we plot the result against the window midpoint, connecting the resulting points by line segments. When our operation is the mean or a quantile, we may plot the summary on the same plot as the data. For a range or standard deviation, we would plot the ranges or sd's against the window midpoint on another plot because these values lie on a scale different from the original data. We typically pick the window width just large enough to give us enough data to make a decent estimate but not so wide that the estimate becomes meaningless.

Figure 2.1 illustrates two windows of width 3 days, one from 1.5 to 4.5 and one from 14.5 to 17.5. The two boxes in the plot enclose all of the observations in the two windows. The mean of the observations in the left window is 260 mg while the mean of the observations in the right window is 810 mg. The standard deviations are left window 61 mg and right window 120 mg. We reasonably conclude that both the population mean and sd are larger around time $t = 16$ than around $t = 3$. An issue is how big the window should be. Around time $t = 3$, the means are increasing rapidly, and taking a wide window may cause us to overestimate the standard deviation. With the Big Mice data, we have enough data to keep the window width down to a width less than 1. We would then take the mean and sd at each time point, and plot them against that time point. These two plots are illustrated in figure 2.6. We see in figure 2.6(b) that the sd at time $t = 3$ is between 40 and 50, and because our earlier window was wider than necessary, it did indeed overestimate the standard deviation.

Inspecting the two plots, we conclude that the mice means increase smoothly over time. The increase is not quite linear, with a slight acceleration in the beginning, and then a slight deceleration after day 10. The sd's also increase in a smooth but somewhat curvilinear pattern. The sd's bounce around more from day to day than do the means. In general, standard deviations are harder to estimate than means, and this is reflected in the greater variability of the standard deviations over time.

Often we do not formally draw figures such as 2.6 or decide on a specific window width. For example, in figure 2.5(a) we see that the minimum response, the maximum response, the average observed response, and the range of the responses seem to be nearly constant over time. We identify these statistics as a function of time by, for example, looking at the subset of observations in the window between the times $t = 1$ and $t = 2$ and comparing that set of observations to the set of observations between for example $t = 5$ and $t = 6$. The maximum value in these two time intervals is nearly identical and come from the same subject. The minimum values

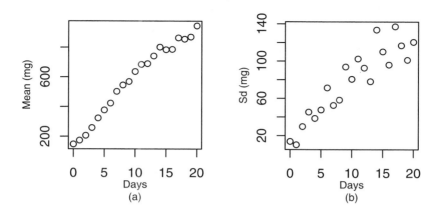

Figure 2.6. Big mice data. (a) Sample means by day. (b) Sample standard deviations by day.

are slightly different and come from different subjects. But this difference is small and is attributable to sampling variability and the fact that the lowest valued subject did not provide any observations before $t \approx 2.5$. Because the time trend within subjects seems flat, and the overall impression of the sample average time trend seems flat, we reasonably hypothesize that the trend of the population mean over time is flat.

Both the minimum and the maximum in figure 2.5(b) appear to be increasing linearly over time. However, the range $=$ max $-$ min is roughly constant over time. The distribution of observations between the max and min is fairly uniform and we conclude that the *population variance* of the responses is constant over time.

The population sd over time of the responses increases in figure 2.5(c). Ignoring the outlier, the range of the 5 responses taken at around $t = 1$ is less than $1/2$ of a unit, from just above zero to less than .5. At time $t = 6$, the observations range from a minimum of around .5 to a maximum near 4. We conclude that the population mean, sd, and range are increasing over time. A rough estimate of the standard deviation is range/4; the range appears to be linearly increasing with time, and we conclude that the range and the sd are increasing in an approximately linear fashion.

We do these sample mean and sd calculations as steps to a further end: the development of a model for the responses as a function of time. For the Big Mice data, we have learned that any model must allow for a population mean and sd that are increasing smoothly with time.

2.3.2 Skewness and the Pediatric Pain Data

Pain is a difficult subject to study because it is hard to design formal experiments if one does not wish to inflict pain on humans; rats cannot tell

42   2. Plots

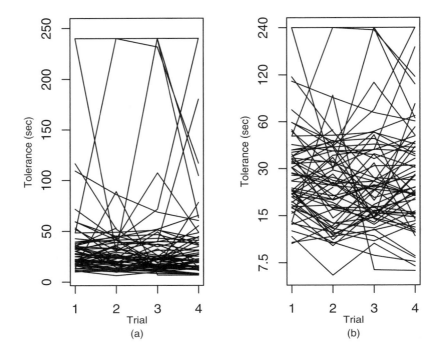

Figure 2.7. Profile plot of the Pediatric Pain data. (a) Original scale. (b) Log scale.

us where it hurts. Most human pain studies are observational. The Pediatric Pain data are unusual in being the result of a designed experiment. The data consist of up to four observations on 64 children aged 8 to 10. The response is the length of time in seconds that the child can tolerate keeping his or her arm in very cold water, a proxy measure of pain tolerance. After the cold becomes intolerable, the child removes his or her arm. The arm is toweled off and no harm is caused. The procedure was considered enjoyable by the children. No one declined to participate in the experiment initially, and no one dropped out for reasons related to the experiment although there is some missing data due to absences and broken arms, common occurrences in children and unrelated to the experiment. Two measures were taken during a first visit followed by two more measures during a second visit after a two-week gap.

During the first visit, the children were classified into one of two groups, *attenders* (A) or *distracters* (D) according to their style of coping (CS) with the pain. The children were asked what they were thinking about during the trials. Those who were thinking about the experiment, the experimental apparatus, the feelings from their arms and so on were classified as attenders. Those who thought about other things, such as the wall, homework from school, going to the amusement park, all unrelated to the experiment, were classified as distracters.

A treatment (TMT) was administered prior to the fourth trial. The treatment consisted of a ten-minute counseling intervention to either attend (A), distract (D), or no advice (N). The N treatment consisted of a discussion without advice regarding any coping strategy. Interest lies in the main effects of TMT and CS and interactions between TMT and CS. Interactions between TMT and CS were anticipated.

The data are plotted in figure 2.7(a), circles for individual observations are omitted. Time is the trial number, ranging from 1 to 4. We see a large mass of observations at times under 50 seconds and relatively sparse data at larger times. This suggests that the data are skewed and that we should consider a transformation. Figure 2.7(b) is the same data on the log scale, labeled with a logarithmic scale on the y axis. The profiles are evenly distributed throughout the range of the data. The data are perhaps slightly more sparse near the top than the bottom, and while we could consider a slightly stronger transformation, the amount of skewness seems minor.

In linear regression, we often use a histogram of our responses to determine if we should transform the data. For longitudinal data, it is not correct to pool all observations into a single histogram. One could draw histograms of the data at a single time point. If we have random times, then we could plot a data set consisting of but one observation per subject, all taken from a narrow window of time. Would it make sense to plot all of the Big Mice data in a single histogram? Two mice would contribute 21 observations and the rest would contribute 7. Consider data like in 2.5(c), with fixed intercept and with the random slopes ranging from 0 up to some positive value. If we had no outlier, and if the study had continued on longer, a histogram of the entire data set would look skewed, yet it would not be correct to transform the data. Histograms of the data in a reasonably narrow window about any time would correctly indicate no need to transform the data.

In figure 2.7(a), there seemed to be a lot of univariate outliers; all the observations above approximately 75 seconds or so. In figure 2.7(b), these high observations seem much less troublesome. Still, a few outliers are visible. The subject with the lowest times at trials 3 and 4 has an unusually high pain tolerance at trial 2. This high trial 2 value causes the line segment between trials 1 and 2 and also between trials 2 and 3 to travel in directions very different from the other line segments between these times. This indicates that the (Y_{i1}, Y_{i2}) pair and the (Y_{i2}, Y_{i3}) pairs for this subject are bivariate outliers. We identify this subject as an overall outlier.

In figure 2.7(a) and also (b), we also see that there appears to be a fixed maximum above which no child scores. Inspection of the data shows that these values are 240.00 seconds, or 4 minutes exactly. Sometimes the investigator may not mention this to the statistician; graphics help us discover these features of the data. The investigators felt that if immersion lasted

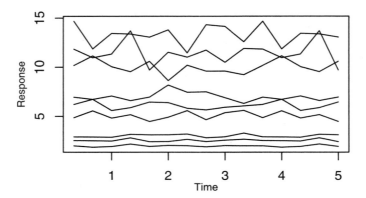

Figure 2.8. Example illustrating different within subject variability. Observations from a single subject have a random intercept and no time trend. Bottom three subjects have observations with the least amount of within-subject variability. Middle three subjects have a middle amount of variability within subject, and the top three subjects have the most within-subject variability.

past 4 minutes, there was no extra information to be gained from allowing the child to keep their arms in longer. This happened in 11 observations out of the 245 total number of observations. The investigators recorded 240.00 seconds as the response in these trials. This *censoring* should be taken account of in the modeling although we do not do this in the analyses presented here.

Figure 2.7(a), is not a beautiful plot; one would never publish it in a medical journal as part of reporting an analysis of this data. Still, this is potentially the single most important step in the analysis of this data! We learned (1) that we should transform the response, (2) that there was possible non-constant variance, (3) that there were some outliers, and (4) that there was a maximum value imposed on the data by the investigators.

2.3.3   Within-Subject Variability

In the Big Mice data, we saw different marginal variances at different times. This was summarized in figure 2.6(b), which plotted standard deviations across-subjects within-time. We also can think about *within-subject, across-time* variability. Figure 2.8 illustrates. The nine subjects each have their own (random) intercept and profiles that have no trend over time. The three subjects with the smallest responses have observations that vary around their means in a tight pattern without much variance. The three subjects in the middle have greater variability around their means, and the three subjects with the highest means have observations with the highest variability around their individual means. The range of the observations within person is low for the subjects with the lowest values; it is middling for the

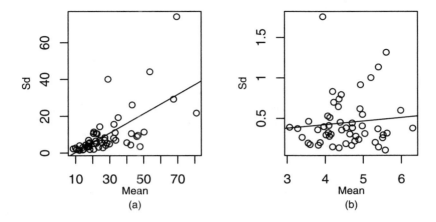

Figure 2.9. Pain data. Plot of standard deviations (y) against means (x) for subjects with 4 observations and no values of 240 seconds. (a) original data, (b) log base 2 data.

subjects with the middle means, and it is highest for the subjects with the highest means.

The Pediatric Pain data in figure 2.7(a) appear to have higher within-subject variability for subjects with higher means, and subjects with lower means appear to have smaller variability. It is somewhat hard to be absolutely certain. Inspection of figure 2.7(b) suggests that the log transformation was useful in eliminating the non-constant within-subject variance along with the skewness. Skewness of the response and non-constant variance are often associated, and it is not surprising that the one transformation does a good job of reducing both.

When profiles are flat or linear, that is, they exhibit a random intercept and do not have a time trend, there is a plot that can help clarify whether within-subject variability increases with the subject-specific mean. For each subject, we can calculate the mean of the n_i observations and the standard deviation of the n_i observations. Then we plot the n standard deviations against the means. Figure 2.9(a) plots, for the Pediatric Pain raw data, the within-subject standard deviations of the four observations on the vertical axis versus the means of the four observations on the horizontal axis. The line is a least squares line drawn through the points without particular regard to assumptions. We see that the standard deviations do definitely increase with the mean. Figure 2.9(b) shows the same plot for the log base 2 data. It shows little if any correlation between the within-subject standard deviation and the subject-specific mean. For both plots, subjects with less than four observations or with a measurement of 240 are not included.

For our Pediatric Pain analyses, we take a log base two transformation of the responses before analyzing the data. The base two log transformation

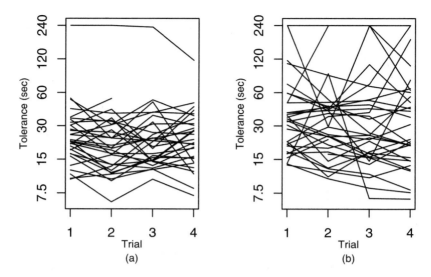

Figure 2.10. Profile plots of the Pediatric Pain data on the log scale separately by coping style: (a) attenders; (b) distracters.

is no different from a base 10 or base e transformation in terms of getting rid of skewness. However, powers of two are much easier to do in our head than powers of e to invert the log transformation and allow us to interpret what a particular log measurement signifies on the original seconds scale. Powers of 10 are also easy to do, but are only useful for data with a wide dynamic range covering several powers of 10.

2.4   Elaborations of Profile Plots

The profile plot is a useful all-purpose tool for understanding longitudinal data. Any truly useful tool develops many variations that are helpful in different circumstances. In this section, we discuss some modifications to the basic profile plot for (i) data sets with covariates, (ii) data sets where the range of the response across subjects obscures the trend of the profiles within subject, and (iii) two kinds of empirical residuals that can be helpful for understanding our data.

2.4.1   Covariates

So far we have plotted entire data sets in our profile plots. It is not required that we include the entire data set in a single plot, and we may use our creativity to determine when it might be helpful to look at a subset of the data in a single plot, or to look at all of the data but spread across several different profile plots.

When we have a discrete covariate, we may plot subjects with different values of the covariate in separate plots. Figure 2.10 illustrates this for the Pain data and the coping style covariate. The left-hand plot shows subjects who are attenders and the right-hand plot shows distracters. Compared with figure 2.7(b), there are fewer subjects plotted on the same plot, and we can more easily distinguish individual profiles of subjects.

Figure 2.10 reveals several interesting features of the data. In figure 2.10(a), most of the subjects range uniformly on the log scale from approximately 7.5 seconds up to slightly below 60 seconds. There is one exceptional subject that we identify as a high outlier. In figure 2.10(b), the distracters range uniformly from around 7.5 seconds up to 240 seconds. A number of distracters have observations over 60 seconds. The average of the attenders appears to be approximately half way between 15 and 30 seconds. Because this is a logarithmic scale, that point is at $\sqrt{(2)} \times 15 \approx 1.4 \times 15 = 24$ seconds. The average of the distracters appears to be around one quarter of the way between 30 and 60, which puts it at around $2^{1/4} \times 30 \approx 36$ seconds. We conclude that distracters have greater average pain tolerance than attenders.

We could have used different line types for the two groups and plotted all subjects on the same plot. This works reasonably well when the groups are well separated in their responses or if there are very few subjects. The Pain data have a bit too many subjects for separate line types to be helpful.

With a continuous covariate, we might slice the covariate into a small number of intervals and create separate profile plots for subjects with covariate values that fall into each interval. A common way of slicing continuous variables is called a *median split*. Subjects with covariate values above the median form one group, and those with values below the median form a second group. All subjects who fall at the median, if they exist, may go all together into either group. We do a median split when there is no particular scientific rationale for splitting the covariate at some other value.

2.4.2   Ozone Data

Ozone is an invisible pollutant that irritates the lungs and throat and causes or exacerbates health problems in humans. Crops may grow less if exposed to excess ozone, and chemical products such as paint may degrade when exposed to ozone. The Ozone data set records ozone over a three-day period during late July 1987 at 20 sites in and around Los Angeles, California, USA. Twelve recordings were taken hourly from 0700 hours to 1800 hours giving us $20 \times 12 \times 3$ ozone readings. Measurement units are in parts per hundred million. Table 2.1 gives the four-letter abbreviation for the sites, the full names of the sites, and the longitude, latitude, and altitude

Site abbr	Site name	Long	Lat	Altitude	Valley
SNBO	San_Bernadino	117.273	34.107	317	SG
RIVR	Riverside	117.417	34	214	SG
FONT	Fontana	117.505	34.099	381	SG
UPLA	Upland	117.628	34.104	369	SG
CLAR	Claremont	117.704	34.102	364	SG
POMA	Pomona	117.751	34.067	270	SG
AZUS	Azusa	117.923	34.136	189	SG
PASA	Pasadena	118.127	34.134	250	SG
BURK	Burbank	118.308	34.183	168	SF
RESE	Reseda	118.533	34.199	226	SF
SIMI	Simi_Valley	118.685	34.278	310	SF
ANAH	Anaheim	117.919	33.821	41	No
LAHB	La_Habra	117.951	33.926	82	No
WHIT	Whittier	118.025	33.924	58	No
PICO	Pico_Rivera	118.058	34.015	69	No
LGBH	N_Long_Beach	118.189	33.824	7	No
LYNN	Lynwood	118.21	33.929	27	No
CELA	Central_LA	118.225	34.067	87	No
HAWT	Hawthorne	118.369	33.923	21	No
WSLA	West_LA	118.455	34.051	91	No

Table 2.1. The Ozone data: general information about sites. The first five columns are the site abbreviation, full name, longitude, latitude, and altitude. Valley is whether the site is in either the San Fernando or Simi Valleys (SF) or San Gabriel (SG) valley. Other sites are adjacent to the ocean without intervening mountain ranges. Abbreviations in names: N North; LA usually means Los Angeles; except La Habra is La Habra.

of each site. Also given is a valley indicator to indicate whether the site is in the Simi or San Fernando Valleys (SF) or San Gabriel Valleys (SG). The remaining sites are adjacent to the ocean or otherwise do not have mountain ranges between them and the ocean. The data was originally collected to compare to output of computer simulations of atmospheric chemistry in Los Angeles. Figure 2.11 shows a map of the site locations. The San Gabriel Valley is on the right or east on the plot. Simi Valley is the left most or western most site and is adjacent to the San Fernando Valley sites of Reseda and Burbank. Each night ozone returns to a baseline value, and we treat the data as having $60 = 20 \times 3$ subjects with 12 longitudinal measures each.

Figure 2.12 plots ozone profiles for the sites separately by day. We see that on day 1 ozone generally increases monotonically up to a peak between 2pm and 4pm before beginning to decrease slightly. There are a range of ozone levels. The ozone peaks appear to be increasing from day 1 to day 2 to day 3. It may be that the peaks are slightly later on day 3 than on day

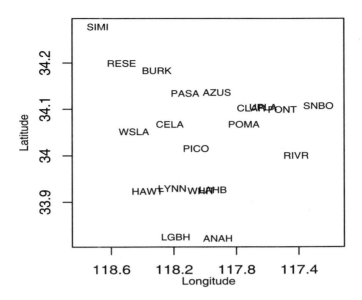

Figure 2.11. Map of Ozone data sites. CLAR overlaps UPLA, which overlaps FONT, and WHIT overlaps LAHB. To place west on the left, longitude increases right to left.

1. The lowest ozone levels appear to be similar across the three days. From this plot we do not know if the same sites are lowest on each of the three days.

Figure 2.13 plots the profiles by site. Sites are ordered by the maximum ozone value over the three days. The individual plots are arranged starting at the bottom left and moving left to right and then from bottom to top. We notice that the sites with the largest ozone concentrations are all in the San Gabriel Valley. The next two sites with high ozone are in the San Fernando Valley. One site not in a valley has higher ozone than Simi Valley, the last valley site. We lumped Simi Valley with the San Fernando Valley to avoid having only one site in that category, but technically it is a different valley from the San Fernando Valley. The sites with the lowest ozone values appear to be rather similar over the three days, while the middle and higher ozone sites have different peak ozone levels over the three days.

We can even plot profiles a single profile per plot and look at as many subjects as we have patience for. In olden times, statisticians would print out a single subject's profile on separate pages, then shuffle the pieces of paper around on a table top like a jigsaw looking for similar patterns among different subjects. In figure 2.14, we print 18 of the 60 cases, with the other 42 on additional pages that are not shown. We now can inspect the patterns

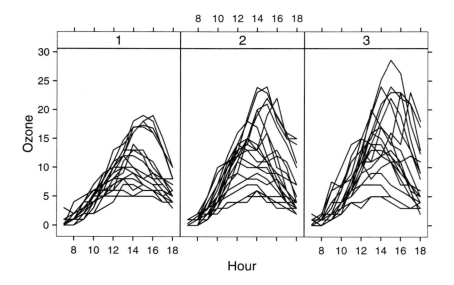

Figure 2.12. Profile plots of the Ozone data separately by days. Days are ordered 1, 2, and 3 from left to right.

of ozone over day as well as over time within a site. In these six sites, we see that the ozone peak is generally increasing from each day to the next, and we again suspect that the peak is moving later within the day from day 1 to day 2 to day 3.

2.4.3   Weight Loss Data and Viewing Slopes

The Weight Loss data consist of weekly weights in pounds from women enrolled in a weight loss trial. Patients were interviewed and weighed the first week and enrolled in the study at the second week. The data from 38 women are plotted in figure 2.15. There are from 4 to 8 measurements per subject. Weights range from roughly 140 pounds to 260 pounds.

Study protocol called for the subjects to visit the clinic at weeks 1, 2, 3, and 6 and weigh themselves on the clinic scale. At weeks 4, 5, 7, and 8, study personnel called subjects at home and asked them to weigh themselves on their home scales and report the measurement. Week 1 was a screening visit; participation in the actual weight-loss regimen did not start until week 2.

Figure 2.15 presents a profile plot of the Weight Loss data. Unfortunately little structure is visible, except for the numerous parallel profiles. This indicates the not surprising result that each woman has her own average weight and her weight varies around that weight over time; this data illustrates a random intercept. We see one slightly heavy subject and one

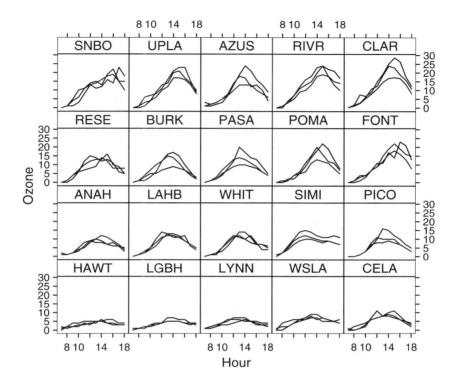

Figure 2.13. Profile plots of the Ozone data separately by site. Sites have been ordered from left to right and bottom to top by their maximum ozone reading over all hours and days.

slightly light subject. It is not obvious whether women are losing weight or not based on this figure.

The problem with figure 2.15 is that we cannot see slopes of individual profiles. Suppose someone lost 5 or even 10 pounds over the 8 weeks; that slope would barely be visible in the plot; a slope of $-1/2$ to -1 pound per week is scientifically quite high yet it would be nearly indistinguishable from a flat profile with slope 0. We want to see slopes of magnitude $-1/2$ to -1, and the question is how to draw the plot so that we can actually see slopes of reasonable magnitude. If we made the figure taller, then a 5 or 10 pound difference would become physically larger on the printed page, and we will be more likely to be able to see it in the plot. The second thing we can do is to make the figure narrower! By narrowing the x axis, we increase the angle of the slopes, so that a slope of -1 pound per week appears steeper on the plot.

Figure 2.15 is square, this is the default shape produced by most statistical software. Instead, if we give instructions to the software to make the

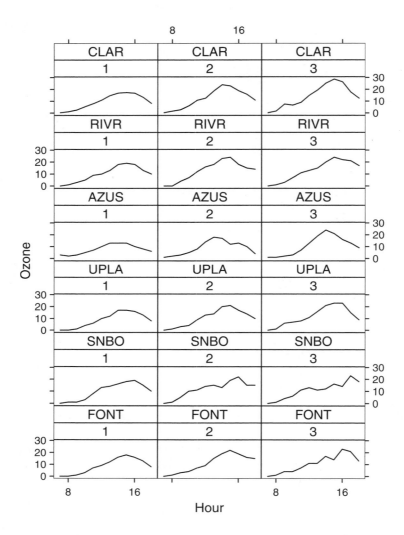

Figure 2.14. Profile plots of Ozone data by day and site. Each row shows profiles from the same site with day increasing left to right. The 6 sites with the highest ozone levels are shown. This is slightly less than 1/3 of a larger display (not presented).

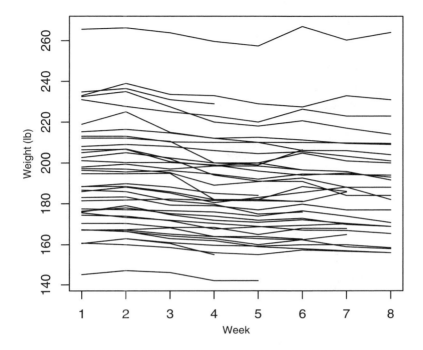

Figure 2.15. Profile plot of the Weight Loss data.

plot taller and narrower, something like either half of figure 2.16 results. In fact, because the taller, narrower figure is larger than the printed page, I broke the y axis in half and plotted the lower half of the data from below 140 pounds up to over 200 pounds in one plot on the left and the upper half of the data from approximately 200 pounds to the maximum on the right. Some data has been plotted in both figures, roughly in the range from 198 to 210 pounds. The *shape* of the plot has been modified so that we can see the trends in the weights. Now we can see that people are losing weight; the observation at week 8 generally appears to be lower than that subject's corresponding observation at week 1 or 2. Another feature of the data is also slightly visible. A number of profiles take a steep drop around weeks 4 and 5 then return to a higher level at week 6, and this seems more pronounced for heavier subjects.

Changing the shape of the plot is often necessary when plotting longitudinal data. When the range of the response *across* all of the subjects is large, but the range *within* subjects is small in comparison, then we often have difficulty seeing the time trends of individual subjects on a plot like 2.15. In contrast to the Weight Loss data, the Big Mice data have a large response range within subjects that is almost the same as the response

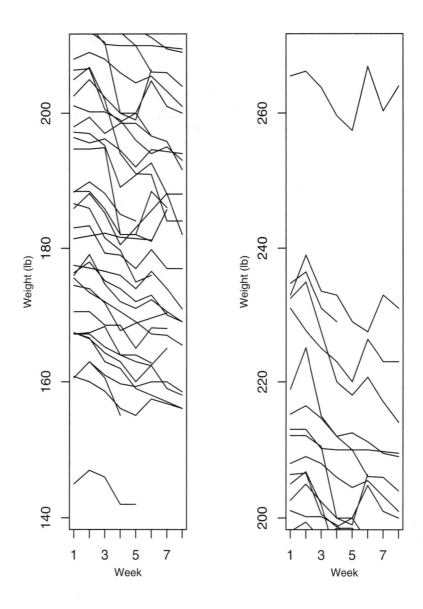

Figure 2.16. Weight Loss data with appropriate shape parameter.

range across subjects. Unfortunately for the Weight Loss data, we would like to make the plot impractically taller. The next subsection provides another solution to the problem of viewing the slopes of individual profiles.

2.4.4   Empirical Within-Subject Residuals

We always begin data analysis with a figure like 2.15. However, this plot was not particularly successful in terms of viewing slopes over time because of the large range in the responses. Reflection suggests that we are not very interested in the absolute weights of the subjects. Rather, we are interested in changes in weight over time. What could we do to focus on the weight changes while not worrying about absolute weight levels?

Ignoring figure 2.16 for now, and just looking at figure 2.15, we see that each subject appears to have her own average weight, and weekly observations vary around these averages. If we estimate the average weight, we can calculate the individual deviations around the average and then consider plotting those in a profile plot. A simple estimator of the intercept for each subject might be the subject's average response. In chapter 9, we will learn about models that produce better estimates of the intercept for each subject. Until then, we consider the subject average $\bar{Y}_i$ as an estimator of the subject's average weight. The difference between the jth observation and the subject mean

$$R_{ij} = Y_{ij} - \bar{Y}_i$$

is an empirical within-subject residual.

Figure 2.17 plots the R_{ij} in a profile plot. At weeks 1 and 2, we see that most residuals R_{ij} are greater than zero and some are as large as 10 pounds, that is, most subject's weights are above their average weight. At weeks 7 and 8, most subject's weights are below their average. Thus we see that yes, subjects do lose weight over the course of the study.

From week 1 to week 2, no weight is lost, if anything a little weight is gained. From week 2 to week 3, the first week of the weight loss treatment, the subjects lose quite a lot of weight. At weeks 4 and 5 they continue to lose weight with a few losing a substantial amount between weeks 3 and 4. At week 6, suddenly, weight is gained, presumably because patients are weighed under supervision with a properly calibrated scale. At weeks 7 and 8, they continue to lose weight. The overall weight loss indicates that a fixed time effect is needed in the model. It is not clear whether the weight loss is strictly linear, an issue we leave for later.

We could consider subtracting off other subject relevant weights such as subject baseline rather than subject average weight from each observation and plot the resulting changes from baseline in a profile plot. Exercise 18 explores what the Weight Loss profile plot looks like if we subtract off the baseline weight measurement instead of the subject average. Exercise 20

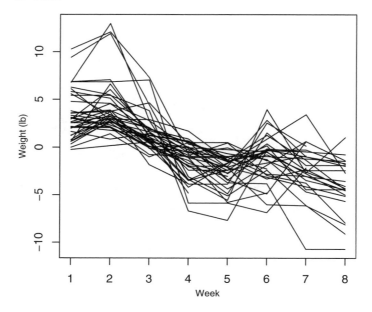

Figure 2.17. Profile plot of empirical within-subject residuals for the Weight Loss data.

explores what happens to the Weight Loss profile plot if we plot differences from one time to the next $Y_{ij} - Y_{i(j-1)}$.

2.4.5 Empirical Population Residuals

Sometimes we wish to see the variation across subjects within a particular time. If there is a large change in the average response over time, then it may be hard to view the individual subject profiles, for example to see if profiles are parallel, or to see if the marginal variance is increasing. In figures 2.1, 2.2, and 2.3, the marginal variance grows quickly initially then appears to stop growing. We can get a better look at the profiles by subtracting off an estimate of the mean at each time point to look at the deviations from the mean. For the Big Mice data, define the sample mean $\bar{Y}_{.j}$ at time j as the mean of all observations at day j and define the empirical population residuals

$$U_{ij} = Y_{ij} - \bar{Y}_{.j}.$$

In figure 2.18, we can now see the bulk of the data better, and we can better see individual profiles. The range of the y axis is approximately 500 mg rather than 1200 mg of figure 2.3. It is easier to tell relative high or low and by how much within a time and to follow the paths of individual

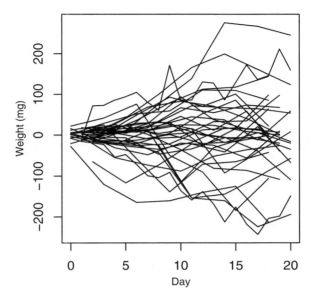

Figure 2.18. Profile plot of empirical population residuals for the Big Mice data.

mice. We lose the population time trend in this plot, the y axis represents differences from the population mean. Positive residuals with a decreasing trend means that the mouse is getting closer to the population mean, not that the weight is decreasing.

2.4.6   Too Many Subjects

Profile plots go by several other names including *spaghetti plots*, *parallel plots*, and *connect-the-dots plots*. The spaghetti plot name describes what profile plots look like when too many profiles are plotted in a single plot. The ink density destroys many features of the data. This can be overcome in several ways. Using a higher quality graphing package and printer can do wonders. Alternatively, a subset of subjects may be plotted. The subset may be a randomly selected subset or, as we did in section 2.4.1, subsets may be specified according to the values of some covariate or, as with the ozone data, we might plot all subjects separately.

2.5   Inspecting Correlations

Longitudinal data are different from linear regression because observations are correlated within subjects. The correlations among observations needs

	Trial			
Trial	1	2	3	4
1	1	.73	.84	.60
2	.73	1	.72	.66
3	.84	.72	1	.76
4	.60	.66	.76	1

Table 2.2. Correlation matrix for the Pediatric Pain data.

to be modeled, and it helps to have summary measures and graphics that help us decide on the covariance model.

The correlations $\hat{\rho}_{jk}$ and variances s_j^2 form many patterns in different data sets; there is not one pattern that describes all data sets. We want to inspect the correlations among our observations to help determine the type of model for the correlations that we will use. A simple and useful summary of the correlations among our longitudinal observations is a table of those correlations.

Table 2.2 gives the correlations among the Pediatric Pain observations. Correlations of observations with themselves are 1, so 1's go down the long diagonal. The correlation between observations at trial 1 and trial 2 is $\hat{\rho}_{12} = \hat{\rho}_{21} = .73$. The pain correlations vary from .60 to .84, with the lowest correlations $\hat{\rho}_{14} = .60$ and $\hat{\rho}_{24} = .66$ being trial 4's correlations with the trial 1 and trial 2 observations. Correlations between trial j and k were calculated by using all subjects that had both observations at times j and k.

Although somewhat different, these 6 correlations are not wildly different, and we might initially consider a model where all correlations among observations are the same. Seeing no other pattern, as an alternative model we might consider a model where all the correlations are different.

An estimate of the uncertainty in a correlation can help with judging the differences in the correlation values. The estimated standard error of a simple correlation is

$$\text{SE}(\hat{\rho}) = \frac{1 - \hat{\rho}^2}{(n-3)^{1/2}}$$

where n is the number of subjects contributing pairs of observation to the computation. For the Pain data, the number of pairs of observations contributing to each correlation ranges from 58 for all correlations involving trial 3 to 62 for $\hat{\rho}_{12}$. The range of standard errors is .04 to .08 for the correlations in table 2.2. We do not have a simple test for the equality of two correlations, but it seems reasonable that differences in correlation less than .1 are not very important. Still, the largest difference in correlations is .24 and that may be significant, suggesting that the six correlations in table 2.2 may be different.

For the Big Mice data, we can calculate sample correlations for any pair of days provided that either group 1, group 2 or group 3 mice were

	2	5	8	11	14	17	20
2	1	.92	.57	.36	.23	.23	.38
5	.92	1	.77	.54	.45	.41	.55
8	.57	.77	1	.86	.80	.76	.81
11	.36	.54	.86	1	.93	.92	.87
14	.23	.45	.80	.93	1	.96	.89
17	.23	.41	.76	.92	.96	1	.92
20	.38	.55	.81	.87	.89	.92	1

Table 2.3. Correlation matrix for the Small Mice data.

measured on both of the days. The group 4 mice do not give us enough observations to calculate correlations for other days. Table 2.3 gives the sample correlations for the Small Mice data based on 14 observations. In inspecting correlation tables like this, we look at the rows beginning at the long diagonal and at the columns also beginning with the long diagonal. We look along diagonals parallel to the long diagonal, looking for simple patterns in the correlations.

Each diagonal away from the long diagonal corresponds to a given *lag* between observations. The first off-the-main diagonal gives the lag 1 correlations; observations being correlated are consecutive observations. The correlation between day 2 and day 5 observations is $\rho_{12} = .92$, indicating a strong relationship between weights at those two early days. The second long diagonal, beginning with correlations .57, then .54 and ending with .89 are the lag 2 correlations; .54 is the correlation between day 5 and day 11 observations. And it continues, until the correlation .38 is the sole lag 6 correlation, the correlation between observations at day 2 and day 20.

To begin more detailed analysis of table 2.3, we inspect the longest and first off diagonal, the lag one diagonal, with correlations of .92, .77, .86, .93, .96, and .92. Although these are not all exactly equal, they are all quite similar, possibly excepting the .77, which is a tad lower than the others. The standard error (se) of .92 is .04 while the SE of .77 is .12. We possibly hypothesize that the lag 1 correlations are all equal, or, approximately equal. Next we go to the diagonal two away from the long diagonal, hoping that this pattern of near equality continues. Here the values start out lower, .57 and .54, then abruptly increase, ranging from .86 to .96, so that either we have increasing correlations along the lag 2 diagonal, or we have two low then three high correlations. The third off diagonal starts low at .37, then steadily increases. The first two values are low, the last two are high, at .76 and .87.

Continued inspection suggests perhaps two groups of observations, the early observations and the late observations. The first four observations at times 2, 5, 8, and 11 have a pattern that has high lag 1 correlations, middling values of lag 2 correlations, and a low lag 3 correlation. The last four observations from day 11 to day 20 have all high correlations,

	7	8	9	10	11	12	1	2	3	4	5	6
7	.63	.27	-.02	-.17	-.24	-.26	-.26	-.27	-.31	-.30	-.26	-.24
8	.13	.73	.53	.33	.03	-.10	-.10	-.19	-.20	-.13	-.07	.03
9	-.02	.56	1.44	.75	.49	.32	.24	.22	.29	.38	.39	.47
10	-.20	.45	2.02	1.87	.79	.59	.39	.25	.33	.37	.37	.47
11	-.37	.05	1.71	3.61	2.43	.86	.60	.45	.51	.49	.51	.54
12	-.54	-.25	1.52	3.61	6.86	3.26	.83	.70	.68	.64	.59	.51
1	-.67	-.30	1.43	3.00	5.95	11.1	4.08	.89	.81	.73	.62	.50
2	-.90	-.75	1.66	2.48	5.86	12.1	19.5	5.35	.94	.86	.77	.63
3	-1.18	-.87	2.55	3.73	7.53	13.5	20.1	30.3	6.07	.93	.84	.75
4	-1.11	-.54	3.18	4.01	7.03	12.3	17.5	26.7	32.9	5.85	.91	.81
5	-.82	-.25	2.81	3.54	6.28	9.75	12.8	20.7	25.7	26.8	5.05	.93
6	-.59	.08	2.68	3.48	5.15	6.61	8.07	13.2	17.9	18.6	18.5	3.94

Table 2.4. Correlation/covariance matrix for the Ozone data. Above the long diagonal are the sample correlations, below the diagonal are sample covariances, and along the diagonal (boxes) are the sample standard deviations.

ranging from .87 to .96. The cross-correlations between the early and late observations generally follow the pattern that the closer in time, the higher the correlation, but that all the later observations have, roughly speaking, similar correlations with any given early time. Day 8 has high correlations with the later observations, but has that mildly lower correlation with day 5, so day 8 may be the dividing day between the early and the late observations.

The lower and upper half of the correlation matrix are the same, and sometimes we omit the lower or the upper half of the matrix. Instead of the format of table 2.2 or 2.3, we can pack information into the table by placing the sample standard deviations down the long diagonal, covariances below (or above) the long diagonal, and correlations in the other half of the table. This is illustrated in table 2.4 for the Ozone data.

The Ozone data standard deviations, correlations and covariances are more complex than the Pain data. The sample standard deviations begin at a very low level in the morning and steadily increase until 4pm in the afternoon, then decrease slightly by 5pm or 6pm. The lag one correlations start low at .27 then increase rapidly to .53, .75, and so on to over .9, and stay high and roughly constant through the end of the data. The lag two correlations start even lower at −.02, then increase to correlations in the .8's, not quite as high as the lag one correlations. The higher lag correlations exhibit a similar pattern, except that the correlation between the first and second observations with the remaining observations remain modest and negative. Along rows, from the row for noon and rows for later times, we see strictly decreasing correlations as the lag increases. For morning rows, at 11am we see a high lag one correlation, then decreasing to a constant correlation. For 9am and 10am, the correlation starts high, decreases to a lower constant correlation and then creeps up slightly at the end, for 7am and 8am, the correlation decreases to negative(!) correlations for most of the day, before starting to creep back up at the end. To summarize, for constant lag, the correlation starts low, then increases to some maximum.

We illustrate two common types of correlation matrices in tables 2.5 and 2.6. In table 2.5, the correlations with constant lag are the same. The lag one correlations are .90, the lag two correlations are all .81, and the lag three

| | Trial | | | |
Trial	1	2	3	4
1	1	.9	.81	.73
2	.9	1	.9	.81
3	.81	.9	1	.9
4	.73	.81	.9	1

Table 2.5. Example correlation matrix illustrating banded correlations.

| | Trial | | | |
Trial	1	2	3	4
1	1	.85	.84	.86
2	.85	1	.87	.87
3	.84	.87	1	.84
4	.86	.87	.84	1

Table 2.6. Example correlation matrix illustrating approximately equal correlations at all lags.

correlation is .73. Constant correlation for a given lag is called a *banded* correlation matrix; many important correlation structures are banded. As the lag increases, the correlations decrease, and in this particular example, they decrease in a nearly geometric fashion, with $.81 = .9^2$, and approximately $.73 \approx .9^{|4-1|} = .9^3$. We see a decrease in correlation with increasing lags in the mice data and in the Ozone data, but neither example appears to illustrate *banding*.

Table 2.6 illustrates a correlation matrix with approximately equal correlations for all pairs of observations. This would, at least approximately, be called an *equicorrelation* correlation matrix. An equicorrelation correlation matrix says that the lag does not matter in calculating the correlations; no matter how distant in time two observations are, they have the same constant correlation. Equicorrelation is of course a special case of banding, but usually we intend the term banded to mean correlations that are not constant for different lags.

2.5.1 Scatterplot Matrices

A scatterplot of two variables is a graphical illustration of the correlation between the two variables. Additionally it shows whether the relationship between the variables is linear, and whether there are outliers, clusters or other deviations from normality in the data. When we have multiple variables to plot, there are many scatterplots to look at; for J variables, we have $J(J-1)/2$ pairs of variables and in each pair either variable may be on the vertical or horizontal axis. A *scatterplot matrix* organizes all of the pairwise scatterplots into a compact arrangement. For longitudinal data,

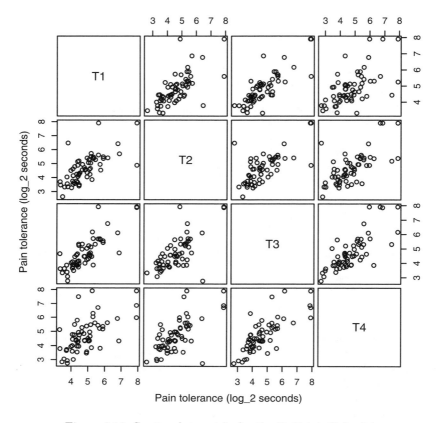

Figure 2.19. Scatterplot matrix for the Pediatric Pain data.

we require balanced or balanced with missing data to conveniently produce a scatterplot matrix.

Figure 2.19 is a scatterplot matrix of the Pain data with responses on the log base two scale. There are 12 scatterplots in the figure. Each plot in row j counting from the top has the response from trial j on the vertical axis. The three plots in a row share the same axis tick marks and tick labels, given either on the left- or right-hand side of the figure. Similarly each plot in a column shares the same variable on the horizontal axis and the same tick marks and labels given either above or below the column. The (j, k)th plot and the (k, j)th plot plot the same data, but with the vertical and horizontal axes reversed. The figures are arranged with the long diagonal going from the upper left to lower right, the same as our correlation matrix in table 2.2. Sometimes the long diagonal goes from lower left to upper right. All i subjects contribute one point to each plot unless they are missing either the jth or kth trial.

The tick mark of 3 $\log_2$ seconds corresponds to $2^3 = 8$ seconds, and the tick marks of 4, 5, 6, 7, and 8 on the log scale correspond to 16, 32, 64, 128, and 256 seconds, respectively. Different rows do not share exactly the same scales or ranges on the axis, although they will be similar because the range of the data at each trial are nearly the same.

A key feature of the scatterplots in figure 2.19 is that the relationship among the pairs of responses is linear. This is important, as it is a major part of the normality assumption that we will use in our analyses. We also see that the observations are generally elliptically distributed, but that there is a halo of points scattered mostly at the higher values; there are some outliers in this data. The correlations in figure 2.19 look approximately equal to us; with experience one can develop the ability to accurately estimate correlations from bivariate normal data to within a few percent. Under closer inspection, it appears perhaps that the lag one correlations are definitely all similar, and that the lag three correlation is lower than the lag 1 correlations. The lag two correlations are difficult to determine, but perhaps, matching table 2.2, the plot of trial 4 against trial 2 is also of lower correlation, whereas trial 3 against trial 1 is of equal or slightly higher correlation to the lag 1 plots.

Figure 2.20 gives the scatterplot matrix for the Small Mice data corresponding to the correlation matrix in table 2.3. The data set is quite small and we often have trouble with identifying both absolute and relative correlations with small sample sizes. Still we see that the lag 1 correlations are all quite high except the plot of days 5 against 8, which seems lower than the other lag 1 plots. Among later days, 8–20, the higher lag plots still have fairly strong positive correlation although the correlation does decrease with increasing lag. Observations from the early days 2 and 5 have low correlations with the observations at later days. We cannot tell the exact values of these correlations, and a correlation less than .4 can be hard to distinguish from independence without actually formally calculating it.

We can also create a scatterplot matrix from randomly spaced data by *binning* the times into convenient intervals and identifying all observations in the same bin as coming from a single *nominal* time. For observations nominally scheduled for every three months, the actual times may vary around the nominal date. We might still use nominal times for plotting in a scatterplot matrix, with the understanding that the variability in times may affect the figure somewhat.

2.5.2   Correlation in Profile Plots

We can identify correlations from profile plots as well. It is easiest to identify the correlation between neighboring observations.

The set of line segments $i = 1, \ldots, n$ between consecutive observations $Y_{i(j-1)}$ and Y_{ij} show the correlation between observations at consecutive times. Figure 2.21(a) illustrates a range of positive correlations between

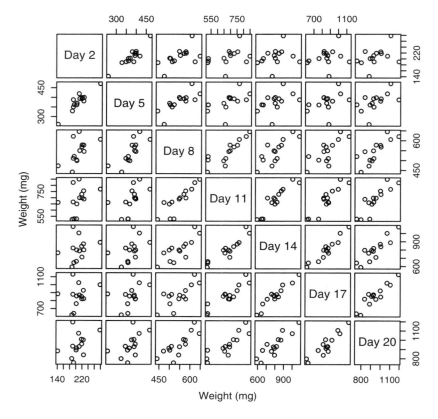

Figure 2.20. Scatterplot matrix for the Small Mice data.

consecutive observations ranging from $\rho = .99$ at the left to $\rho = 0$ at the right, and 2.21(b) shows a range of negative correlations from $\rho = -.99$ at the left to $\rho = 0$ at the right. Both figures 2.21(a) and 2.21(b) show profiles of a sample of 20 subjects observed at times $t = 1, 2, \ldots, 8$. The population mean is zero and the population standard deviation is 1 at all times. The population correlation between time t and $t + 1$ is given at the bottom of the plot. In figure 2.21a, the correlation between observations at $t = 1$ and $t = 2$ is .99. The line segments between $t = 1$ and $t = 2$ rarely cross, and this is indicative of a high positive correlation. The correlation between $t = 2$ and $t = 3$ is lower at $\rho = .95$ than the correlation between the first two times, and there is more crossing of the profiles. As t increases, the correlation between the observations at t and $t + 1$ decreases, and the amount of crossing of the profiles increases from left to right. The last pair of observations are uncorrelated with $\rho = 0$ between observations at $t = 7$ and $t = 8$, and the profiles are at their most haphazard.

In figure 2.21(b), the correlation decreases in absolute value from left to right, but this time the correlations are negative. The crossing of the line segments between observations increase from the right-hand side to the left-hand side. The difference from right to left is that the crossing is less and less haphazard and becomes more and more focused in a smaller and smaller region as the negative correlation increases. Between $t = 1$ and $t = 2$, the correlation is highest, and the line segments all intersect in a very narrow region near a point $(1.5, 0)$. Exactly where this point is in general depends on the mean and variances of the two observations but the intersection point is between the two times when the observations are negatively correlated and is outside the two times when the correlation is positive. If the points were perfectly negatively correlated, then all line segments would intersect exactly at a single point.

When the correlation between consecutive points is positive, the line segments between observations tend not to intersect. The stronger the correlation the fewer the intersections. The line segments will not be parallel unless the variances at the two times are equal. Generally for positively correlated data, the line segments would intersect if we were to extend the lines out toward the direction of the time with the smaller variance. The closer the variances, the farther we must extend the lines to see the intersections. And if the variances are equal, the lines are parallel and the intersection points go out to infinity.

2.5.3   The Correlogram

For equally spaced data, an empirical correlogram is a plot of the empirical correlations ρ_{jk} on the vertical axis against the lag $|j - k|$ on the horizontal axis. If our observations are balanced but not equally spaced, we might instead plot ρ_{jk} against $|t_j - t_k|$. As when we wish to draw a scatterplot matrix, we must bin the data for randomly spaced data to create a correlogram. Various enhancements to the basic correlogram plot are possible. Figure 2.22 shows a correlogram for the Ozone data. Correlations ρ_{jk} whose j are equal are connected by line segments. This correlogram tells us the same information as the correlation matrix in table 2.4, in a different form. It is easy to see that correlations decrease with increasing lag, and that they tend to level out once the lag reaches 3 or 4, and that the correlations may even begin to increase for still greater lags.

2.6   Empirical Summary Plots

An empirical summary plot presents information about the average response over time. One simple way to estimate the mean at a given time is to take the average $\bar{Y}_{\cdot j}$ of all observations at a given time t_j. We can plot

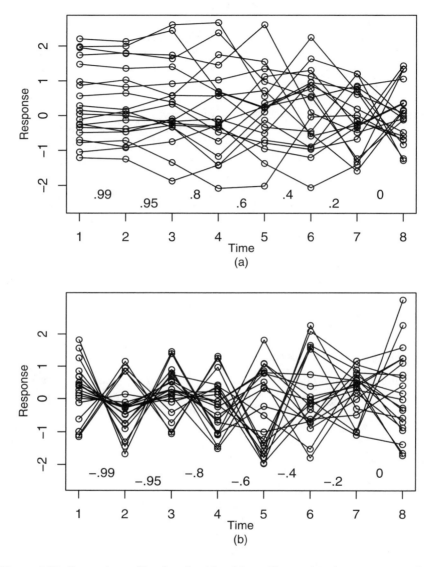

Figure 2.21. Example profile plots for 20 subjects illustrating decreasing correlations between consecutive observations. Part (a) shows positive correlations .99, .95, .8, .6, .4, .2, 0 between consecutive times. Part (b) shows negative correlations −.99, −.95, −.8, −.6, −.4, −.2, and 0.

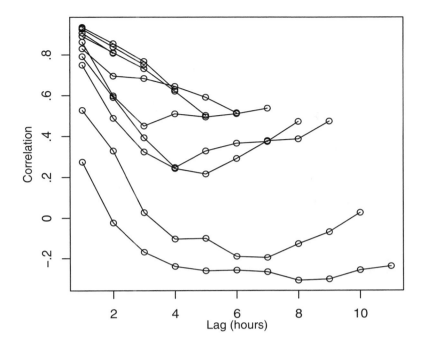

Figure 2.22. Correlogram for the Ozone data.

the $\bar{Y}_{.j}$ against t_j; usually we connect the dots between consecutive time points, as with the profile plots. When we plot the means $\bar{Y}_{.j}$, we often want to show a measure of uncertainty. When we average n independent observations with sample standard variance $s_{jj}^2 = (n-1)^{-1} \sum_{i=1}^{n} (Y_{ij} - \bar{Y}_{.j})^2$, then the standard error of the mean is $\mathrm{SE}(\bar{Y}_{.j}) = n^{-1/2} s_{jj}$. We may plot the means along with error bars that illustrate plus and minus 1 standard error to show the size of the standard error. More often we plot plus and minus 2 standard errors to show approximate 95% confidence intervals around $\bar{Y}_{.j}$, which we call an *empirical summary plot*. A third possibility is to show an interval that covers most of the data. We might show error bars that are plus and minus 2 sample standard deviations, $\pm 2s_{jj}$. This is an approximate 95% prediction interval and we call this plot an *empirical prediction plot*.

Figure 2.23(a) gives an empirical summary plot, and 2.23(b) gives an empirical prediction plot for the Big Mice data. The prediction intervals are much wider than the inference intervals; predictions address the prediction of a new observation with all the variability an individual observation has.

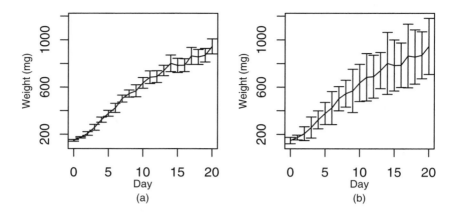

Figure 2.23. (a) Empirical summary plot and (b) empirical prediction plot for the Big Mice data.

The empirical summary interval is making an inference about the *average* response at a given time.

We often wish to distinguish between subjects with different covariate values when we plot empirical summary plots. We may plot the mean profiles from different covariate values either in different plots, or on the same plot, slightly offset from each other so that neither set of intervals obscures the other. Figure 2.24(a) illustrates inference profiles for the attenders and distracters on a single plot. The error bars for attenders and distracters are slightly offset from each other to avoid overplotting. The units are log base two seconds. The means and standard deviations at each time are calculated using either attender or distracter observations at the given time.

Figure 2.24(b) is a back-transformed version of the Pain data empirical summary plot. The means and the interval endpoints have been transformed back to the original seconds scale prior to plotting. The log and back-transformed plots look similar visually. The advantage of the original seconds scale is that we can read off numbers in convenient units for the centers and endpoints of intervals.

For a continuous covariate, we might do a median split before creating our empirical summary plots. Then we create two of the desired plots, one for subjects above the median and one for subjects below the median.

When we have substantial missing data, we must be careful in drawing conclusions from empirical summary plots; we need to try to confirm that observations are not missing differentially in one group or another, or that high or low observations are not differentially missing. Chapter 12 discusses this at length. Fitting a statistical model to the data and then

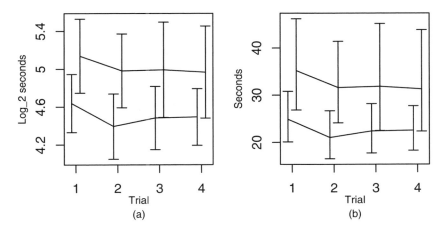

Figure 2.24. Empirical summary plots on the (a) log base two seconds scale and (b) back-transformed seconds scale for the Pain data, separately by coping styles. The upper intervals are for distracters.

plotting an inference plot based on the fitted model can sometimes, but not always, overcome the problems caused by missing data. Thus the empirical summary plot is something we draw early in a data analysis to help us understand the general population time trend and what the effects of covariates might be on the mean. It is not necessarily a good source of conclusions.

2.7   How Much Data?

With longitudinal data, we need to understand how much data we have. How many observations do typical subjects have? For otherwise balanced data except for some missing observations, we tabulate the number of missing (or observed) observations at each time point. An observed observation sounds redundant. Here, observation refers to data we intended to collect by design, and observed means that we actually collected that data from the subject. Similarly, a missing observation is an observation we intended but failed to collect. For randomly observed data there are no missing or observed observations, rather, there are just the observations that we managed to collect. For data with actual times that can be different from the nominal times, we compare the two sets of times to see how they differ. A histogram of actual times can be helpful to give an idea of the times of observations.

Round	# obs
1	543
2	510
3	509
4	497
5	474

Table 2.7. At each round of the Cognitive data, the number of subjects with a Raven's score.

n_i	# obs
0	7
1	8
2	21
3	16
4	40
5	455

Table 2.8. Number of Kenya subjects with from 0 to 5 Raven's observations.

2.7.1   Cognitive Data: Raven's

The Cognitive data is from a school lunch intervention in rural Kenya. The school lunch intervention began at time $t = 0$ in 9 out of 12 schools in the study. Students at the other three schools formed a control group. A number of different measurements were taken. Here we study a particular Cognitive measure called Raven's colored matrices®, a measure of cognitive ability. Up to 5 rounds of data were collected on children in the first form (first grade) in the schools. Round 1 data is baseline data collected in the term before the onset of intervention. Round 2 was taken during the term after the intervention started, rounds 3, 4, and 5 were during the second, fourth, and sixth terms after intervention started. We explore here how much data was collected and when it was collected.

Table 2.7 gives the number of observations taken at each round. We see a steadily decreasing number of observations. This is a common pattern in longitudinal data as subjects drop out of the study or get tired and decline to answer questions or supply information. There are 547 subject identification numbers (ids) in the data set, but only 540 have data in the Raven's data set. How did we get 543 observations at baseline? Further inspection of the data shows that only 530 subjects had baseline data. There are 13 subjects with a second observation before $t = 0$. Other than those 13 subjects, no subject had two Raven's observations during a single round.

Table 2.8 gives the numbers of subjects with from 0 up to 5 observations. Most subjects (83%) have a full 5 observations. The average number of observations per subject is $4.6 = (7 \times 0 + 8 \times 1 + 21 \times 2 + 16 \times 3 + 40 \times 4 + 455 \times 5)/547$.

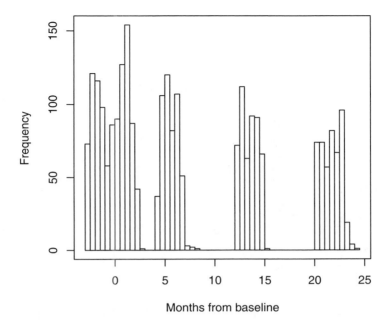

Figure 2.25. Histogram of the actual times of observations for the Kenya Cognitive data.

Inspection of the times of the actual observations can be useful. Figure 2.25 is a histogram of all actual times of observations. We see round 1 observations with times less than zero continuing smoothly into the second round of observations from zero to three months. The third round is clearly separated in time from the second round with observations taken between $t = 4$ up to $t = 8$ months. The fourth and fifth rounds are also clearly separated from each other and from round three.

Figure 2.26 plots the actual times of observations for a selection of subjects. Plotting all subjects requires several figures to create an adequate display, and so figure 2.26 shows 80 subjects. Vertical lines are drawn at months 0 and 3 to show breaks between different rounds of data collection. In inspecting these plots, I redrew them several times adding various vertical lines to aid in drawing conclusions about the times. We see that observations appear to have been taken in clusters. Most of the round 1 observations were taken between -3 and -2 months, with fewer observations taken between -2 and -1 months, and about 11 observations taken between -1 and 0 months. Most of the round 2 observations in this set of subjects were taken right at 1 month. The remainder were taken a bit after month 2. Round 3 has two observations taken right at month 4 but most observations were taken after month 5, with the remainder taken after

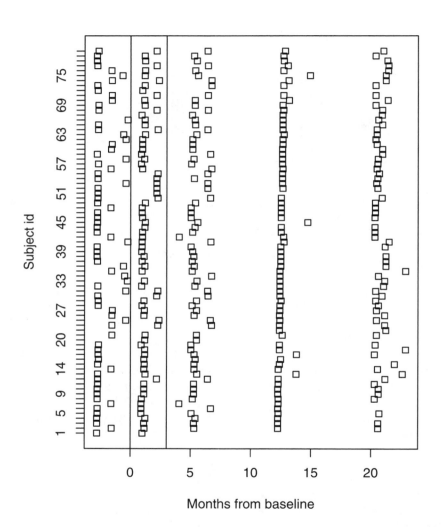

Figure 2.26. Plot of times of observations against ID number for Cognitive data.

month 6. Round 4 has observations taken between month 12 and 13, but there are several straggler observations taken almost to month 15. These comments apply only to the 80 subjects we see here. Figure 2.26 is an *event chart*.

2.8   Discussion

Our plots so far are exploratory; they are designed for investigating the basic distribution and characteristics of the data. Profile plots and scatterplots plot the raw data, empirical correlograms and empirical summary plots plot simple summaries of the data.

Our primary inferences from longitudinal data are about the average response over time and how it varies as a function of covariates. Our models for longitudinal data require us to specify how profiles vary over time and as functions of covariates; profile plots and empirical summary plots can help us with the initial model specification. We also need to specify models to describe the variances of the observations and the correlations among observations. Based on the fitted values from our model, we may plot inferred correlations in a correlogram of the model fit, and we may plot fitted means as functions of time and covariates in inference and prediction plots. These plots may be compared informally to the empirical correlogram and empirical summary plots that we drew in this chapter for model checking. Our model based inferences will usually be more accurate than the empirical summary inferences, but if our model assumptions are incorrect, then the empirical plots can show us how to fix them.

2.9   Problems

1. Consider finding the number of times where a mouse's weight decreases between consecutive measurements.

   (a) In figures 2.1 and 2.2, for how many observations can one be absolutely certain that there was a decrease in weights between one observation and the next for any mouse?

   (b) How about in figure 2.3?

   (c) Inspect the raw data and identify the mouse and the times at which a mouse weight decreases from one observation to the next. How many of these observations can be found in the first three plots of this chapter?

   (d) How difficult is it to identify all circumstances of consecutive measurements with decreasing mouse weight by hand? How long does it take? Is it easier if the Big Mice data were in long or wide format?

(e) Write a program to find the observations where weight decreases from one measurement to the next and present the results.

(f) Compare the difficulty of doing the task by hand as compared to finding all decreases by looking at a figure like 2.3.

2. For the Ozone data,

   (a) How often does ozone decrease from one hour to the next during the morning? What locations and what times?

   (b) Construct a scatterplot of ozone versus time. Compare it to the profile plot. Which shows more information? What can we figure out about the data using the profile plot that we can't from the scatterplot?

3. Assume balanced data and exactly two time points. Show that the average of the individual slopes is the slope of the line segment connecting the sample means.

4. Does the within-subject average of γ_{ij} from equation 2.1 estimate anything of interest? What is it? To answer these questions:

   (a) First assume equally spaced observations and answer the questions.

   (b) Second consider unequally spaced and answer the questions.

   (c) For unequally spaced observations, is there a weighted average that estimates something similar to what we get for equally spaced observations? Are the weights interpretable?

5. Is the mean over all observations Y_{ij} equal to the mean of the individual averages $\bar{Y}_i$? Explain under what conditions it is and when it is not.

6. Two profiles are said to have similar patterns over time when $Y_{ij} - \bar{Y}_i = Y_{lj} - \bar{Y}_l$, assuming the two subjects' times are the same $t_{ij} = t_{lj}$. Assuming same times, show that this is equivalent to $\gamma_{ij} = \gamma_{lj}$, where γ_{ij} is the slope between times t_{ij} and $t_{i(j-1)}$ for subject i, defined in equation 2.1.

7. For each of the following statements, state whether it is true or false, and come up with a rationalization (or proof) if true or a counterexample if false.

   (a) Suppose that all the individual profiles are flat. Then the sample average will be flat over time.

   (b) Suppose that all the individual profiles have the same non-zero slope. Then the empirical summary plot slope will be equal to the slope of the individual profiles.

   (c) Suppose the data are balanced and the individual profiles are flat. Then the sample average is flat.

(d) Suppose the data are balanced and the individual profiles have the same pattern over time. Then the empirical summary plot pattern over time will be equal to the individual patterns over time.

8. For the data sets in figures 2.5(a) and (c), calculate the mean, sd, min, and max of the observations in each window of width 1 beginning at $t_L = 1$ and increasing t_L in steps of .5 up to $t = 5$. Plot each summary statistic against the midpoints of the window, connecting the dots. For (c), do the calculations both with and without the outlier. In 2.4(a) the means vary quite a lot, even though from looking at the plot we think that they should be approximately constant across time. Explain why the means vary so much in this plot. Explain why dropping the outlier makes the plots for (c) smoother.

9. Create data sets that illustrate the following points. Use 8 subjects with 5 observations per subject unless otherwise specified. Data sets may be sketched with paper and pencil or generated using a statistics package with a random number generator.

    (a) Illustrate a data set where each subject has its own intercept and slope.
    (b) Continuing from the previous example, have the subjects with higher intercepts have the higher slopes also.
    (c) Continuing, have the overall population slope be negative.
    (d) Illustrate subjects who all start low, grow high with separate growth rates, then level off at different heights. You may wish to have more than 5 observations per subject.
    (e) Illustrate subjects who start low, go up to a subject-specific high, then come down low again.
    (f) Invent two more patterns. Plot sample data, and describe the patterns in a sentence.

10. Dropout. Consider three hypothetical weight-loss studies. The first study has only one group of subjects, all treated the same with some intervention that begins immediately after the baseline measurement. The second study is a double-blinded randomized chemical weight loss intervention study. The treatment group gets a diet pill, whereas the control group gets a placebo pill. The third study is also randomized. It is a study of a behavioral weight-loss intervention. The treatment group gets regular meetings with an attitude control specialist, group therapy, and weekly phone calls from a nurse practitioner. The control group gets a pamphlet on weight loss.

    (a) For the three studies, if the treatment is successful, what results would we expect to see? Sketch empirical summary plots for the groups in the study. Describe the plots in one sentence. Do not include error bars, just the mean is fine for these problems.

(b) In general, in a weight loss study, who would be more likely to drop out, those who lose a lot of weight or those who do not lose weight? What effect will this have on a profile plot of weights we drew?

(c) Study 2. Suppose that the pill does not work. What will the profile plots for subjects in the two groups look like?

(d) Study 2, cont. Suppose the pill works. Consider drawing conclusions from the empirical summary plot. Would the apparent conclusions be stronger if there was dropout as compared to if there was no dropout?

(e) Study 3. Which group is likely to stay with the trial longer, which is likely to drop out sooner?

(f) Study 3, cont. Assume both groups lose weight equally, and assume that subjects who don't lose weight drop out differentially in the two groups. Which group, will appear from the empirical summary plot to have better results?

(g) Study 3, cont. Suppose that the treatment group loses weight, whereas the control group does not. Suppose dropout is solely related to the treatment group but not to the amount of weight loss. Will the empirical summary plots make it look as if one group is doing better than it really is?

11. Plot histograms of the Pain data for each trial.

(a) Does the original scale appear to be skewed? Try various transformations to improve the normality of the data. What is your preferred transformation?

(b) Does the time point you choose to plot affect the choice of transformation?

(c) Without plotting histograms of the mice data (or you may if you want to!), describe the differences between two plots, one of which has the data from a specific time as opposed to another that includes data from all trials.

12. Draw profile plots of the Pain data using different line types for the attenders and distracters. Can you tell that the distracters have greater pain tolerance on average? Try using different colors for the lines instead and answer the question.

13. Plot the Pain data with one subject per plot. Make sure that the y axis has the same log base 2 scale for all subjects. Do you observe that low average subjects seem to have less within-subject variability than high average subjects?

14. Take averages of all the observed log base 2 Pain data responses for (a) the attenders, (b) distracters, and (c) attenders omitting the high outlier subject. Transform the averages back to the mean scale.

How well do our eyeball judgments compare with the estimates from subsection 2.4.1? Explain any discrepancies.

15. On a logarithmic scale, suppose that two tick marks are labeled c and d with $c < d$, and you estimate that an observation is $x(100)\%$ of the way from c to d. Show that the observation value is at $c(d/c)^x$. If the proof isn't easy, try plugging in some values for c and d and x. For the Pain data, we already have $d/c = 2$, which is why we used 2 as the power in those calculations.

16. Draw pictures of the Weight Loss data, one profile to a plot.

   (a) There are several choices to be made.
      i. One may construct the y axis of each plot so that they all cover the entire range of the data.
      ii. One may draw each profile's plot so the y axis covers just the range of the given profile.
      iii. One may draw each profile's plot so that the range of the y axis is the same for each profile but is as small as possible.
      For each choice of y axis, is the plot most akin to (i) figure 2.15 or (ii) figure 2.16 or (iii) neither? Use each of the three answers exactly once!

   (b) One can also plot our empirical within-subject residuals $Y_{ij} - \bar{Y}_i$ instead of Y_{ij}. Is there any advantage to the residual profiles one to a plot instead of the original Y_i?

17. In subsection 2.4.4, we plotted profiles of the empirical within-subject residuals $Y_{ij} - \bar{Y}_i$. Would this be of much value for the (i) Big Mice data? (ii) How about the Ozone data? (iii) The Pediatric Pain data? Calculate the empirical within-subject residuals and draw the plots. What do you learn, if anything?

18. In subsection 2.4.4, instead of plotting the empirical within-subject residuals $Y_{ij} - \bar{Y}_i$, suppose that we instead subtract off the baseline measurement and define $W_{ij} = Y_{ij} - Y_{i1}$.

   (a) Plot all of the W_{ij} in a profile plot.
   (b) From this plot, what characteristics of the plot tell you
      i. that subjects are losing weight over the duration of the trial?
      ii. that subjects lose weight at different rates?
      iii. that there is something odd going on from trial 5 to 6?
   (c) Is it easier or harder to detect these three items in this plot as compared to the plot of empirical within-subject residuals? Which plot is better?

19. How could you produce an estimate of a single subject's intercept that is better than the mean $\bar{Y}_i$??? By better, I mean closer to the true value on average.

20. For the Weight Loss data, suppose that we took each observation Y_{ij} for $t_j > 1$ and subtracted off the previous observation $Y_{i(j-1)}$ giving consecutive differences.

$$W_{ij} = Y_{ij} - Y_{i(j-1)}$$

   (a) For a fully observed Y_i, what is the length of $W_i = (W_{ij})$ the vector of all W_{ij} for subject i?
   (b) Plot the W_{ij} in a profile plot.
   (c) What features do you see in the profile plot?
   (d) What do these features imply about the original data Y_{ij}?
   (e) If the Y_{ij} have a random intercept, i.e., $Y_{ij} = \mu_i + \epsilon_{ij}$, what will the W_{ij} profiles look like?
   (f) If the Y_{ij} fall on a subject-specific line $Y_{ij} = a_i + b_i j + \epsilon_{ij}$, then what will the W_{ij} profiles look like?
   (g) If the Y_{ij} follow a subject-specific quadratic $Y_{ij} = a_i + b_i j + c_i j^2 + \epsilon_{ij}$, what will the differences look like?
   (h) What is the problem with this differences plot if (a) There is missing data in the middle of the times? (b) The data are observed at random times?

21. The empirical population residuals were of some use for the mice data, allowing us to look at the individual observation to observation variation. In contrast, it seems implausible that the empirical within-subject residuals would be useful for the mice data.

   (a) For the Pain data, without calculating the two types of residuals and without drawing the plots, one of the residual plots is very unlikely to show us interesting structure, and one might or might not show us interesting structure. Which is which, and briefly, why?
   (b) Answer the same question for the Ozone data.
   (c) Draw both residual plots (empirical within-subject, and empirical population) for the Pain data and illustrate your conclusion from problem part 21(a).
   (d) Draw both plots for the Ozone data and illustrate your conclusion from problem 21b(b).

22. Occasionally, the profile plot plan of connecting the dots may obscure the actual trends in the data. This tends to happen when there is a combination of rapid changes in responses over time and missing data. Table 2.9 presents the Vagal Tone data. Vagal tone is supposed to be high and in response to stress it gets lower. The subjects in this study were a group of 21 very ill babies who were undergoing cardiac catheterization, an invasive, painful procedure. The columns in the table give the subject id number, gender, age in months, the duration of the catheterization in minutes, up to 5 vagal tone measures, and a measure of illness severity (higher is worse). The first

Id	Gender	Age (m)	Dur (min)	Pre1	Pre2	Post1	Post2	Post3	Med sev
1	F	24	150	1.96	3.04		3.18	3.27	4
2	M	8	180						13
3	M	3	245				0.97		20
4	M	14	300						18
5	F	10	240	3.93	3.86		3.59	3.27	12
6	M	23	240		2.24		2.55	2.27	20
7	M	5	240	2.57	2.52	3.92	3.22	1.28	12
8	M	4	210	4.44	3.07	1.7	3.43	3.43	3
9	M	15	180				1.15	1.03	19
10	F	6.5	330	1.74		1.23	1.49	0.92	5
11	M	15.5	180			2.14		4.92	22
13	F	11	300		3.06	1.94	2.6	1.18	7
14	F	5.5	210						11
15	M	10	300			2.97	5.23		4
16	M	19	330	4.4	3.92	0			29
17	M	6.5	540	1.96	1.95		2.51	1.25	23
18	F	9	210		3.51	1.88	2.14	2.42	13
19	M	15	120	3.63	3.11	1.87	4.46	3.7	15
20	F	3	80	2.91	2.91	0.61			5
21	F	23	65		5.03				5

Table 2.9. Vagal Tone data. Columns are subject number, gender, age in months, length of time in minutes of cardiac catheterization procedure, five vagal tone measures, and a medical severity measure. Blank indicates missing measurement. Subject number 12 has no data.

two measures are before the catheterization, the last three are after. The first measure was taken the night before, the second measure the morning before, then the catheterization; the third measure was taken right after the catheterization, the fourth was taken the evening after, and the last measure was taken the next day. There is a substantial amount of missing data; blanks in the table indicate missing data; subject 12 is missing all variables.

Figure 2.27 shows the Vagal Tone profile plot drawn in two ways. In 2.27(a) we draw the usual plot and connect the dots between all observations within a subject, even if they are not consecutive observations; in 2.27(b), points are connected only if they are consecutive observations from the same subject.

(a) Describe the impressions one gets from the two plots. How are the impressions different?

(b) Which plot do you prefer?

23. Plot the Weight Loss data one profile to a plot. What fraction of subjects appear to be losing weight?

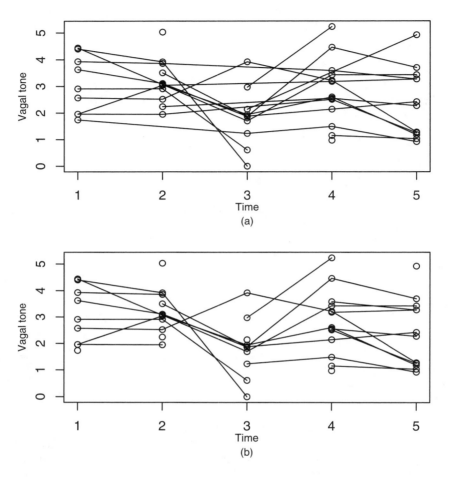

Figure 2.27. Vagal Tone data, plots of vagal tone by measurement. (a) Profile plot. (b) Profile plot, but non-consecutive observations within subject are not connected.

24. Inspect the profile plot of the Weight Loss data.

    (a) Make a rough guess of the correlation between any two observations. Does your guess depend on the specific times that you choose?

    (b) Calculate the correlation matrix for the Weight Loss data.

    (c) Plot the scatterplot matrix of the data.

    (d) Describe your conclusions about the correlations.

    (e) Will the Weight Loss residuals have a greater, lesser, or equal variety of the correlations as compared to the raw Weight Loss data?

(f) Look at figure 2.17. Between what sets of observations do you expect to find

    i. a strong positive correlation,

    ii. a strong negative correlation, and

    iii. a low or zero correlation?

    Briefly explain your reasoning.

(g) Calculate the correlation matrix and draw the scatterplot matrix for the Weight Loss residuals. Describe your findings.

25. In the Pain data, draw separate scatterplot matrices for the attenders and distracters. How do the scatterplots differ? What conclusion do you draw?

26. Dental data. The Dental data set is a classic data set for longitudinal data analysis. The response is the length in millimeters from the center of the pituitary gland to the pteryomaxillary fissure for 11 girls and 16 boys. The measurements were taken every two years at ages 8, 10, 12, and 14. There is a single covariate, gender. The purpose of this analysis is to correctly describe the important characteristics of the data.

(a) Create a profile plot of the data. Use separate line types (or colors or plots) for the boys and the girls.

(b) Briefly report your findings. What is the overall pattern? Are boys and girls different? In what ways?

(c) Calculate the correlations among the observations, and draw a scatterplot matrix. Use separate plotting characters for boys and girls.

(d) Report any additional findings.

(e) Draw an empirical summary plot, and repeat separately for boys and girls. Is there a difference in level between boys and girls?

(f) Is there a difference in average slope between boys and girls?

(g) Inspect the profile plot of empirical within-subject residuals. What do you learn about the data? There are four important items to identify about this data set. What are they? You may or may not have seen all of them in the original profile plot.

27. Draw a correlogram for the Small Mice data. Interpret the results.

28. Calculate the correlation matrix and draw a correlogram for the Dental data. What are your conclusions about the correlations?

29. Draw a correlogram for the Pain data. Draw it separately for the attenders and distracters. Describe your conclusions.

30. The standard deviation of an estimated correlation when the true correlation is zero is $\mathrm{SE} = (n-3)^{-1/2}$. Often we add horizontal lines

Subject no.	Gender	\multicolumn{4}{c}{Age at measurement}
		8	10	12	14
1	Girl	21	20	21.5	23
2	Girl	21	21.5	24	25.5
3	Girl	20.5	24	24.5	26
4	Girl	23.5	24.5	25	26.5
5	Girl	21.5	23	22.5	23.5
6	Girl	20	21	21	22.5
7	Girl	21.5	22.5	23	25
8	Girl	23	23	23.5	24
9	Girl	20	21	22	21.5
10	Girl	16.5	19	19	19.5
11	Girl	24.5	25	28	28
12	Boy	26	25	29	31
13	Boy	21.5	22.5	23	26.5
14	Boy	23	22.5	24	27.5
15	Boy	25.5	27.5	26.5	27
16	Boy	20	23.5	22.5	26
17	Boy	24.5	25.5	27	28.5
18	Boy	22	22	24.5	26.5
19	Boy	24	21.5	24.5	25.5
20	Boy	23	20.5	31	26
21	Boy	27.5	28	31	31.5
22	Boy	23	23	23.5	25
23	Boy	21.5	23.5	24	28
24	Boy	17	24.5	26	29.5
25	Boy	22.5	25.5	25.5	26
26	Boy	23	24.5	26	30
27	Boy	22	21.5	23.5	25

Table 2.10. The Dental data. Columns are subject number, gender, and then the four repeated measurements. Responses are the length in millimeters from the center of the pituitary gland to the pteryomaxillary fissure on each subject. Measurements were taken at ages 8, 10, 12, 14.

at $\pm 2(n-3)^{-1/2}$ to a correlogram to identify correlations that are not significantly different from zero.

(a) Suppose you were to add these lines to a correlogram of the Pain data. Would it change any conclusions? Explain why you can answer this question without actually drawing the correlogram.

(b) Add these lines to the Ozone data correlogram. What does it suggest on an individual correlation basis? Still there are many correlations all of similar size, so perhaps all of those correlations are not equal to zero.

31. Draw an empirical summary plot for the Ozone data. Then draw one separately for valley and non-valley sites. Finally, draw a third for each of the three days and briefly summarize your conclusions.

32. Suppose that some subjects drop out of your study early. Could this cause the empirical summary plot to be misleading? Consider the following examples. For each, (a) sketch and describe how the empirical summary plot will look as compared to how it would look if you had full data, (b) whether the empirical summary plot is misleading, and (c) if it is misleading, how it would be misleading.

    (a) Subject profiles follow a random intercept pattern. All subjects have a 25% chance of not appearing for any given observation.
    (b) Subject profiles follow a random intercept pattern. Subjects who are below average are much more likely to drop out than those above average.
    (c) You have two groups. Subject profiles all start from similar starting points and have different, random, slopes. Subjects who score too high are cured and then tend to drop out permanently. Assume the average slope is the same in both groups. You are interested in either the trend over time or the differences in trend between the two groups; how are these inferences affected by the dropout?
    (d) The same situation as the previous part, but now, subjects in group 1 have a higher slope than subjects in group 2.

33. In the construction of the empirical summary plot, we plotted plus and minus two standard errors of the mean, or plus and minus two sample standard deviations to make an empirical prediction plot. Assuming normally distributed data, how might you improve on these two plots to show (a) an exact 95% confidence interval for the mean, and (b) an exact 95% prediction interval for future data? (Hint: we used the number 2 in constructing our plots. What number should you use instead?) Draw your improved plots for the mice data, can you tell the difference between your plots and figure 2.23?

34. Sketch by hand how your empirical summary plots would look like in figures 12.1(a)–(d). From data that looked like those in (b) and (d), how might you figure out that subjects with low responses were dropping out more than subjects with high responses?

35. The data in figures 12.1(b) and (d) seem troubling. However, approximately, what can happen when we fit a model is that the model first estimates intercepts and slopes for each subject, then averages subjects' intercepts or slopes to get a estimate of the population intercept and slope. Explain why this might be sufficient to get your inference plot from the fitted model in these two figures to look more like the desired empirical summary plot from 12.1(a) and (c).

36. For each of the following data sets, produce a table of the number of observations at each nominal time of observation.

    (a) Small Mice
    (b) Big Mice
    (c) Pediatric Pain
    (d) Weight Loss
    (e) BSI total
    (f) Cognitive
    (g) Anthropometry weights

    Describe each data set observed pattern in a few words.

37. For each of the following data sets, produce a table of the number of subjects with each possible number (i.e., 0 up to J) of observations.

    (a) Small Mice
    (b) Big Mice
    (c) Pediatric Pain
    (d) Weight Loss
    (e) BSI total
    (f) Cognitive
    (g) Anthropometry weights

38. For each of the following data sets, produce a histogram of the actual times t_{ij} that observations were taken. On your plot, mark the nominal times that observations were taken.

    (a) Weight Loss
    (b) BSI total
    (c) Cognitive
    (d) Anthropometry weights

3
Simple Analyses

'Tis a gift to be simple
'Tis a gift to be free

– Shaker Hymn

Doth my simple feature content you?
Your features! Lord warrant us! What features?

– William Shakespeare

Overview

- Why researchers summarize longitudinal data before analyzing it
- How we summarize longitudinal data
- Inferences of interest from longitudinal data
- How we analyze those summaries
  - Paired t-test
  - Two-sample t-test
  - Difference of differences design
- Subsetting data sets

In discussing the analysis of longitudinal data with other researchers, we occasionally meet up with otherwise intelligent researchers who prefer to use simple statistical methods to analyze fairly complex designs. Having spent thousands if not millions of dollars to collect a complex data set, they then want to use a simple and cheap statistical tool to analyze their data. We need to understand what these simple statistical methods are, why people may want to use them, and why they are not the best methods to analyze longitudinal data. In this chapter, we discuss some of the more common simple statistical methods that researchers use to analyze longitudinal data.

When faced with complex data sets, data analysts have traditionally looked for ways to simplify the data into a data structure that permits analysis using a familiar statistical technique. In longitudinal data, simplification may result in ignoring all but one or two observations per subject, or we may compute simple summaries such as the mean response for each subject or the change in response from beginning to end of the study. These simplifications allow the analyst to use familiar analyses such as paired or two-sample t-tests, analysis of variance (ANOVA), or regression models in place of unfamiliar, unknown, and certainly more complex analyses. The extent of the simplification depends on available statistical software, time, and the data analyst's sophistication. In other words, the analysis depends on available tools, resources, and knowledge.

For longitudinal data, the problem is that repeated observations within a subject are correlated. Most, if not all, analyses taught in the first year or two of statistical training require observations to be independent of each other. Suppose we have J observations on each of $n/2$ subjects in each of two groups. We may not pool the $Jn/2$ observations in each group and do a two-sample t-test with $Jn - 2$ degrees of freedom (df) for estimating the error. This will massively overestimate the precision of our conclusion. Similarly, we cannot run a regression with Jn cases in the sample when we have covariates to regress upon; this again will overestimate the precision of our conclusions. The correlated observations within a subject violate the assumptions required for a two-sample t-test or linear regression analysis. If there were exactly two observations per subject, then a paired t-test may be appropriate, but we commonly have three or more observations per subject and often unequal observations per subject.

A common way to simplify longitudinal data is to reduce each multivariate response Y_i to a univariate summary and then apply a familiar analysis. If need be, we can do this several times in an attempt to answer all of the questions we wish to ask of the data.

The following sections discuss summaries of longitudinal data and the associated analyses. With some trepidation, example analyses are given. We need to know about these analyses, although they are often of lower quality than the longitudinal analyses we will learn about in the remainder of the book. This chapter also introduces some of the basic inferences that researchers wish to make with longitudinal data. Having first set the reader

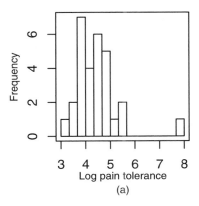 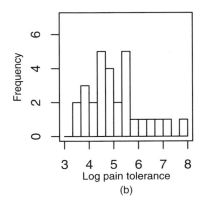

Figure 3.1. Histograms of the baseline average log pain tolerance: (a) attenders, (b) distracters.

up with examples of simple analyses, these analyses are critiqued in the next chapter.

3.1   Summarizing Longitudinal Data

3.1.1   Mean

Perhaps the simplest univariate summary of longitudinal data is the average of the responses from a single subject

$$\bar{Y}_i = \frac{\sum_{j=1}^{n_i} Y_{ij}}{n_i}.$$

The average $\bar{Y}_i$ is treated as a single response per subject and analyzed as the response variable with a familiar statistical technique, whether a t-test, ANOVA, or linear regression. A common variation is to average a subset of the observations within a subject. For example, if nearly every subject has four observations, but a few have five or six, perhaps we average only the first four observations per subject. A question is what to do for subjects that have only three observations: ignore them, average the three observations and continue, average the three observations and perform some form of weighted analysis, or still something else?

The Pediatric Pain data provides examples for this chapter. Suppose we wish to test for a difference between the attender average response and the distracter average response. We might average all observations for subjects and perform a two-sample t-test between attenders and distracters. In this data, we have three longitudinal measures taken before the treatment intervention and one after treatment. Because the fourth observation is different

Group	Name	Sample size	Mean	Sd	t-test
1	Attenders	29	4.47	.88	$t = -2.3, p = .025$
2	Distracters	29	5.08	1.09	

Table 3.1. Pediatric Pain data. t-test summary statistics: sample size, mean, and standard deviation (sd) of the average of the first three log responses. The t-test compares the average of the first three responses between attenders and distracters. The reported p-value is two-sided on 56 degrees of freedom. The grand mean and pooled standard deviation are 4.77 and .99.

from the first three, perhaps it should be omitted from the averages. A few subjects are missing some observations, and we omit those subjects from all analyses in this chapter.

Thus we select only subjects with four observations, and we then average the first three observations. As discussed in chapter 2, we perform all analyses on the log base two scale. Figure 3.1 shows histograms of these values for the two groups. Inspection suggests that distracter averages are higher than attender averages; attenders have a single subject with $\bar{Y}_i$ over $5.5 = \log 45$, distracters have 6. Table 3.1 presents summary statistics of the two groups and the results of a two-sample t-test of the null hypothesis H_0: no difference in means. The means are significantly different. Of course, we see a high outlier in figure 3.1 in the attender group. As compared to the data set without the outlier, the outlier tends to inflate the pooled variance, makes the means more similar and decreases the statistical significance of the result.

To help interpret these results, table 3.2 presents the original seconds scale and the corresponding log base two seconds scale. Not surprisingly, most people find the original seconds scale easier to interpret. Conversions are given from the mean log times of 4.47 and 5.08 and for the difference $.60 \approx 5.08 - 4.47$. We see that the difference between the two group means is about $34 - 22 = 12$ seconds. Alternatively, the distracters' average pain tolerance is approximately 52% longer than the attenders, where $.52 = 2^{(5.08-4.47)} - 1 = 33.7/22.2 - 1$.

The times 22.2 and 33.8 seconds are the back-transformed values of 4.47 and 5.08; the .88 and 1.09 are the standard deviations on the log seconds scale. If the data are symmetrically distributed about their mean on the log seconds scale, then the exponentiated values (i.e., $2^{4.47} = 22.2$) are estimates of the population medians in the original seconds scale.

3.1.2 Slope

When we collect longitudinal measures, we are frequently interested in change over time. The simplest summary of change, the difference of two measurements, is studied in the next section. Here we summarize each profile by a slope. We calculate the slope for subject i by regressing Y_{ij} against t_{ij} with n_i data points for subject i, and we repeat the regression

$\log_2$ seconds	seconds
.60	1.52
.88	1.84
1.09	2.13
4.47	22.2
5.08	33.7
3	8
4	16
5	32
6	64
7	128
8	256

Table 3.2. Back-transforming from the log base two seconds scale to the original measurement scale in seconds.

separately for all n subjects. Define $Y_{ij}^* = Y_{ij} - \bar{Y}_i$ and $t_{ij}^* = t_{ij} - \bar{t}_i$, where $\bar{t}_i$ is the mean of the observation times for subject i

$$\bar{t}_i = \frac{\sum_{j=1}^{n_i} t_{ij}}{n_i}.$$

The formula for the slopes can be written

$$\tilde{\beta}_i = \frac{\sum_j t_{ij}^* Y_{ij}^*}{\sum_j t_{ij}^* t_{ij}^*}.$$

The n slopes from these regressions are treated as regular data and are entered into the data set as another variable. We can now perform t-tests, ANOVAs, and regressions with the slopes as the response for each subject. We may compare the average slope to zero in a one sample t-test to see if there is any trend in the responses over time. We may compare the average slopes in two groups to each other using a two-sample t-test. If there are more than two groups, we can use ANOVA to compare the group mean slopes to each other. Regression might be used to see if subject covariates (age, gender) predict differences in slopes.

In the Pediatric Pain data, we might wonder if learning occurs with the repeated trials, and whether this learning is different between attenders and distracters. One way to assess possible learning by subjects is to consider the subject-specific slopes of log pain tolerance regressed on time where time is represented by the trial number. We calculate the slopes for the 58 subjects with complete data. Figure 3.2 gives separate histograms for attenders and distracters. We see that the shapes are akin, and the centers of both distributions are similar and near zero. The summary statistics in table 3.3 and the two-sample t-test also do not show any important differences in the groups nor are the average slopes in each group separately significantly different from zero.

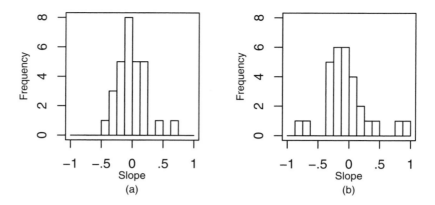

Figure 3.2. Histograms of the log pain tolerance slopes: (a) attenders, (b) distracters.

		Sample			Test	
Group	Name	size	Mean	sd	t	p
1	Attenders	29	-.017	.24	$-.39$	.7
2	Distracters	29	-.046	.37	$-.66$	.5
Attenders = Distracters					.34	.7

Table 3.3. Summary statistics for several t-tests for the slopes. Columns headed test and t and p are the tests of H_0 that the mean slope for attenders and for distracters is zero. The last line is the t-test of no difference in the mean slopes of the two groups. The reported p-values are two-sided on 28 or 56 degrees of freedom. The pooled standard deviation is .217.

Some basic calculations can help us interpret these slopes. Over times from 1 to 4, a slope of $-.046$ would lead to an increase of $(4-1) \times -.046 = -.138$ units, (i.e., a decrease). On the unlogged scale, this is $2^{-.138} = .91$, a 9% drop. The change for the attenders group, $2^{(3\times(-.017))} = .965$ for a 3.5% drop, is even closer to zero. Neither decrease seems of practical, much less of statistical, significance, and we conclude that there is certainly no evidence of an increase in pain tolerance with practice in this data set.

Further analysis is possible. Subjects were randomized to one of the three treatments. Although we might not expect the two coping style (CS) groups to differ significantly in terms of slopes, we might expect children given an effective treatment (TMT) to increase their pain tolerance while the others stayed constant. Figure 3.3 presents boxplots of the slopes for the 6 CS×TMT groups. The picture is uninspiring, and we are not surprised that the six groups are not significantly different from each other when we do an ANOVA. However, testing each group separately with a one sample t-test, the average slope for the DD group is significantly greater than zero

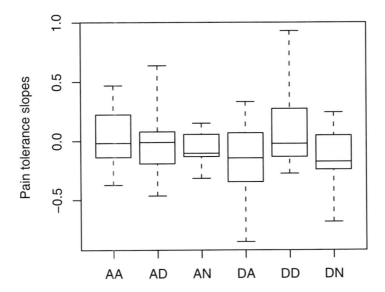

Figure 3.3. Pain data: Boxplots of slopes for the Pediatric Pain data for six groups. First letter abbreviates the coping style, and the second letter abbreviates the treatment.

($p = .008$) and the slope for the DN group is significantly less than zero ($p = .03$).

3.2   Paired t-test

Many longitudinal studies are designed to be analyzed using a paired analysis. A paired t-test can be a reasonably appropriate early analysis. Suppose we have $n_i = 2$ for all subjects, with Y_{i1} as the *before* measure and Y_{i2} is the *after* treatment measure. The differences

$$d_i = Y_{i2} - Y_{i1}$$

are the amount of increase from the first observation to the second. If we expect the response to decrease over time, we may well subtract the second observation from the first to save the inconvenience of dealing with negative numbers. We proceed to analyze the d_i using a one sample t-test.

Using the 58 complete data observations from the Pediatric Pain data, let us take differences between the fourth and second observation. These observations are comparable; in each trial children used the same hand

Group	Name	Sample size	Mean	Sd	t	p
1	AA	10	.18	.61	.80	.4
2	AD	10	.18	.50	.80	.4
3	AN	9	.03	.38	.12	.9
4	DA	10	-.24	.51	-1.0	.32
5	DD	11	.72	.67	3.3	.008
6	DN	8	-.68	.93	-2.7	.04

Table 3.4. Summary statistics for paired t-tests for each of the six groups. The first letter of the name stands for CS, the second is for TMT; A is attender, D, distracter, N, none. The reported p-value is two-sided with df equal to the sample size minus 1.

and both were the second trial of a session following a practice trial. Table 3.4 presents a summary of the results of the paired t-test analysis for each group. We see that no attender result is significantly different from zero; the treatments neither helped nor hurt the attenders. In contrast, distracters taught to distract had significantly longer immersion times on trial 4 than on trial 2. The distracters given the null counseling intervention decreased their times a surprising amount, but the sample standard deviation is quite high due to a single low outlier. Deleting the outlier does not change the significance of the results, as the mean would increase toward zero even as the sd shrinks substantially. Figure 3.4 presents side by side box plots of all 6 sets of differences. The three attender groups are more homogenous while the three distracter groups seem to be more different from each other. The DD group in particular seems unusually high compared with the other groups, having the highest lower quartile and highest minimum value of the 6 groups.

3.3   Difference of Differences

The difference of differences (DoD) design generalizes the single sample paired t-test design to two groups instead of a single group. The DoD design is more common than a single group paired comparison design. To understand the need for the DoD design, first consider a hypothetical simplification of the Pediatric Pain study. Suppose that only the distract treatment is given to distracters, and we see a significant improvement in $\log_2$ seconds for trial four over trial two using a paired t-test analysis. The problem with this design is that we do not know if the increase in pain tolerance is due specifically to the distract treatment. Two alternative explanations are (1) the increase is due to the extra practice the subjects have had and (2) it is due to the passage of time. All three reasons could cause the fourth trial outcome to be longer than the second trial outcome. See problems 2 and 3 for another reason for using the DoD design.

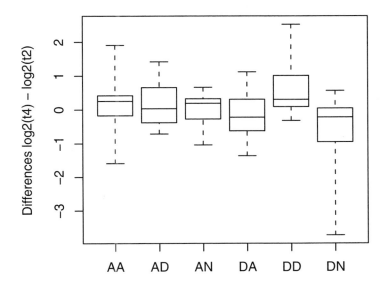

Figure 3.4. Pain data: Boxplots of differences of trial 4 minus trial 2 for the six groups.

Intervention	Control	Mean difference	SE	t	p
AA	AN	.15	.34	.45	.6
DD	DN	1.39	.53	2.65	.017

Table 3.5. Two difference of difference analyses for the Pediatric Pain data, comparing the AA group to the AN group and the DD group to the DN group.

The DoD design provides a solution to this issue by following a second group of subjects with a separate treatment. The experimental design now includes a second group with treatment which may be a null treatment as with a control or placebo group or there may be an active second treatment. Otherwise the second group must experience the same pattern of measurements and events over time as the original treatment group. If the improvement in time in the original group is due to extra practice or the simple passage of time, then the second group will also show an improvement in time. We compare the average improvements in the two groups. If they are similar, the improvement in the first treatment group is not due to the differences in treatments. If there is an improvement in the first treatment group not seen in the second group, then we attribute the improvement to the difference in treatments that the two groups receive.

We formally define the DoD using notation specific to this paragraph. Define notation with observations Y_{kij} indexed by three subscripts, k for group, i for person within group, and j for observation within person. The first subscript $k = 1$ or 2 for group 1 or 2; subscript i runs from 1 up to n_k, which is the number of subjects in the kth group, and subscript $j = 1$ for the first observation and $j = 2$ for the second observation on subject i. The analysis takes differences

$$d_{ki} = Y_{ki2} - Y_{ki1}$$

within subject as in the paired t-test analysis. Now we do a regular two-sample t-test on the d_{ki}. The two-sample t-test numerator is then a difference of mean differences

$$\bar{d}_1 - \bar{d}_2,$$

with

$$\bar{d}_k = \frac{\sum_{i=1}^{n_k} d_{ki}}{n_k}$$

for $k = 1$ and $k = 2$. The denominator of the DoD t-test uses a pooled standard deviation

$$s_{\text{pooled}} = \left[\frac{\sum_{k=1}^{2} \sum_{i=1}^{n_k} (d_{ki} - \bar{d}_k)^2)}{n_1 + n_2 - 2} \right]^{1/2}$$

times the square root of the sum of the inverse sample sizes. The full formula is

$$t_{\text{DoD}} = \frac{\bar{d}_1 - \bar{d}_2}{s_{\text{pooled}}} \left(\frac{1}{n_1} + \frac{1}{n_2} \right)^{-1/2}.$$

3.3.1   Pediatric Pain Example

The Pain data was designed specifically to be a DoD analysis. The null (N) treatment was intended to mimic the D and A treatments but without the educational component. Two DoD examples are given in table 3.5. The first is for the AA group compared to the AN group, and the second is for the DD group compared with the DN group. The DD group is significantly different from the DN group. This design strengthens the conclusions from the paired t-test analysis. In contrast, the AA group is not significantly different from the AN group. We conclude once again that the distract treatment was helpful to the distracters, whereas the attend treatment was not helpful to the attenders. We do not show the mis-matched treatments AD and DA comparisons to the AN and DN treatments, respectively, as these are not of great interest.

3.4 Using a Subset of the Data

In practice, it is very common to analyze a subset of the data. There are several reasons for this

1. To eliminate subjects or observations not in the population of interest.

2. To ignore variables of little or no interest.

3. To produce a data set of manageable size.

4. To produce a subset of the data for which a particular known analysis may be applied.

Eliminating subjects not in the population of interest is not just scientifically valid, it is necessary. In a study of female breast cancer survivors, we would not include the rare males with breast cancer. In a study of melanoma, we do not include bone cancer patients. And in that same study, we might not wish to study subjects with additional severe medical problems, so a subject who suffers a heart attack might be dropped from the study. All human studies have inclusion and exclusion criteria. An ineligible subject may sometimes be accidentally enrolled, and when this is discovered, that subject is removed from further analyses.

In most all studies, we ignore variables we are not interested in for practical reasons. If we were required to pay attention to everything, we would go crazy under the burden of too many analyses most of which we would not care about and most of which would show no significant results. It is, however, easier and cheaper to collect extra information than to discover that we missed an important variable and to then go back and collect more information from our subjects, even assuming that the subjects are still living and willing. Thus we have a tendency to collect large numbers of variables on our subjects. We select the most important and interesting variables for further followup and analysis.

The third reason above is a valid practical reason for subsetting a data set. Many data sets are quite large, and it can be difficult in practice to work with every observation. Some studies may have hundreds of thousands if not millions of observations. Working with a randomly selected subset of the subjects makes for an easier time of data management, may make it possible to draw interpretable plots that are not a blur of ink, make it possible to execute exploratory analyses, and may substantially decrease the computer time to execute an analysis. As computing power increases, as software flexibility and power increases, the need to analyze a subset of a data set will decrease over time.

The fourth reason for subsetting data is strictly utilitarian. It is preferable to have some analysis than no analysis. It may be we are familiar with a particular analysis, or we may own software that can produce that analysis, for example a paired t-test or a one-way ANOVA. It may be that subjects with missing data make it difficult to analyze data in our software, so the software may remove them from the analysis. Analysts may take subsets of the longitudinal responses, subsets of cases, or both. Unlike the first three reasons, this fourth reason is due to ignorance of the methods that we discuss in later chapters. Sometimes it is the case that no one knows how to analyze a particular data set well; then we all do our best with the tools at hand. The Pediatric Pain data set falls in between; not everyone knows how to analyze it, but a number of good methods are available, and we present them in later chapters.

Averaging responses within subject, then analyzing only the averages, is a form of data subsetting. In many data sets there are additional trends or patterns in the profiles not caught by averages. These patterns are buried in the residuals, the differences $Y_{ij} - \bar{Y}_i$. Unless we analyze additional data summaries, we will not learn about the other trends or patterns in the data.

3.4.1   Pediatric Pain Data

The Pediatric Pain analyses in this chapter analyzed various subsets of the data. Only subjects with a complete set of four responses were included in any analysis. If subjects with missing data had been included, then an additional burden would have been placed on the analyst, as analyses and software that allowed for missing data would have been required, and these may be unfamiliar or software may be unavailable.

The data were subset in several ways. In the six paired analyses presented in table 3.4, we used only the second and fourth observations within person, ignoring information from the first and third observations. These six analyses only used the observations from subjects in the group being analyzed. The degrees of freedom of the tests were equal to the number of people in that group minus one. In the DoD design, we only included the AA and the AN groups in that analysis; other groups did not contribute to, for example, the variance estimation.

3.5   Problems

1. Table 3.3 for the slopes has 2 additional t-tests in it as compared to table 3.1 for the means. What hypotheses would these two tests test if they had been done in table 3.1? Are these tests sensible? Explain in one sentence.

2. Use the Dental data for this problem. We will look at analyses of some subject summaries.

   (a) Calculate averages for each child. Plot these 27 numbers, distinguishing boys from girls. Describe the distributions of the two samples (one sentence). Is there a difference between groups?

Present (i) a formal inference (i.e., a test and p-value) and (ii) a plot of the data illustrating the conclusion. Do not present a plot with just two points on it!

(b) Repeat the previous item for the slopes for each child.

(c) Are the variances of the intercepts the same for boys and girls? How about the slopes?

(d) Compare this analysis to the one in problem 2.26. Which analysis shows more about the data? In a short itemized list, list what we learn in each analysis and whether we can learn it from both analyses or just from one analysis.

3. Redo the Pediatric Pain baseline two-sample t-test in several different ways.

(a) Do the test without the outlying attender. Verify the statements: As compared to the data set without the outlier, (i) the outlier tends to inflate the pooled variance, (ii) makes the means more similar and (iii) decreases the statistical significance of the result.

(b) Another way to do the two-sample t-tests comparing attenders to distracters is to compare only a single observation such as the first, second or third observation. Do the three possible two-sample t-tests again, using one observation at a time and as many of the subjects as possible for each analysis.

(c) Of all the baseline two-sample t-tests, comparing attenders to distracters, which analysis is best?

4. Use the Pediatric Pain slopes for this problem.

(a) Do an F-test to see if the slopes are significantly different among the 6 CS×TMT groups.

(b) Do an F-test among three treatment groups just for the distracters and a second time just for the attenders. Report the results.

(c) Are the slopes in any of the 6 groups significantly different from zero?

5. Is a slope a reasonable way to summarize a subject's responses in the Pediatric Pain data?

6. Use the Weight Loss version 1 data for this problem. Calculate the first weight minus the last weight for each subject $d_i = Y_{i1} - Y_{in_i}$ and use that in the following problems.

(a) Plot the differences d_i in a histogram and describe the resulting histogram.

(b) Plot the d_i against n_i. Does any pattern emerge? Describe two theoretical patterns that might be seen, depending on characteristics of the data.

(c) Is the mean of the d_i significantly different from zero? Should we do a one-sided or a two-sided test? Draw a one-sentence conclusion.

7. Use the Weight Loss version 1 data for this problem. Calculate slopes for each subject.

(a) Plot the slopes in a histogram and describe the resulting histogram.

(b) What is the average slope? What is its standard deviation?

(c) Test if the average slope is equal to zero.

(d) Use a regression program to calculate the slope for several people with different amounts of data. How different are the standard errors? Critique the t-test analysis.

(e) Given what you learned in chapter 2 about the patterns of weight loss, critique the use of the slope as a summary of an individuals weight loss.

8. Could the Ozone data be summarized by a difference d_i equal to last observation minus first observation? Is there possibly a different difference that might make for a better summary measure?

4
Critiques of Simple Analyses

He that will not apply new remedies must expect new evils; for time is the greatest innovator.

— Francis Bacon

Caveat emptor.

— John Fitzherbert

Overview

- Omitting subjects is bad because it leads to increased standard errors.

- Omitting subjects with missing data can cause bias.

- Analyzing a subset of times is bad because it ignores relevant information and causes increased standard errors.

- Summaries ignore the complexity inherent in longitudinal data.

- To combat prejudice against complex statistical analyses, one must know why simple analyses are not optimal and how complex statistical analyses improve on them.

The analyses presented in the previous chapter are easy to implement, especially as compared to analyses in upcoming chapters. Chapter 3 analyses are covered in introductory statistics classes, and many software packages exist to handle the required calculations. They are analyses we have likely used frequently in the past, and we can produce them relatively quickly when faced with a new data set. Chapter 3 analyses are also more familiar to non-statisticians than the analyses from later in this book; they therefore may be easier to communicate and easier to sell to non-statisticians.

Statistical methodology is an advanced technology. It has grown in parallel with the growth of data collection and computing power. Unlike many technologies where there often seems to be a blind prejudice in favor of advanced technology, in disciplines where statistics is applied, we sometimes find that "better, more sophisticated" technologies are the subject of discrimination. This is likely due to ignorance of the advantages of the technologies. The need for advanced statistical technology and the lack of familiarity is driven in part by traditionally non- or not-very-quantitative disciplines turning quantitative. Biology is an example of a discipline that turned quantitative in the 1990s. In contrast, psychology and economics are fields that have strong quantitative traditions, notably psychometrics and econometrics. Medicine has a less strong quantitative tradition and there we occasionally find problems of prejudice against advanced statistical techniques. Also, lack of proper marketing on the part of statisticians has undoubtedly contributed to this prejudice.

Unfortunately all of the analyses presented in chapter 3 have serious drawbacks, if not outright deficiencies. There are three sets of key issues:

Efficiency loss results from not using all of the relevant data and information available to us.

Bias results from selectively omitting or ignoring data.

Over-simplification of summaries causes us to overlook the richness of information inherent in longitudinal data.

In the following discussions, efficiency loss is easily described. Bias and over-simplification blend into each other as topics. For all three topics, we present simple yet suggestive examples but do not present definitive mathematical results.

4.1   Efficiency Loss

Experimental design is the part of statistics that studies the trade-offs between cost of data collection and resulting gain in information. The efficiency of a design is the ability of the design to maximize the information gain for the cost. For example, we may decide to sample J observations on each of n subjects. If we make J smaller, we may be able to increase n or conversely. What choice of J and n is best?

Similarly, when analyzing data, the statistical analysis itself must be designed. Design of statistical analyses is a favorite task of statisticians. A good statistical analysis maximizes the information that we extract from the data for given resources. A good statistical analysis produces estimates and inferences that are *efficient*, whereas a poor analysis loses efficiency compared to the more efficient and therefore better analysis. Resources that affect the design of a statistical analysis are the availability of time, money, data, computing power, and statistical knowledge.

We often measure information by the standard errors of our estimates; the smaller the standard errors, the greater the information. Similarly, power is the ability to reject a null hypothesis when an alternative hypothesis is true. Appropriate modeling of our data allows us to achieve the greatest power for hypothesis tests and smallest standard errors for parameter estimates. Essentially, efficiency means minimizing variability and maximizing information. Ignoring data relevant to a hypothesis reduces the efficiency of the analysis, causing increased standard errors and decreased power. There are several ways that we may ignore relevant data in the analysis of a longitudinal data set.

4.1.1 Omitting Subjects

We want to analyze data from all the subjects that we have in a study. When we analyze a subset of the subjects in our study, we lose efficiency. It seems silly to even consider dropping subjects from an analysis much less discussing it. We might even propose analyzing all subjects as a general principle of data analysis. Yet we did drop 6 subjects in our Pain data analyses in the previous chapter. We did not know how to (easily) incorporate subjects with missing data into the analysis and consequently omitted them from the analyses. Was that a good idea? Well, if all we know how to do are two-sample and paired *t*-tests then it was necessary to get to an analysis. But the principle to never omit subjects from the analysis seems to be almost unquestionable. The longitudinal models that we learn about in later chapters will have no trouble including subjects with missing data. If there is any information in cases with missing data that is relevant to our conclusions, longitudinal data models will use that information to draw conclusions.

4.1.2 Omitting Observations

Omitting observations is another way to omit relevant information in our analysis.

We took various subsets of the observations in the Pediatric Pain data analyses of chapter 3. This was done to manipulate the data into a structure that we knew how to analyze with various *t*-tests. Is this a good reason to ignore the first and third observations? Do we know that the first and third observations are very different from the second and fourth observations? They are a little different, as they were done with different hands, but do we know that subjects keep the dominant and non-dominant hands for different lengths of times in the cold water? The hand could well be unimportant for this experiment.

Let us consider in detail the paired t-tests presented in table 3.4. The purpose of taking the difference $Y_{i4} - Y_{i2}$ is to compare the post-treatment response to the pre-treatment response. We used Y_{i2} as representative of the pre-treatment response: same arm, same number of previous practice trials in the same session. However, Y_{i1} and Y_{i3} are also pre-treatment measures. Are there reasons to use those observations instead of Y_{i2}? For example, Y_{i3} was taken on the same day as Y_{i4}. Should we do three different paired t-tests, one for each baseline? Hopefully, each would give the same answer. In analyzing this data, I have not seen any systematic differences among the Y_{i1}, Y_{i2} and Y_{i3}. Perhaps all should be used to estimate the baseline response? We might calculate

$$\bar{Y}_{i,\text{baseline}} = \frac{1}{3}(Y_{i1} + Y_{i2} + Y_{i3})$$

as a summary of the three baseline responses. Then our paired t-test could take differences

$$d_i^* = Y_{i4} - \bar{Y}_{i,\text{baseline}}.$$

Averaging all three baseline measures makes better use of our data and makes for a more efficient estimate of the baseline response for each person.

When we did the paired t-test in table 3.4, we averaged the d_i over the subjects in the group g we are studying. Let N_g be the sample size of this group.

$$\bar{d}^* = \frac{1}{N_g} \sum_i d_i^*$$

where the sum is over subjects i in the group under study. Then $\bar{d}^*$ is the numerator of the paired t-test. Equivalently, this is the mean of the time 4 observations minus the mean of the baseline responses

$$\bar{d} = \frac{1}{N_g} \sum d_i = \frac{1}{N_g} \sum_i y_{i4} - \frac{1}{N_g} \sum_i \bar{y}_{i,\text{baseline}}.$$

But at baseline, there is no difference between the three treatment groups in the Pain data. So to estimate the baseline difference, we really ought to combine the data for the three groups. This will give us a more efficient estimate of the baseline mean, meaning that it would have a lower variance than the estimate based solely on the subjects in group g. The problem is that then we will not have a paired t-test analysis.

We removed subjects with missing data from the analyses. But they have information about the average baseline response for everyone in the coping style group they belong to. Can't we include them in the analysis somehow also?

Finally, we have to estimate the standard deviation of the differences. Could this standard deviation be the same across different groups? If so, perhaps we could pool sums of squares from different groups to estimate it.

This discussion was mainly about the paired t-test analyses. We found that we might be able to do a better job estimating the baseline average response than just using the Y_{i2} observations from the group under study, and we suggested, but did not explicitly identify, the possibility that there may be ways to improve our estimate of the variance. These problems exist for all of the analyses discussed in the previous chapter, including the DoD design. The methods of our later chapters will use all of our data appropriately and allowing us to efficiently estimate baseline and treatment differences of interest.

4.1.3   Omitting Subjects and Observations: Some Special Cases

There are a few situations where we do omit or appear to omit subjects or observations from our analysis.

When there are too many subjects to comfortably manipulate the data and to analyze using present-day computer programs and computing power, we may randomly sample a subset of the data to create a manageably sized data set. This lets us produce an appropriate analysis of the data. Too much data can well happen in practice, particularly in industrial and business applications or in large scale surveys where there may well be hundreds of thousands or millions of subjects in a data base. Essentially the design of the data collection has an additional step where we subsample the original data set.

An example where apparently omitting data can be appropriate and useful occurs when an analysis is performed separately on two disjoint and exhaustive subsets of a larger data set. In psychological studies, we often analyze men and women separately because men and women respond differently on many outcomes. As long as every subject contributes to one (and only one!) analysis, then we have fully analyzed the data set. Typically we analyze groups defined by levels of a discrete covariate. In statistical terms, we say that we have allowed for interactions between the covariate and all parameters in the model. Suppose we were performing regression analyses of a psychological outcome regressed on treatment. Analyzing men and women separately allows for a separate intercept and slope and residual variance for men and for women.

4.2   Bias

Bias can be introduced into an analysis at many stages and in many different forms. Bias can be introduced into our analysis (i) through bad design, (ii) by subjects who may drop out for reasons related to our study, and (iii) by the analyst through mis-analysis of our data. We talk about each of these in turn.

4.2.1   Bias By Design

Bias can enter if we do a poor job of subject selection. For example, if we select very healthy subjects and assess measures of health, we may generalize our outcomes only to the very healthy. We err if we attempt to generalize our study to the population as a whole. The selection of healthy subjects may be done by design or it may have been done by accident. If by design, then hopefully we intended to generalize to the population of healthy subjects and there will not be a problem. Sometimes we acquire a healthy or unhealthy population unintentionally. If, for example, we select random people entering a workout gym, versus if we pick random people entering a hospital, we will get very different samples. If we pick random people accompanying the people entering the hospital, we might find an acceptable mix of healthy and sick people.

If we do a bad job of randomization, we may end up with healthier subjects in, say, the treatment group, causing their outcomes to be better than the control group, and leading to erroneous inference about the treatment. In particular, we might over-estimate its effectiveness. This will introduce bias into our conclusions no matter how we analyze our data.

4.2.2   Bias From Bad Analysis

Bias is introduced into an analysis when we analyze a non-random subset of otherwise well collected data. Our subjects may help us introduce bias into our study by dropping out when things go poorly (or too well!). Problems 2 and 3 present an important example.

Unfortunately, many reasons for subsetting longitudinal data introduce bias into the analysis. Particularly problematic is any analysis that omits subjects with missing data. Removing subjects with missing data changes the population we are studying. It eliminates anyone who for any reason has produced missing data. If they are missing for innocuous reasons unrelated to our study, then this does not change our population and does not change the desired inference, although it may lose some efficiency. But suppose that sicker people, or poorer, or less intelligent, or more active subjects miss data collection visits and so miss observations. Generally we can say that subjects with missing data are poor compliers; if they have trouble complying with data collection, they may well also have trouble complying with the intervention also. Dropping these people alters the population we are studying and therefore changes the inferences that we make. Dropping poor treatment compliers will over-estimate the treatment effect in the general population. Dropping subjects with some missing observations cannot be recommended, and analyses that permit missing or unbalanced observations are preferred to those that require balanced data.

4.3   The Effect of Missing Data on Simple Summaries

It is easy to assume that a particular summary is appropriate when it is not. Unbalanced or randomly spaced measurements can cause apparently reasonable summary measures to become unreasonable by introducing bias and variance into the simple analyses.

To illustrate, consider two subjects, one observed at times $t = 1, 2, 3$, and the second observed at times $t = 4, 5, 6$. Suppose that observations from both subjects follow the straight line $Y_{ij} = t_j = j$. If we summarize subjects by the average response, then the first subject will have an average of 2, while the second subject will have an average of 5, and we might erroneously conclude that the two subjects were quite different in their responses, when in fact the difference is due to the different times of measurement, and had nothing to do with the subjects at all!

More generally, suppose we want to compare two groups on their average levels. Suppose all subjects' responses are increasing linearly over time and the rate is similar on average between the two groups. Now suppose one group tends to be observed early and one group tends to be observed late, we will see a difference in the groups, and this difference will be due solely to the times of observation and not due to the actual differences in average response between the groups. Differential times of measurement in our subjects can lead to bias of different forms, depending on how we summarize the data.

We have similar problems if we summarize subjects' observations by the slope, when the profiles do not follow a linear trend. Again consider two subjects, both of whose profiles follow a quadratic trend, $Y = (t - 3)^2$ as a function of time. If subject one is observed at times $t = 0, 1, 2$, and 3, while subject two is observed at times $t = 3, 4, 5$, and 6, the estimated slope for subject one is -3, whereas the slope for subject two is $+3$. The problem is similar to the previous example where we used means to summarize profiles that had a time trend. There are two problems: (i) we used a summary measure that did not acknowledge the complexity of the data, and (ii) the two subjects are measured at different times.

Different summaries have different problems in various scenarios. We give a number of examples in the following sections.

4.3.1   Summarizing Profiles by the Mean

Consider averaging the Y_{ij} responses, as in subsection 3.1.1. Consider various special cases.

4.3.1.1   Balanced Data, Flat Profiles

By flat, we mean profiles with no slope or trend over time. Averaging the Y_{ij}'s is not such a horrible thing to do. Inferences will be valid except for a slight loss of efficiency in analyzing means as opposed to the random intercept models that we learn about later.

4.3.1.2   Unbalanced Data, Flat Profiles

Now a problem occurs. The $\bar{Y}_i$ no longer have equal variances. A proper analysis must take into account the differences in variances. Unfortunately, the correlation model of the observations within a subject causes the unequal variances to be complex functions of n_i, the times of observation and the parameters of the covariance matrix of the Y_{ij}. Even for the simplest covariance model for longitudinal data, the random intercept model, the variance of $\bar{Y}_i$ is still a complicated function of n_i.

4.3.1.3   Linear Profiles

Suppose the data profiles fall along subject-specific lines. Is an average what we want to summarize the profiles by? If observations are randomly spaced, then the quantity that the average estimates depends on the t_i vector. For example, if

$$Y_{ij} = \beta_{i1} + \beta_{i2} t_{ij}$$

where β_{i1} and β_{i2} are the intercept and slope of the ith subject's observations, then

$$\bar{Y}_i = \beta_{i1} + \beta_{i2} \bar{t}_i,$$

where

$$\bar{t}_i = n_i^{-1} \sum_{i=1}^{n_i} t_{ij}.$$

Unless $\bar{t}_i$ is a constant across subjects, this mean is a different quantity for each subject. A t-test of equality of means for two groups may give misleading results if the two groups have strongly different missing data patterns or different times of observations.

4.3.1.4 Nonlinear Profiles

Consider summarizing a nonlinear profile by the mean response. What is the interpretation of the mean? It varies with the times of observations. Even if our study has balanced data, the next study will not likely have the same times of observations, and our estimates will not generalize to a new study. With random or unbalanced times, the interpretations of the mean will be different across subjects.

4.3.2 Slopes

Consider summarizing longitudinal data by individual slopes. If the data are not balanced, then each slope has a different variance depending on the pattern of the observed times. Combining the slopes to estimate the grand mean of the slopes needs to take the variances of the slopes into account. Unfortunately, like taking the mean of unbalanced data, the variances are complex functions of the t_{ij} and the covariance matrix of the observations.

If profiles are nonlinear, then the slope will depend on the distribution of the t_{ij}'s. The interpretation of the slope depends on the t_{ij}'s and is problematic.

4.3.3 Weighted Least Squares Analysis

Sometimes people do a *weighted least squares* analysis to analyze data where there is known differences in the accuracy (i.e., variability or standard errors) of the responses for different subjects. This requires that the standard errors be known for each subject's response. Unfortunately, the standard errors reported by regression software are not correct for input into the weighted least squares analysis, because of the correlation among the observations within a person. As indicated for the unbalanced data, flat profiles example above, even under the simplest models for longitudinal data, the variances of the mean or slope estimates are not simple functions of the times and numbers of observations. Doing the correct analysis using weighted least squares is unlikely to be easier than the longitudinal analyses we suggest in later chapters, which naturally adjust parameter estimates and hypothesis tests for unbalanced numbers and times of observations. Farther into the future, the longitudinal analyses will generalize naturally and easily to more complex data structures; while weighted least squares analyses and similar analyses will require additional work.

4.4 Difference of Differences

The DoD t-test is a one number summary of the four means: baseline and followup means for the treatment group and for the control group. The four means can take many forms yet lead to the same statistically significant test result. Many of these patterns prevent unequivocal conclusions. Proper analytical and graphical supplementation is needed before drawing a final conclusion.

Figure 4.1 presents six possibilities. In each subfigure, we plot the baseline $t = 1$ and followup $t = 2$ group means for the treatment ($+$) and control ($\circ$) groups, connecting each group's means by a line segment (dashed: treatment; solid; control). In looking at these pictures, let us suppose that the control group was intended not to change from baseline to followup, and that the treatment group was expected to increase substantially over time. Also, let us suppose that, because of randomization, the two groups are supposed to be comparable at baseline.

Figure 4.1(a) exemplifies the prior expectations about the results. The two groups have the same baseline mean, implying comparability at baseline in terms of the average response. The control group did indeed remain constant, whereas the treatment group response was much larger at the end of the study. Assuming that this increase is both significant and of substantive interest, then 4.1(a) represents the canonical outcome for a difference of differences experiment.

Figures 4.1(b)–(f) represent five alternative sets of means for the DoD study. In figures (b)–(e), the difference in mean differences is one unit; that is, the treatment group improves by one unit more than the control group improves. All could potentially have the exact same t-statistic for testing the null hypothesis that the treatment group improves the same amount as the control group.

Figure 4.1(b) depicts equality between treatment and control groups at baseline, both groups improve but the treatment group improves more than the control. Issues to decide here are, are the groups significantly and importantly different at followup? Why is there an improvement in the control group? Is the group difference in improvement caused by a single subject or subjects? Occasionally, a transformation such as the log of the responses Y_{ij} can eliminate much of the difference in the groups at followup, so it is necessary to look at the individual observations Y_{ij} at baseline and at followup and to decide that we are analyzing the data on the correct scale.

In figures 4.1(c) and (d), the groups are different at baseline. In (c), the treatment group is worse off at baseline and better off at followup. The fact that treatment started worse, yet ended better is probably a plus for the evaluation for the treatment. Still, there will always be a nagging worry. The two groups were not comparable at baseline, and this may be the cause of the relative improvement in the groups. In figure (d), treatment is better at baseline. In evaluations of treatments for various diseases, non-randomized studies are often considered unreliable for evaluating the treatment. The worry is that the treatment group will be picked, even if unconsciously, to perform better than the control group. The results of the study can never be considered definitive since the groups are not comparable. In our example, we are assuming that the two groups were randomized. Randomization can sometimes result in groups that are not comparable and when this happens, we are unable to draw definitive conclusions. Observational studies can often show results like figure (d), and while we can find a significant result suggestive, the result cannot be considered definitive.

Figure 4.1(e) looks like a success for the treatment, but recall that we were expecting the treatment group to improve and the control group to stay constant. That both groups declined has to be explained before determining that the treatment works in treating the disease. Often with this outcome, as many questions are raised by the study as are answered.

Figures 4.1(c), (d), and (f) begin with non-comparable groups at baseline. Any differences in changes in the two group means may be ascribed either to the treatment or to the fact that the two groups are not comparable at baseline.

In figure 4.1(f), we see a disappointing result, but a conclusion may be tentatively drawn that the treatment causes no increase in response, as we see no change in the treatment group. Figure 4.1(f) shows why we additionally need baseline measures on subjects. If we only look at the followup treatment observations, the treatment group looks to be better than the control. However, neither group improved over time and the treatment group was just as much better than the control group at baseline.

A concern is to ask whether transforming the response scale to another scale eliminates the difference in improvements. Many difference of difference response scales may not have an absolute meaning. Suppose for example that the response is an educational or a psychological test. A difference of one unit, say, on the response scale at different scores in the exam may not have the same meaning. That is, a change from 88 to 98 on a test may be more or less important than a change from 8 to 18. Transforming individual scores can eliminate the interactions that were seen in all figures of 4.1 except figure (d).

For example, suppose that an increase as seen in 4.1(a) or (b) or especially (d) is due to a few subjects with very large increases and everyone else is remained similar between baseline and followup. Taking a log or square root transformation of the response will bring these extreme subjects in toward the group mean while keeping the rest of the subjects similar. This can eliminate the significance of differences in these examples. When results depend on the transformation interpretation of the conclusion is again problematic.

4.5 The Richness of Longitudinal Data

Occasionally, in our haste to *shoehorn* our data into a *t*-test, we may completely miss the boat on key issues. The vagal tone data of table 2.9

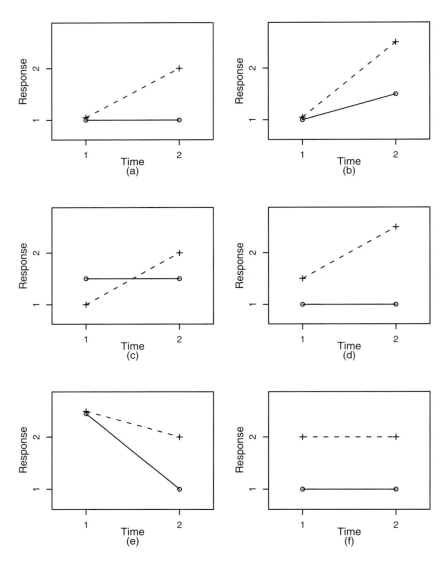

Figure 4.1. Examples of baseline and followup mean configurations in the difference of difference design. The two groups are + (treatment) and ∘ (control), with their baseline ($t = 1$) and followup ($t = 2$) group means connected by lines: solid – control, dotted – treatment. (a), (b), (e) Group means are equal at baseline; (c) tmt group is worse at baseline, (d) and (e) tmt group is better at baseline. (a), (b), (c), (d), (e) Tmt group improves 1 unit more than control group. (c), (d), (e) The groups are different at baseline, so a firm conclusion is not possible. (f) Treatment did not improve outcome, but groups differed at baseline.

illustrates this point. These data have two pre-cardiac catheterization measures (pre1 and pre2) and three post-catheterization measures (post1, post2 and post3). If we wish to do a paired t-test, we need to settle on a particular baseline measure and a particular followup measure. Because vagal tone drops after the catheterization and then hopefully returns to normal some time after the stressful event, we would prefer to use the first post1 measure as our followup measure. As the closest observation before the catheterization, we would use pre2 for our pre-measure. Unfortunately, there are 10 missing post1 measures out of 20 infants and a further three missing pre2 measures for the 10 observed post1s. We could waste a substantial amount of time setting up rules of when we might substitute post2 or post3 or averages of the two in place of post1 when it is missing. Similarly, we might use pre1 for our pre- measure rather than pre2 when pre2 is missing.

However, this mostly misses the point. Inspection of figure 2.27(b) suggests that there is a separate level for each infant, that the response at the first two times is approximately the same, the response decreases at post1, and then increases back to baseline or above at post2 and post3. A better approach would allow us to use all of this information in a single statistical method. The method should allow us, for example, to use either, both or neither pre- measures as appropriate for estimating the pre- average response. Similarly, we would like to use all of the observed post1 measures to estimate the immediate post- response. A t-test will never allow us all this flexibility. A statistical model is required, not a single hypothesis test.

We have other problems with the simple summary approach. Longitudinal data are rich and the statistical methods need to acknowledge this richness. A single summary measure can be flawed in many examples. Consider again the first example of section 4.3, where one subject had observations at times 1, 2, and 3, while subject two had observations at times 4, 5, and 6, and the response at each time t_j was equal to j. If we take a bivariate summary, namely the intercept and slope, we get that subject one has an intercept and slope of 0 and 1 respectively as does subject two, and we recognize that both subjects have the same pattern of response. This need to assess both the intercept and slope is ubiquitous in longitudinal data, and is explicitly allowed for in the methods we develop later.

Even multiple summary measures, analyzed independently can miss interesting structure in our data. Supposing that observations fall about subject-specific lines, then we have the set of intercepts and slopes. It is not enough to summarize the various intercepts and the various slopes. Confer figure 4.2. In all three figures, profiles fall along subject-specific lines. The population average response at all times is 25. In all three figures, the mean intercept is 25, and the standard deviation of the intercepts is 4. In all three figures, the standard deviation of the slopes is 1.4. But each set of profiles is different. In 4.2(a), slopes and intercepts are positively correlated and over time the profiles spread apart. In figure 4.2(b), slopes and intercepts are uncorrelated and the profiles spread apart but not nearly at the same rate

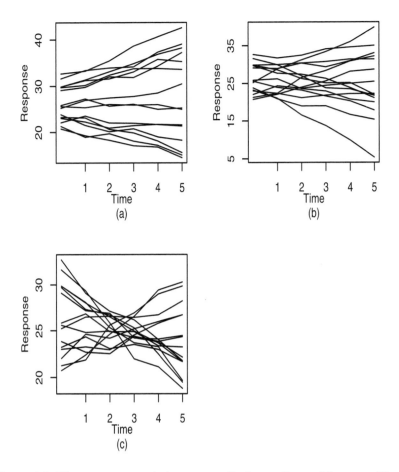

Figure 4.2. Three constructed data sets, all observations with mean 25 at all times. Intercepts have mean 25 and standard deviation 4. Slopes have mean zero and standard deviation 1.4. (a) Strong positive correlation between slopes and intercepts; (b) zero correlation between slopes and intercepts; (c) strong negative correlation between slopes and intercepts.

as in figure (a). In figure 4.2(c), slopes and intercepts are highly negatively correlated, so that subjects who start low rise up over time, whereas those who start high drop over time. We will miss this if we analyze the slopes and intercepts separately.

A serious problem with univariate summaries is that to provide a simple and practical methodology, the methodology must ignore the covariances among the observations. If subjects have unbalanced or random times of observation, then the problem of adjusting the univariate methodology for the covariances is problematic; a longitudinal analysis is much easier.

Univariate summaries do not allow us to predict future outcomes, either within subject or for new subjects. Our description of subjects' observations falling around subject-specific lines for figure 4.2 lets us make future predictions for subjects where we already have enough data to estimate their line. If we add in the knowledge about the correlation between the intercepts and slopes, we can begin to make predictions for subjects with only a single baseline observation.

Another, more subtle problem, is that there are better ways to estimate individual profile summaries than by only using the data from the individual alone. These methods are called empirical Bayes estimators methods and are discussed in chapter 9.

Summarizing subject profiles, no matter how cleverly, involves information loss. Using a reduced data set is inefficient. In the Pain data, we have three pre-treatment measures, and the loss of efficiency is potentially severe when we use only a single measure to represent all three. Perhaps the biggest deficit of simple analyses is that they hide information. Summarizing profiles first, then analyzing them loses information about the complexity of the profiles. If we summarize profiles by intercepts and slopes, and then only analyze those, we cannot discover many features in the profiles. We won't be able to see nonlinearity in the profiles nor non-constant variance except for non-constant variance of the intercepts or slopes.

4.6 Discussion

The process of adapting data to familiar statistical tools rather than adapting our tools to the data is so ubiquitous that it has a name: the *shoehorn*. We *shoehorn* our data into the shoe we have rather than getting a larger shoe as our data grows. Another relevant adage is the one about when we own a hammer, all problems tend to look like a nail. When all we know about are *t*-tests, then all analyses are solved with *t*-tests.

The term *shoehorn* is disparaging; the disparagement is well earned. The insidious flaw with the shoehorn is that in adapting the data to the tools rather than vice versa, we limit the scientific questions that we ask to those answerable by the tools we use.

Paired and two-sample *t*-tests are seemingly ubiquitously applicable, but they provide yes-no reject/do-not-reject answers to binary yes-no questions. There is much more to research than asking yes-no questions. Anyone that has played the game 20 questions knows how limiting a yes-no question can be when you want to ask a *what* or a *how much* type question.

It is reasonable to begin a data analysis with plots first and then with simple summaries of the data. This gives us an idea of what the data look like. But analyses of summary measures throw away information that more holistic models can incorporate and use to improve efficiency and reduce bias in the conclusions. It is important to move on to a full longitudinal data model. The longitudinal models we discuss starting in the next chapter better accommodate issues with our data and with our science, for example missing data and complex hypotheses.

Our approach in this text is to develop a complete statistical model of our data, including models for the mean over time, and the covariance among the observations and the effects of covariates on the mean and possibly the covariance. The statistical model approach to data analysis develops a description of the data and by inference of the world. This answers a complex question not easily summarized by yes-no reject/do-not-reject answers. We may then summarize our model in any way we desire. For any simple summary of our data, there is a corresponding summary at the end of the longitudinal analysis. The longitudinal model summary will use the data as efficiently as possible to answer the scientific questions of interest.

4.7   Problems

1. This problem continues problem 2 from chapter 3. Suppose that children's responses follow straight lines.

   (a) When we calculate an average for one child, what does it estimate (i.e., the child's response at what time)? Is this quantity the same time for each child?

   (b) Suppose we have various patterns with 1 up to 3 missing observations for some children. We take averages $\bar{Y}_i$. Suppose that the individual profiles follow a linear trend plus residual error. Exhaustively list all possible times $\bar{t}$ at which the average $\bar{Y}_i$ potentially could estimate the child's response at.

   (c) Are slopes and averages reasonable summary measures of this data set? (Yes/No) Are they all that is needed? (Yes/No, + one sentence).

2. Consider a hypothetical diet study where we measure subject's weight at baseline and every 2 weeks for 10 weeks. Complete data for 20 subjects is given in table 4.1. The bold entry in some rows indicates that a subject's weight increased to four or more pounds over their baseline weight.

   (a) Do a paired t-test using the weight at week 10 and the baseline weight using data for all 20 subjects.

   (b) Suppose that subjects with observations in bold font drop out of the study just BEFORE that first bolded observation. Now do a paired t-test (week 10 minus baseline) using the 10 complete observations only.

Subject	Weight (pounds)					
number	0	2	4	6	8	10
1	219	222	221	215	214	211
2	167	161	165	161	158	159
3	152	144	144	147	145	146
4	194	193	193	189	191	189
5	235	231	235	236	235	231
6	200	199	195	199	195	197
7	181	183	177	175	180	180
8	211	204	210	208	211	211
9	250	249	247	251	251	250
10	210	209	209	212	212	213
11	142	140	143	142	**147**	*144*
12	192	195	193	193	**198**	*201*
13	217	217	220	219	**224**	*226*
14	187	189	189	**193**	*195*	*202*
15	188	**196**	*194*	*191*	*193*	*191*
16	187	**191**	*187*	*187*	*181*	*179*
17	184	**189**	*184*	*188*	*189*	*189*
18	197	**201**	*200*	*198*	*201*	*204*
19	200	**205**	*205*	*206*	*209*	*210*
20	219	**223**	*225*	*222*	*226*	*231*

Table 4.1. Constructed data for problem 2 and 3. Boldface entries indicate the first observation where a subject's weight increased to four or more pounds over their starting weights. Thereafter, observations are italicized.

(c) For subjects with at least two observations, do a paired t-test using the last recorded (i.e., last regular typeface) observation.
(d) Suppose that subjects drop out just AFTER the boldface observation. For all subjects, use the last available observation for the "after" measurement and baseline for their before measurement. Repeat the paired t-test again.
(e) Compare the conclusions for the three t-tests. Is the diet effective? Is it possible that a naive researcher might conclude differently? Draw a moral from the analyses.
(f) For those subjects who drop out, how does their last observation compare to the observation at week 10? Is the 10-week observation always higher?
(g) Explain how this example provides an argument for the DoD design over the single sample paired t-test.

3. Continue with the data from problem 2 and table 4.1. Calculate weight differences at week j as week j weight minus baseline weight for each subject for $j = 2$ through 10. There are three versions of the data set: (i) the complete data, (ii) those differences that do not

use italicized data, and (iii) those differences that omit both italicized and bold font data. For each of the three data sets, draw the following plot.

(a) Draw a profile plot of the differences.
(b) Draw an empirical summary plot.
(c) For the two sets of three plots, decide if the plot suggests that there is a diet effect or not.
(d) Draw a conclusion.

4. Suppose the story underlying figure 4.1 had been that subjects should get worse, and that the treatment is considered successful if subjects stay the same.

(a) Sketch a figure like those in figure 4.1 to illustrate what we expect to happen.
(b) Explain what conclusions one would draw for each plot of figure 4.1. Which plots lead to unequivical conclusions?
(c) Suppose the story underlying figure 4.1 had been that subjects should get worse, and that the treatment is considered success-ful if subjects decline less slowly. Now answer the previous two questions.
(d) The figures in 4.1 mostly assume that the treatment gets better. Many additional possibilities are possible. Create 6 more possi-ble figures. Identify which of them suggest that the treatment works.

5
The Multivariate Normal Linear Model

I'm very well acquainted too with matters mathematical,
I understand equations, both the simple and quadratical,
– Gilbert & Sullivan

Not everything that can be counted counts, and not everything
that counts can be counted.
–Albert Einstein

Overview

- The multivariate normal model

- Needed elaborations to the multivariate normal model

- Parameterized covariance matrices: an introduction to the
    - compound symmetry
    - autoregressive
    - random intercept and slope

  covariance models.

- Multivariate linear regression models
    - Covariates
    - Parameter estimates
    - Tests and confidence intervals for regression coefficients
    - Contrasts
    - Inference plots

5.1 Multivariate Normal Model for Balanced Data

We model our response Y_{ij} on subject i at time t_j with a normal distribution

$$Y_{ij} \sim N(\mu_j, \sigma_{jj}) \qquad (5.1)$$

with mean μ_j and variance σ_{jj}, for observations $j = 1, \ldots, J$. We initially work with the model where the mean μ_j and variance σ_{jj} depend only on j and where all subjects have $n_i = J$ observations taken at the same times. The difficulty with the univariate specification (5.1) is that we have multivariate data

$$Y_i = \begin{pmatrix} Y_{i1} \\ \vdots \\ Y_{iJ} \end{pmatrix}$$

collected longitudinally on subjects. Longitudinal data have non-zero, usually positive, correlations among observations within a subject. We must incorporate these correlations in our modeling. It is inconvenient to write model (5.1) and then separately specify the covariance σ_{jl} between Y_{ij} and Y_{il} where l is a subscript that runs from 1 up to J. For example

$$\mathrm{Cov}(Y_{ij}, Y_{il}) = \sigma_{jl} = \sigma_{lj} = \mathrm{Cov}(Y_{il}, Y_{ij}) == \mathrm{E}[(Y_{ij} - \mu_j)(Y_{il} - \mu_l)]. \quad (5.2)$$

These two symbols σ_{jl} and σ_{lj} are equal; they are a single parameter with two names. A more compact notation allows us to combine equations (5.1) and (5.2) in terms of the full vector of observations Y_i and a multivariate normal model

$$Y_i \sim N_J(\mu, \Sigma). \qquad (5.3)$$

The subscript J on N signifies that Y_i is a J dimensional multivariate normal random variable. The mean μ is a J-vector of means

$$\mu = \begin{pmatrix} \mu_1 \\ \vdots \\ \mu_J \end{pmatrix}$$

with jth element $\mu_j = E[Y_{ij}]$. The *expected value* or *average value* or *expectation* of Y_i is

$$E[Y_i | \mu, \Sigma] = \mu.$$

The vertical bar "|" is read "*given*" and the quantities to the right of the bar, here μ and Σ, are assumed known.

The covariance matrix of Y_i is a J-by-J matrix Σ with elements $\sigma_{jl} = \mathrm{Cov}(Y_{ij}, Y_{il})$

$$\Sigma = \begin{pmatrix} \sigma_{11} & \cdots & \sigma_{1J} \\ \vdots & \ddots & \vdots \\ \sigma_{J1} & \cdots & \sigma_{JJ} \end{pmatrix}.$$

The covariance matrix Σ has many names. In words, it is called the *variance matrix* of Y_i and the *variance-covariance matrix* of Y_i. Using notation, we have $\Sigma = \text{Var}(Y_i)$ and $\Sigma = \text{Cov}(Y_i)$. The formula for the covariance is

$$\sigma_{jl} = \text{E}[(Y_{ij} - \mu_j)(Y_{il} - \mu_l)].$$

In matrix form we write the covariance as

$$\text{Var}[Y_i] = \text{E}[(Y_i - \mu)(Y_i - \mu)'] = \Sigma.$$

Matrix Σ has J variance parameters and $J(J-1)/2$ covariance parameters for a total of $J(J+1)/2$ unknown parameters.

The correlation of Y_{ij} and Y_{il} is

$$\text{Corr}(Y_{ij}, Y_{il}) = \rho_{jl} = \rho_{lj} \equiv \frac{\sigma_{jl}}{(\sigma_{jj}\sigma_{ll})^{1/2}}.$$

The two correlation parameters $\rho_{jl} = \rho_{lj}$ are equal and as before with $\sigma_{jl} = \sigma_{lj}$, are really a single parameter with two names. If we wish to talk about the correlation matrix separately from the covariance matrix, we define the correlation matrix R with elements ρ_{jl}

$$\text{Corr}(Y_i) \equiv R \equiv \{\rho_{jl}\}.$$

The notation $R \equiv \{\rho_{jl}\}$ means that the matrix R has elements ρ_{jl} and the $\equiv$ sign is read as "is defined to be," or "equivalently."

Observations from the same subject are correlated, whereas observations from different subjects are independent. The model (5.3) says that each Y_i has the same distribution. We say that subjects are *independent and identically distributed*, abbreviated *iid*. This is the *multivariate normal model for balanced data*. Later we will introduce regression models to allow subject means to vary from subject to subject according to covariate values.

There are advantages to writing our model as in (5.3) instead of in (5.1) and (5.2). One advantage is the compactness of model (5.3); observation means and variances and correlations among observations are explicitly acknowledged. We can communicate information more quickly using the more compact notation. The second advantage relates to the development of algorithms to estimate the unknown parameters μ and Σ from sample data Y_i, $i = 1, \ldots, n$. Algorithm development requires a knowledge of matrix algebra, calculus and numerical analysis. The mathematical operations required flow much more easily using the multivariate model specification (5.3) rather than the separate specifications (5.1) and (5.2).

5.1.1 Estimating μ and Σ

Model (5.3) has unknown parameters μ and Σ to make inference on. We will use θ, a vector of length $J(J+1)/2$, to refer to the unique parameters in Σ

$$\theta = (\sigma_{11}, \sigma_{12}, \ldots, \sigma_{1J}, \sigma_{22}, \sigma_{23}, \ldots, \sigma_{2J}, \ldots, \sigma_{JJ})'.$$

This lists the parameters in Σ starting with the entire first row, then the $J-1$ unique parameters in the second row, and so on, ending with σ_{JJ}. The prime symbol $'$ indicates the transpose of the preceding vector or matrix.

We assume basic familiarity with the univariate bell-curve shape of the normal distribution. Normal observations Y_{ij} are most likely to appear close to their mean μ_j. The standard deviation $\sigma_{jj}^{1/2}$ is a ruler that measures how far from the mean observations are likely to fall. We know that 68% of observations fall within one standard deviation of the mean $(\mu_j - \sigma_{jj}^{1/2}, \mu_j + \sigma_{jj}^{1/2})$ and that 95% of observations fall within two standard deviations of the mean $(\mu_j - 2\sigma_{jj}^{1/2}, \mu_j + 2\sigma_{jj}^{1/2})$. The multivariate normal also has a bell-curve shape in higher dimensions; under model (5.3), Y_i is most likely to be found in the neighborhood of its mean μ; this is where the density of Y_i is highest. As Y_i gets farther and farther from μ, the density gets lower and lower.

Given a sample Y_i, $i = 1, \ldots, n$, the average of the observations is

$$\bar{Y} = n^{-1} \sum_{i=1}^{n} Y_i$$

where $\bar{Y}$ has J components $\bar{Y}_j$. The mean $\bar{Y}$ is an unbiased estimate of μ

$$\mathrm{E}[\bar{Y}] = \mu.$$

The estimate of μ is $\hat{\mu}$, and because $\bar{Y}$ estimates μ, we write

$$\hat{\mu} = \bar{Y}.$$

Recall that the *hat* " $\hat{\ }$ " on a parameter indicates an estimate of that parameter.

The J by J *sample covariance matrix* is

$$S = (n-1)^{-1} \sum_{i=1}^{n} (Y_i - \bar{Y})(Y_i - \bar{Y})'.$$

It has individual elements $s_{jl} = (n-1)^{-1} \sum_{i=1}^{n} (Y_{ij} - \bar{Y}_j)(Y_{il} - \bar{Y}_l)$. The matrix S provides an unbiased estimate of Σ and is our estimate of Σ

$$\tilde{\Sigma} = S.$$

The tilde " $\tilde{\ }$ " on top of an unknown parameter also denotes an estimator of an unknown parameter. Later, we will distinguish between two types of estimates and use hats to identify one type of estimator and tildes to identify the other.

The covariance matrix of the estimate $\hat{\mu}$ is

$$\mathrm{Var}(\hat{\mu}) = \frac{\Sigma}{n}. \tag{5.4}$$

Because we do not know Σ, we plug in its estimate S giving the estimated variance S/n. The *standard error* of $\hat{\mu}_j$ is $s_{jj}^{1/2}/n$. The standard error is

j	Day	Est	SE
1	2	206.29	7.957
2	5	376.93	12.742
3	8	545.14	15.585
4	11	684.29	27.404
5	14	801.71	35.741
6	17	864.43	36.679
7	20	945.29	32.250

Table 5.1. Estimates and standard errors of μ_j for the Small Mice data.

a type of standard deviation; the term *standard error* instead of *standard deviation* is used when we are talking about a standard deviation of a parameter estimate. For $0 < a < 1$, we can construct a $(1 - a)100\%$ confidence intervals for μ_j

$$(\bar{Y}_j - t_{n-1}(1 - a/2)s_{jj}^{1/2}, \bar{Y}_j + t_{n-1}(1 - a/2)s_{jj}^{1/2}) \qquad (5.5)$$

where $t_{n-1}(1-a/2)$ is the $(1-a/2)$ quantile of the t distribution with $n-1$ degrees of freedom (df). As $n-1$ increases, the t distribution approaches the normal distribution and the quantile $t_{n-1}(1 - a/2)$ approaches $z(1 - a/2)$, the $(1 - a/2)$ quantile of the normal distribution. When $a = .05$ giving a 95% confidence interval, we often approximate $t_{n-1}(.975)$ or $z(.975) \approx 1.96$ by 2.

5.1.1.1 Small Mice Data

The Small Mice data have 7 observations on each of 14 mice. Observations are taken every three days starting at day 2. We label observations within subject by j starting at $j = 1$ up to $j = 7$. The actual day t_j corresponding to observation j is $t_j = 3j - 1$. We fit model (5.3) to this data. Table 5.1 presents estimates and standard errors for the μ_j. To more easily interpret these results, figure 5.1 presents the results graphically. A line connects the means at each day, illustrating the average growth of the mice over time. Error bars at each time depict a 95% confidence interval for each estimate.

This data set is balanced, point estimates of μ and Σ, and standard errors for μ are all easily calculated with simple algebraic formulas. What will happen for the Big Mice data, which does not have balanced data? Looking at figure 5.1, the average weight trend does not appear to be linear in time, but it might well be quadratic. Can we determine if a quadratic polynomial in time is sufficient to describe the trend? Or might we need a cubic polynomial? The Small Mice does not have covariates like age or gender, but most data sets do. How do we include covariates in model?

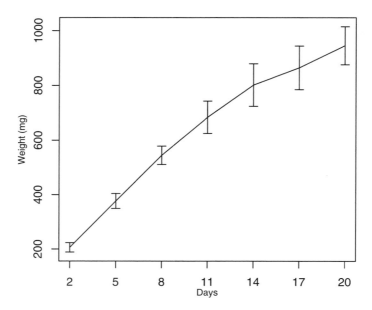

Figure 5.1. Plot of inferences for the Small Mice data. The increasing curve connects the means at each time point, and the error bars at each day give a 95% confidence interval for the mean. This graphically represents the results of table 5.1.

5.1.2   Elaborations to the Multivariate Normal Model

The multivariate normal model (5.3) for balanced data featured *closed-form* estimates and covariances of the parameters, meaning that we can write down an algebraic formula for the estimates and covariances as functions of the data. Test statistics for testing hypotheses of interest and for constructing confidence intervals have known distributions. This is all incredibly convenient. Unfortunately model (5.3) is not flexible enough to analyze most longitudinal data. In making it flexible enough, we will lose the ability to have closed-form parameter estimates and standard errors and known distributions for our test statistics.

Model (5.3) has three limitations that need relaxing to make it applicable for more longitudinal data analyses. The first limitation is that there are a lot of parameters in θ, the vector of unknown covariance parameters in Σ. This can cause problems with estimation, as the number of parameters can overwhelm the ability of the data set to determine good estimates for the parameters. We introduce *parameterized covariance models* for Σ in the next section. These reduce the number of parameters in Σ, but these parameterized covariance matrices do not usually allow for closed-form parameter estimates and known distributions for our test statistics.

The second limitation is that model (5.3) has no covariates. How covariates affect the response is usually the key scientific issue in longitudinal data analyses. We briefly introduce covariates in section 5.3 and chapter 7 discusses covariate specification at length.

The third limitation involves missing or unbalanced data. Missing data are common in longitudinal data. Missing data means that, while we intended to collect J observations on each subject, some subjects do not provide us with all J observations. For example, in the Pain data, 6 subjects do not have the full 4 observations, and these subjects have a total of 11 missing observations. In the Weight Loss data, 20 of 38 subjects do not have a complete set of 8 observations. In the Small Mice data, the data are balanced, but in the Big Mice data, only two mice have complete data; all the remaining mice have only 7 of a possible 21 observations. When some responses are missing we usually will not have closed-form parameter estimates, even if closed-form parameter estimates were possible with complete data. More worrisome, missing data can contribute to bias in estimates if the missing data are systematically different from the data that is observed. Chapter 12 discusses missing data and bias.

5.2 Parameterized Covariance Models

Model (5.3) assumed an *unstructured* (UN) covariance model for $\Sigma = \mathrm{Cov}(Y_i)$, meaning that there are $J(J+1)/2$ unknown parameters in Σ to estimate. This is a lot of parameters to estimate. We often cannot estimate all those parameters accurately due to a combination of too little data and too many parameters. Additionally, even when there is enough data, our computational algorithms may be imperfect and again we cannot estimate Σ. Even when the unstructured covariance model can be fit to the data, having so many unknown parameters can introduce substantial variability into the estimation of the means and consequent inflation of the standard errors.

A widely used solution is to use a parameterized covariance model for the covariance matrix Σ. A *parameterized covariance model* is a model that determines the value of σ_{jl} as a function of a q-vector of unknown parameters $\theta = (\theta_1, \ldots, \theta_q)'$ where usually q is usually much less than $J(J+1)/2$. We write $\Sigma = \Sigma(\theta)$ to indicate that Σ is a function of θ. Parameterized covariance models can substantially reduce the number of covariance parameters to be estimated, and they increase the number of data sets to which we can fit longitudinal models.

The Big Mice data set provides an example of the problems with the unstructured covariance model. As one issue, there are only two mice measured on both day 1 and day 2. Without heroic efforts we cannot estimate the correlation between observations on days 1 and 2 and this problem is repeated for every combination of days except those that are a multiple of three days apart.

The opposite extreme from the unstructured covariance model is the independent and identically distributed (IND) covariance model. This model assumes that all observations within a subject are independent and that all observations have the same variance. This model is essentially never appropriate for longitudinal data. It is a special case of all other covariance models that we discuss. We will call this the independence model.

We introduce three parameterized covariance models in the next subsections.

5.2.1 Compound Symmetry

Two very commonly used covariance models for longitudinal data are the unstructured (UN) covariance model and the *compound symmetry* (CS) model. In other circumstances, the CS model is sometimes known as the *one-way random effects* model. The CS model specifies the elements σ_{jl} of $\Sigma(\theta)$ as functions of two parameters, a variance τ^2 and a correlation ρ

$$\sigma_{jl} = \begin{cases} \tau^2 & j = l \\ \tau^2 \rho & j \neq l \end{cases}$$

The covariance σ_{jl} between any two observations Y_{ij} and Y_{il}, $j \neq l$ is constant, hence another name for this model is the *equicovariance* model. If we used this model to fit the Big Mice data, we could use data from times that are three (or six or nine, etc.) days apart to estimate the common covariance $\tau^2 \rho$, and then we would apply this same estimate for the covariance between observations that were 1 or 2 days apart.

5.2.2 Autoregressive

The *autoregressive* (AR(1)) model is also parameterized by two parameters τ^2 and ρ, but the covariance between Y_j and Y_l is now a function of the absolute differences in times $|t_j - t_l|$ between the observations

$$\sigma_{jl} = \tau^2 \rho^{|t_j - t_l|}.$$

When $j = l$, then $t_j - t_l = 0$ and $\sigma_{jj} = \tau^2$ is the variance of the observations.

The decreasing correlation with increasing lag is a key difference between the autoregressive and the compound symmetry models. In the CS model, the covariance $\text{Cov}(Y_{ij}, Y_{il}) = \tau^2 \rho$ is constant no matter what $|t_j - t_l| > 0$ is, from ten seconds to one hundred years.

5.2.3 Random Intercept and Slope

A covariance model that we use later in this chapter is the *random intercept and slope* (RIAS) model. The RIAS model says that each subject's data varies from the population mean μ by a subject-specific line. That is, the residuals $Y_{ij} - \mu_j = \epsilon_{ij}$ fall near a subject-specific line. A subject-specific line is characterized by a subject-specific intercept and a subject-specific slope. If the means μ_j are not linear in time, then the data Y_{ij} will not be linear over time, even though the RIAS model still applies. The RIAS model is partially illustrated in figures 2.5(a) and (c). In 2.5(a), the data exhibit subject-specific intercepts; in 2.5(c), the data exhibit subject-specific slopes. Combine the two figures to have data that have subject-specific intercepts and slopes.

Chapter 8 discusses the CS, AR, and RIAS models at length and many other parameterized covariance models as well.

5.3   Regression Models for Longitudinal Data

In model (5.3), we have $\mathrm{E}[Y_i] = \mu$. This model allows the population mean to vary arbitrarily across time, but all subjects i have the same mean across time. In the Pediatric Pain data we have two covariates, coping style and treatment in addition to trial number. We wish to understand how coping style and treatment affect responses. The Ozone data has variables site, day and hour of day. We want to know how the ozone level varies by site and day as well as by hour within day. When we have covariates that vary from subject to subject, each subject Y_i will have different means μ_i indexed by subject i. As we introduce covariates, it is also convenient at the same time to allow for missing or unbalanced data in our data sets. In general, the number n_i of responses is not the same from subject to subject. We relax the restriction that the n_i all equal J and allow the n_i to vary. The mean μ_i of Y_i is

$$\mathrm{E}[Y_i] = \mu_i = \begin{pmatrix} \mu_{i1} \\ \vdots \\ \mu_{in_i} \end{pmatrix}.$$

The mean of Y_{ij} at time t_{ij} is μ_{ij}.

In the linear regression model with scalar response y_i on subject i, covariate K-vector $x_i = (x_{i1}, \ldots, x_{iK})'$, α a K-vector of regression parameters, and residual error δ_i we have

$$y_i = x_i'\alpha + \delta_i$$

for $i = 1, \ldots, n$. Errors δ_i are independent and normally distributed

$$\delta_i \sim N(0, \sigma^2).$$

We often put this into vector form $Y = (y_1, \ldots, y_n)'$, residual errors $\delta = (\delta_1, \ldots, \delta_n)'$ and covariate matrix X of dimension n rows by K columns with ith row x_i' and individual elements x_{ik}. The linear regression model is formulated in matrix form

$$Y = X\alpha + \delta$$

with

$$\delta \sim N_n(0, \sigma^2 I)$$

and the n by n identity matrix I has ones on the long diagonal and zeros elsewhere. The zeros in I on the off-diagonal says that elements of δ are independent of each other. The ones on the long diagonal multiply σ^2 and say that the variance of each element of δ is σ^2.

5.3.1 Covariate Vectors and Covariate Matrices

In longitudinal data, we have observations Y_{ij} at time t_{ij}. Like the linear regression model, covariates will be a K-vector, but the K vector of covariates x_{ij} has two subscripts and is associated with the subject i and the specific time t_{ij}. The covariate vector x_{ij} has K elements

$$x_{ij} = \begin{pmatrix} x_{ij1} \\ \vdots \\ x_{ijK} \end{pmatrix}.$$

The kth covariate measured at time t_{ij} on person i is x_{ijk} and has three subscripts.

We store covariates x_{ij}, $j = 1, \ldots, n_i$ for individual i in a matrix X_i of dimension n_i by K. Individual rows x'_{ij} (lower case x) are indexed by $j = 1, \ldots, n_i$ corresponding to different times. Individual columns X_{ik} (capital X) are indexed by $k = 1, \ldots, K$. Each column X_{ik} for all i contains a particular covariate such as age, race, treatment, gender, current economic status, et cetera. Typically the first column $X_{i1} = \mathbf{1}$ is an intercept. The boldfaced $\mathbf{1}$ indicates a column vector made up entirely of ones. As discussed in chapter 1, covariates can be time-fixed or time-varying. Time-fixed covariates have X_{ik} equal to some scalar G_i times $\mathbf{1}$; obviously G_i varies from subject to subject; otherwise we would have another intercept in the model. Time-varying covariates have at least two unequal elements of X_{ik} for some subject i.

Time t_i and functions of time are frequent components of the X_i matrix. When we want to include a quadratic time term, we write

$$t_i^2 \equiv \begin{pmatrix} t_{i1}^2 \\ \vdots \\ t_{in_i}^2 \end{pmatrix}$$

to mean the element-wise square of the vector t_i.

Suppose we have a randomized treatment that is assigned to half of the subjects. The remaining subjects are in the control group. Treatment begins prior to t_{i1}. Let G_i indicate the group that subject i is assigned to, $G_i = 1$ for subjects in the treatment group, and $G_i = 0$ for subjects in the control group. To include a column X_{ik} in X_i to identify the treatment group, we set $X_{ik} = G_i \mathbf{1}$.

If we take our first measurement at time t_{i1}, then randomize subjects to treatment or control, the treatment/control indicator column X_{ik} might be time-varying, with X_{ik} a vector with a zero in the first position followed by $n_i - 1$ ones if subject i is in the treatment group, and $X_{ik} = \mathbf{0}$ if the subject is in the control group where $\mathbf{0}$ is a column of zeros of the appropriate length. Because treatment is not delivered until after t_{i1}, subject i cannot be in the treatment group until after t_{i1}.

If we want a time-fixed grouping variable (treatment/control or gender) by time interaction, we will write this as

$$t_i \times G_i = \begin{cases} t_i & \text{subject } i \text{ in treatment group} \\ \mathbf{0} & \text{subject } i \text{ in control group.} \end{cases}$$

If we have a time-varying grouping variable $H_i = (H_{i1}, \ldots, H_{in_i})'$, we will write the interaction also as $H_i \times t_i$. Like t_i^2, the multiplication in $H_i \times t_i$ is elementwise with elements $t_{ij} \times H_{ij}$.

One way of describing the covariates in our model is to specify the columns of the X_i matrix. Matrix

$$X_i = (\mathbf{1}, t_i)$$

is a simple covariate matrix corresponding to a model with an intercept $\mathbf{1}$ and time t_i covariates. A more complex example is

$$X_i = (\mathbf{1}, t_i, \text{BAge}_i \times \mathbf{1}, G_i \mathbf{1}, t_i \times G_i),$$

which has an intercept $\mathbf{1}$, time t_i, a column with baseline Age (BAge) $\text{BAge}_i \mathbf{1}$, a grouping variable G_i, and a time by grouping variable interaction $t_i \times G_i$.

What we have learned previously about specifying covariates in the linear regression model carries over to specifying covariates for longitudinal analysis. However there are additional complexities because we frequently have means which change nonlinearly over time. We must model the time trends and how covariates affect the time trends. We continue discussing covariates in chapter 7.

5.3.2 Multivariate Normal Linear Regression Model

We use a multivariate linear regression model to model the longitudinal observation Y_i. To make the connection between the covariates X_i and the mean μ_i, we introduce the coefficient vector α of length K. When we multiply $X_i \alpha$ we have an n_i vector of means $\mu_i = \mathrm{E}[Y_i] = X_i \alpha$ for subject i. Each $\mu_{ij} = x_{ij}' \alpha$ is a *linear predictor* and captures the effects of covariates on the ith subject mean at time t_{ij}. The vector of linear predictors $X_i \alpha$

generalizes μ in model (5.3), and the iid model is an important special case of this multivariate regression model. The K-vector of regression coefficients α are called *fixed effects* parameters or *population mean* parameters. The mean $\mathrm{E}[Y_{ij}] = x'_{ij}\alpha$ is the *population mean*, the mean of the population of observations whose covariates are equal to x_{ij}.

Our multivariate regression model is

$$Y_i = X_i\alpha + \epsilon_i. \tag{5.6}$$

The ϵ_i is an n_i-vector of residual errors. Residuals ϵ_i from different individuals are independent. We assume a normal distribution

$$\epsilon_i \sim N(0, \Sigma_i(\theta)) \tag{5.7}$$

and we expect the errors $\epsilon_{ij}, j = 1, \ldots, n_i$ to be correlated because they are all measured on the same subject. If, at the first time point, Y_{i1} is greater than its mean $x'_{i1}\alpha$, we typically expect that the second measurement Y_{i2} will also be greater than $x'_{i2}\alpha$. In other words, if $\epsilon_{i1} > 0$, we expect that quite probably $\epsilon_{i2} > 0$ also. This assumes the covariance σ_{12} is positive, the usual situation in longitudinal data.

The covariance matrix $\mathrm{Var}(Y_i) = \Sigma_i(\theta)$ depends on the q-vector of unknown parameters θ, which are common across all subjects. If we have balanced data, then $\Sigma_i(\theta) = \Sigma(\theta)$ does not vary by subject. If we have balanced with missing data, then $\Sigma_i(\theta)$ depends on i at a minimum because it depends on the number of observations n_i. For example, if $t_i = (1, 2, 3, 4)'$ and $\Sigma_i(\theta)$ is the AR-1 covariance model, then

$$\Sigma(\theta) = \tau^2 \begin{pmatrix} 1 & \rho & \rho^2 & \rho^3 \\ \rho & 1 & \rho & \rho^2 \\ \rho^2 & \rho & 1 & \rho \\ \rho^3 & \rho^2 & \rho & 1 \end{pmatrix}.$$

If subject i has only the first and third observation $Y_i = (Y_{i1}, Y_{i3})'$, their covariance matrix will be

$$\Sigma_i(\theta) = \tau^2 \begin{pmatrix} 1 & \rho^2 \\ \rho^2 & 1 \end{pmatrix}$$

This is the previous matrix missing the second and fourth rows and columns. If subject i had observations $1, 3, 4$, then the covariance matrix is the full matrix missing the second row and second column. In most covariance models, Σ_i will also depend on the times t_i.

5.3.3 *Parameter Estimates*

This section is somewhat technical and parts require more advanced mathematics and statistics than the majority of the book. This is for (1) thoroughness, (2) to inform more advanced readers, and (3) to expose readers who will eventually take more advanced mathematical statistics courses to these formulas for the first time. The technical material here will not be used much later in the text. We give formulas for the estimate $\hat{\alpha}$ and its covariance matrix. We explain weighted least squares. We give estimates for fitted values and predicted values and for their covariance matrices. If the math is too much for you at this time, please skip down to the paragraphs on REML and ML estimation and then go on to the example.

In the linear regression model $Y = X\alpha + \epsilon$, α a K-vector, $\epsilon \sim N_n(0, \sigma^2 I)$, we have closed-form estimates $\hat{\alpha}$ for α, an estimate $\tilde{\sigma}^2$ for σ^2, fitted values $\hat{y}_i$ and the estimated covariance matrix of the $\hat{\alpha}$. Briefly,

$$\hat{\alpha} = (X'X)^{-1}X'Y, \tag{5.8}$$

$$\hat{y}_i = x_i'\hat{\alpha},$$

$$\tilde{\sigma}^2 = \frac{1}{n-K}\sum_{i=1}^{n}(y_i - \hat{y}_i)^2, \tag{5.9}$$

$$\text{Var}(\hat{\alpha}) = \sigma^2(X'X)^{-1}. \tag{5.10}$$

We estimate $\text{Var}(\hat{\alpha})$ by plugging in the estimate $\tilde{\sigma}^2$ for σ^2.

The term $X'X$ shows up in two places in these formulas. We may rewrite $X'X$ as a sum over individual covariate vectors

$$X'X = \sum_{i=1}^{n} x_i x_i'. \tag{5.11}$$

Similarly, the second component of $\hat{\alpha}$ can be written

$$X'Y = \sum_{i=1}^{n} x_i y_i. \tag{5.12}$$

The formula for $\hat{\alpha}$ in longitudinal data uses a generalization of these two formulae.

The estimate $\hat{\alpha}$ is called the *ordinary least squares estimate*, or OLS estimate. The OLS formula is simpler than the corresponding formula for longitudinal data. In the linear regression model, we had *constant variance* σ^2 for all observations, and we had independent observations. In longitudinal data models, we have correlated observations within a subject, and we often have non-constant variance; either condition is enough to require *weighted least squares estimates*. The weighting adjusts the contributions of the observations so that more variable observations contribute less information to our inference and less variable observations contribute more. Two highly correlated observations with the same mean contribute less than two independent observations toward estimating that mean. Two positively correlated observations with different means contribute more to estimating the difference in the two means than do two independent observations. If the two observations are negatively correlated instead, then they contribute less than two independent observations to estimating the difference in means.

Suppose for the moment that θ is known. For correlated normal data, the estimate of α will depend on θ. We write $\hat{\alpha}(\theta)$ to denote this dependence. The weighted least squares estimate of α is

$$\hat{\alpha}(\theta) = \left[\sum_{i=1}^{n} X_i' \Sigma_i^{-1}(\theta) X_i\right]^{-1} \sum_{i=1}^{n} X_i' \Sigma_i^{-1}(\theta) Y_i. \qquad (5.13)$$

Formula (5.13) uses generalizations of formulas (5.11) and (5.12). Each generalization has a $\Sigma_i^{-1}(\theta)$ between the X_i' and the X_i or Y_i. The inverse matrix $\Sigma^{-1}(\theta)$ of $\Sigma_i(\theta)$ satisfies, in matrix notation,

$$I = \Sigma_i^{-1}(\theta)\Sigma_i(\theta) = \Sigma_i(\theta)\Sigma_i^{-1}(\theta)$$

The matrix $\Sigma_i(\theta)$ in $X_i'\Sigma^{-1}(\theta)X_i$ and in $X_i'\Sigma_i^{-1}(\theta)Y_i$ acts as a *weight*; it downweights high-variance observations by multiplying them by the inverse variance. It is a multivariate weight, and contributions from correlated observations are adjusted appropriately.

Our current estimate $\hat{\alpha}(\theta)$ depends on unknown parameters θ. There are two popular estimates for θ, the *maximum likelihood* (ML) estimate $\hat{\theta}$ and the *restricted maximum likelihood* (REML) estimate $\tilde{\theta}$. Software will produce either estimate $\hat{\theta}$ or $\tilde{\theta}$. It will then plug $\tilde{\theta}$ or $\hat{\theta}$ into $\hat{\alpha}(\theta)$ to give the REML estimate

$$\tilde{\alpha} = \hat{\alpha}(\tilde{\theta})$$

or the ML estimate

$$\hat{\alpha} = \hat{\alpha}(\hat{\theta})$$

with no explicitly written dependence on θ.

We present the next set of formulas using ML estimates. The covariance of $\hat{\alpha}$ is

$$\text{Var}(\hat{\alpha}(\theta)) = \left[\sum_{i=1}^{n} X_i' \Sigma_i(\theta)^{-1} X_i\right]^{-1}$$

which we estimate by plugging in $\hat{\theta}$ for θ. Standard errors $\text{SE}(\hat{\alpha}_j)$ of individual elements α_j are the square root of the jth diagonal element of $\text{Var}[\hat{\alpha}(\hat{\theta})]$. Fitted values are

$$\hat{Y}_i = X_i \hat{\alpha}$$

The standard error of the fitted values is

$$X_i' \text{Var}[\hat{\alpha}(\hat{\theta})] X_i. \qquad (5.14)$$

We may make predictions as well. The predicted mean of a new subject with covariate vector X_i is also $\hat{Y}_i$. The predictive covariance involves two components

$$X_i' \text{Var}[\hat{\alpha}(\hat{\theta})] X_i + \Sigma_i(\hat{\theta}). \qquad (5.15)$$

The first component is from the estimation error in $\hat{\alpha}$; the second component $\Sigma_i(\hat{\theta})$ is the covariance from sampling a new subject at X_i.

5.3.3.1   REML and ML

As mentioned in the previous subsection, there are two popular estimates for the covariance parameters θ, the *maximum likelihood* (ML) estimate $\hat{\theta}$ and the *restricted maximum likelihood* (REML) estimate $\tilde{\theta}$. In linear regression, $\tilde{\sigma}^2$ is the REML estimate and $\hat{\sigma}^2 = \tilde{\sigma}^2(n - K)/n$ is the ML estimate; we use the REML estimate in linear regression as a matter of course. We will continue that practice here too and generally use REML estimates, although there are a few circumstances where it is convenient or necessary to use ML estimates. Formulas for ML or REML estimates, predictions and standard errors are the same with the appropriate estimate $\hat{\theta}$ or $\tilde{\theta}$ plugged into the formulas. To get REML formulas in place of ML formulas for the previous subsection, substitute in $\tilde{\theta}$ in place of $\hat{\theta}$.

Except in rare circumstances, ML and REML estimates for θ do not have a closed-form algebraic formula. Various *computational algorithms* have been implemented in software to produce estimates $\hat{\theta}$. Section 6.5 gives an overview of computational issues. Section 6.3 discusses ML and REML estimation in detail.

5.3.3.2   Small Mice Data

We asked whether the Small Mice population means increased linearly, quadratically, or perhaps in a more complex fashion with time. We continue using the unstructured covariance model, and we use REML estimation to answer this question.

We fit three models to the Small Mice data. The first model is the linear model that says that the population mean grows linearly in time

$$E[Y_{ij}] = \alpha_1 + \alpha_2 t_{ij}.$$

Then we try out the quadratic model $E[Y_{ij}] = \alpha_1 + \alpha_2 t_{ij} + \alpha_3 t_{ij}^2$ and the cubic model in time $E[Y_{ij}] = \alpha_1 + \alpha_2 t_{ij} + \alpha_3 t_{ij}^2 + \alpha_4 t_{ij}^3$.

Table 5.2 gives output from the three models. For example, the fitted values from the quadratic model are $89.591 + 64.190 t_{ij} - 1.111 t_{ij}^2$, and the standard error of the day×day quadratic term is .095. The linear term is very significant in the linear model, so we know that we need the linear term. Similarly, the quadratic term is also very significant in the quadratic model, with a t-statistic of $-1.111/.095 = 12$. So we know that we need the quadratic day×day term. In the cubic model, the cubic day×day×day term is not significant, indicating that we do not need the cubic term, and can settle for the quadratic model to describe the increase in weight over time for the Small Mice data.

The covariance matrix of the parameter estimates is given in table 5.3. The .008954 entry is the covariance estimate of the quadratic coefficient;

Model	Parm	Est	SE	t	p
Linear	Intercept	104.76	4.39	24	
	Day	41.458	1.290	32	
Quad	Intercept	89.591	4.576	20	
	Day	64.190	2.328	28	
	Day$\times$ day	-1.111	.095	-12	
Cubic	Intercept	92.323	5.891	16	
	Day	60.142	5.971	10	
	Day$\times$day	-.418	.945	-.4	.67
	Day$\times$day$\times$day	-.0237	.0322	-.7	.47

Table 5.2. Estimates and standard errors of coefficients for the linear, quadratic, and cubic models for the Small Mice data with REML fitting. Missing statistical significance indicates p-values less than .0001.

	Intercept	Day	Day$\times$day
Intercept	20.9351	-1.8105	.1222
Day	-1.8105	5.4173	-.1833
Day$\times$day	.1222	-.1833	.008954

Table 5.3. The REML covariance matrix of the parameter estimates for the quadratic time model fit to the Small Mice data.

take the square root of this, giving .095; and this is the standard error (SE) of the estimate -1.111 that is reported in table 5.2.

5.3.4   Linear Combinations of Parameters

Very frequently, we need to estimate not just coefficients but also linear combinations of the coefficients in our model. For our quadratic mice model, we might wish to predict the estimated mean weight at each day from day 2 up to day 20. More generally, we may wish to estimate a linear function of the α_js

$$\sum_k b_k \alpha_k \equiv B'\alpha \qquad (5.16)$$

where $B = (b_1, \ldots, b_K)'$ is a K-vector of known coefficients. Linear combinations of the coefficients are most commonly used to estimate the mean response at various combinations of the covariates. When the elements of B sum to zero, $\mathbf{1}'B = \sum_{k=1}^{K} b_k = 0$, we call B a *contrast* in the means. Contrasts and linear combinations can be used to estimate quantities of interest. Contrasts are usually used to compare different means for groups. Linear combinations are more general, but the algebra is the same for both. The estimate of $B'\alpha$ substitutes in $\hat{\alpha}$ for α, giving $B'\hat{\alpha}$. The estimated variance of $B'\hat{\alpha}$ is a linear combination of the covariance matrix of the parameter

Day	Est	SE	Est	SE
2	214	6	206	8
5	383	10	377	13
8	532	14	545	16
11	661	18	684	27
14	771	20	802	36
17	860	23	864	37
20	929	27	945	32

Table 5.4. Estimated population means and standard errors of the means over time. Columns 2 and 3 are from the quadratic model, whereas columns 4 and 5 are from model 5.3 and are copied and then rounded from table 5.1.

estimates $\text{Var}[\hat{\alpha}(\hat{\theta})]$. In particular,

$$\text{Var}(B'\hat{\alpha}) = B'\text{Var}[\hat{\alpha}(\theta)]B = \sum_{k,l=1}^{K} b_k b_l \{\text{Var}[\hat{\alpha}(\theta)]\}_{kl} \qquad (5.17)$$

where $\{\text{Var}[\hat{\alpha}(\theta)]\}_{kl}$ is the klth element of the matrix $\text{Var}[\hat{\alpha}(\theta)]$. The standard error of $B'\hat{\alpha}$ is then the square root of formula 5.17.

5.3.4.1 Small Mice Data

Table 5.4 presents estimates and standard errors of the means at each day where we had data. The last two columns copy the results from table 5.1 for comparison.

The covariance model is the same for both models, the only difference is in the mean model. We had 7 mean parameters μ_j in model 5.3, whereas the quadratic model has only 3 coefficients, or 4 fewer parameters. The standard errors from the quadratic model are uniformly smaller than the general model, sometimes even 60% smaller. This indicates that the quadratic model is more efficient at estimating the means than the multivariate model 5.3. This is very commonly the case. And in those circumstances where a quadratic or other polynomial in time model does not fit better than the multivariate normal model, we can still use the multivariate normal model since it is a special case of our multivariate regression model $Y_i = X_i \alpha + \epsilon_i$.

5.3.4.2 Weight Loss Data

We are interested in whether the Weight Loss data illustrates actual weight loss. We are interested in the population mean at week 8 minus the population mean at week 1. We can use a contrast to help us determine the estimate of the difference in weights and whether the difference is significant. We fit a model with a mean like that in model (5.3) but with a random intercept and slope covariance model. The estimated mean at week 8 was 185.3 with a standard deviation of 4 pounds. The mean at week 1 was 192.8 with a standard deviation of 4 as well. The estimated difference was -7.4

pounds with a standard error of only 1 pound, indicating a very significant
weight loss over the time of the study.

5.3.5 Inference

Inference on the fixed effects coefficients α is fairly standard. We use software to calculate estimates $\hat{\alpha}_k$ and standard errors $\mathrm{SE}(\hat{\alpha}_k)$. We construct
confidence intervals in the time-honored fashion

$$\hat{\alpha}_k \pm 2\mathrm{SE}(\hat{\alpha}_k)$$

where 2 approximates 1.96 and this gives us an approximate 95% confidence interval. You may substitute for 2 your own favorite z score to create
confidence intervals of varying confidence level. We can construct tests for
hypotheses about α_k as well. The usual test statistic is

$$t = \frac{\hat{\alpha}_k}{\mathrm{SE}(\hat{\alpha}_k)}$$

for a test of $H_0 : \alpha_k = 0$. If $|t| > z_{1-a/2}$ then we reject the null hypothesis.
This is a two-sided test of H_0 with type I error rate $a(100)\%$. The p-value
is the a such that $z_{1-a/2} = |t|$. It is the probability of seeing a test statistic
more extreme than the observed t-statistic, given that the null hypothesis
is true.

One-sided p-values also have a Bayesian interpretation. Suppose for the
moment that the estimate $\hat{\alpha}_k$ is positive. The one-sided p-value $a/2$ is the
probability that α_k is negative given the data. Conversely, $1 - a/2$ is the
probability that α_k is positive, given the data. Similarly, if $\hat{\alpha}_k$ is negative,
the one-sided p-value is the probability that the parameter α_k is positive
given the data, In mathematical notation we write $P(\alpha_k > 0|Y)$. The
Bayesian interpretation of the one-sided p-value applies as well to tests of
contrasts of linear combinations of the α's. The Bayes inference is specific
to the particular data set and linear combination under discussion. This
is in stark contrast to the classical p-value inference which is a statement
about the statistical procedure and not particularly about the data set
being analyzed.

5.3.6 Degrees of Freedom in t-Tests

There are no explicit degrees of freedom (df) given with the formulas in the
previous subsection, for example to allow use of a t rather than a normal z
for tests and confidence intervals. Parameterized covariance matrices $\Sigma(\theta)$
and unbalanced data contribute to the lack of a known distribution like
the t distribution for constructing exact confidence intervals. There are
exceptions; for a few covariance models and balanced data sets with special
mean models, parameter estimates can have known distributions. When we

do have data and a model that allow for an exact t-distribution, it is nice to use it, but this is the exception, not the rule.

In spite of these difficulties, software often estimates the df and presents a t-statistic, and gives a p-value based on that t-statistic. Unless the df are low, this p-value will not be too different from the p-value based on the normal distribution. Use of the normal distribution is asymptotically correct, but slightly liberal (meaning intervals are too small) in finite samples. My practice is to accept the advice of the software and to use the p-values from the output. However, there are frequent circumstances where modest changes to the data set can cause enormous changes to the df produced by software, so it is best not to take the df too seriously. Further, in the rare circumstances where exact calculation is possible, and a specific df is known, software may not recognize this and may get the df wrong!

The problem is perhaps that users are trained in early statistics courses to expect to have a t-statistic in regression or ANOVA problems and expect to see t-statistics and df calculation in places where they do not exist. Software manufacturers respond to this demand by implementing algorithms to approximate the df. When we get to longitudinal data though, there is no simple t-statistic df that is easy to calculate. Various approximations have been formulated and are implemented in software, and often there are a number of choices. The problem is further exacerbated in that the best df approximation must depend on the level a or $1 - a$ of the test or confidence interval; this problem is mitigated by the practice of most data analysts to use only $a = .05$. The df depends on the covariance model and the actual values of the parameters. Different coefficients in the same model can have very different df!

It is important to have a rough idea of what the df should be as a check on software run amok. In a simple t test, either paired or two-sample, the df is the df for estimating the variance, not the number of observations that were averaged to create the mean. In longitudinal data, a rule of thumb for balanced data with unstructured covariance model is that the df should be $n - 1$, the number of subjects minus one. The minus one comes from having a single mean parameter at each time point. Suppose we had two means estimated at each time point? Then the df should be $n - 2$. In general, we can suggest the adjustment to n should roughly be the number of parameters K divided by the number of time points J. Approximately $n - K/J$ df is a rough guess when using the unstructured covariance matrix.

In model 5.3, we use an unstructured covariance matrix Σ with $J(J+1)/2$ unknown parameters. When we have a parameterized covariance matrix with q much less than $J(J + 1)/2$ parameters, we should see some gain in terms of the df or in terms of the distribution of the test statistics, even if it is no longer t. For a covariance model with constant covariance, the df could even be $N - K$ since all observations contribute to estimating the one variance parameter.

An approximation to the df to get approximate t intervals and tests must depend on the values of the covariance parameters θ as well as the sample size. Suppose independent constant variance observations, then $\Sigma = \sigma^2 I$ and we are back in the linear regression framework. The df should be $N - K$, the number of observations minus the number of regression parameters. If Σ has correlations that are all 1, then we really only have one observation per subject, and again we are again back in the linear regression framework but this time with $n - K$ df for our t-statistic. When neither extreme holds, the truth should lie somewhere in the middle.

When I calculate tests, p-values, and confidence intervals myself, I use quantiles of the normal distribution, but more commonly I do use the t-statistic and associated p-value produced by software. If the df from the software appears grossly over-estimated, then the software is doing what I would have done by myself, as the t with many df is a normal. When the df is very small compared to what I think it should be, then one must worry that there may be a problem with the computations or that there are problems with too little data and a consequent lack of power to test hypotheses of interest. Contrary to popular misconception though, if you reject the null hypothesis in spite of a lack of power, that does suggest there is a possibility that you may have a large effect!

For compound hypotheses with a set of several contrasts $B_l'\alpha = 0, l = 1, \ldots, L$ set equal to zero under H_0, the situation is similar. Software will often report an F statistic with numerator and denominator df. The numerator df will be correct, but the previous discussion of df for t-tests applies identically to the denominator df of the F statistic. If faced with a p-value from software, I usually use that directly; if I need to calculate the p-value myself, I compare the F statistic times its numerator df to a χ^2 statistic with df equal to the numerator df of the F statistic. Asymptotically, as the F statistic denominator df gets large, the F statistic times the numerator df approaches the χ^2 distribution.

5.4   Graphical Presentation of Inferences

In section 2.6, we created empirical summary plots based on sample averages and standard deviations. Based on $\hat{\alpha}$ and $\hat{\theta}$, we can draw *inference plots* similar to the empirical summary plots. The difference is that inference plots summarize results from our analysis, whereas empirical summary plots provide exploratory data analysis to be used for model development. Figure 5.1 is an inference plot for model (5.3). To make an inference plot for the multivariate regression model, we must specify a set of covariates and times for the inferences we want. When the design has a set of nominal times, those are the times we use. When we have random times, then one must specify a set of interesting times. Point estimates and standard errors for the means can be plotted in the same way as the empirical summary plots as in figure 5.1. Essentially we graphically display information from table 5.4.

The inference plot estimates a mean $x'_{ij}\hat{\alpha}$ and associated standard error $\mathrm{SE}(x'_{ij}\hat{\alpha})$ at each time point for specified values of the other covariates. The fitted value and standard error are for estimates of the population mean at x_{ij}.

Also possible is to make a *prediction plot*. Predictions refer to potential values of observations from a single subject. The population mean refers to the average response of everyone in the population with a particular covariate value. For normal response data, the point estimate $x'_{ij}\hat{\alpha}$ for prediction will be the same as for the inference plot. However, the associated standard error is much larger for a prediction plot; it is the square root of the jth diagonal element of the prediction variance given in formula (5.15). The remaining construction is the same for the prediction as for the inference plot, extending error bars that are plus or minus two standard errors out from the point estimate.

If we have a number of combinations of covariates, we can pick one or more covariate combinations to illustrate the conclusions from the analysis. Suppose we have subjects of both genders followed every six months for two years, for a total of five measurements. We estimate the average response and associated standard error for each gender at each time point, then plot the point estimates and plus or minus two standard error bars as in figure 2.24(a). If we have four groups instead of two, we make a more complicated picture. If there are too many groups, we might draw several figures or we might drop the error bars to de-clutter the figure. If there are two covariates, gender and treatment, then we might choose one of the treatments to illustrate the effect of gender. Then we draw a second plot, picking one gender to illustrate the effect of treatment. These would be fine if we have no gender by treatment interaction. If we have an interaction, then we need to draw two plots illustrating the effect of treatment, one for each gender. Or we could draw two plots illustrating the effect of gender, one for each treatment.

If we have a continuous covariate, say age at study entry, we might fix that to the median age to do the gender plot and group plot. Then to illustrate the effect of age at study entry, we specify one group and one gender, then pick a young and an old age at study entry and draw an inference plot for subjects of those ages.

5.4.1 Weight Loss Data

For the Weight Loss data, figure 5.2 shows an inference and a prediction plot. The model used has a separate mean at each time, as in model (5.3). The covariance model is the random intercept and slope model. Figure 5.2(a) shows an inference plot. The point estimates of the average weight at

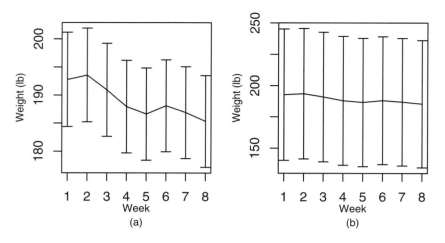

Figure 5.2. (a) Inference and (b) prediction plots for the Weight Loss data.

each time point generally decreases from week 1 to week 8 with exceptions from week 1 to week 2 and again from week 5 to week 6, but there is substantial uncertainty about the level of the mean at each time point relative to the average amount of weight lost. From this plot alone, we might not be willing to conclude that weight is lost from week 1 to week 8, although the point inference indicates weight loss. We need the information from the contrast in the previous section to supplement these plots.

Figure 5.2(b) is a prediction plot. The range of the Y axis is larger than in figure 5.2(a), as prediction variance is much larger than estimation variance.

5.4.1.1 Multivariate Inference and Prediction Plots

Inference and prediction plots display information about the estimate and uncertainty at each individual time point. This is *univariate* information at each time point. Our data and inference is multivariate. We would prefer to display the multivariate uncertainty in the fitted or predicted values.

We can investigate the multivariate uncertainty in the fitted values as a function of time by simulating complete vectors Y_f of fitted mean values from a normal distribution whose mean and variance are the fitted estimate and estimated covariance matrix of the estimate at a selected set X_f of covariates

$$Y_f \sim N(X_f'\hat{\alpha}, X_f'\text{Var}[\hat{\alpha}(\hat{\theta})]X_f).$$

Simulate a number of these, perhaps 20. We plot these in a profile plot. This is a *multivariate inference plot*. These plots are useful for tracking the effects of time on responses, particularly when the time trend is not obvious from the parameter estimates.

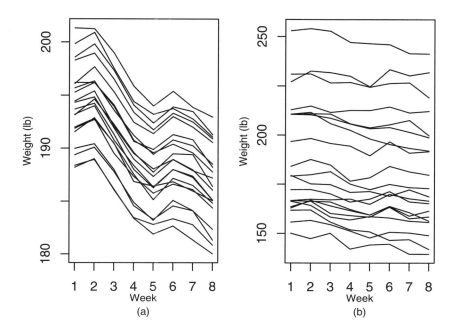

Figure 5.3. Multivariate inference and prediction plots for the Weight Loss data. (a) Multivariate inference plot. (b) Multivariate prediction plot.

We can also simulate new observations from new subjects with a specified covariate value. Then we plot these new observations in a profile plot, called a *multivariate prediction plot*. To simulate complete vectors of observations, we take the parameter estimates $\hat{\alpha}$ and the estimated prediction covariance matrix (5.15) and simulate new future observations Y_{p} at covariate values X_{p} as

$$Y_{\mathrm{p}} \sim N(X'_{\mathrm{p}}\hat{\alpha}, X'_{\mathrm{p}}\mathrm{Var}[\hat{\alpha}(\hat{\theta})]X_{\mathrm{p}} + \Sigma(\hat{\theta})).$$

The multivariate prediction plot can be used as a diagnostic. If the plot looks like our data, then our model provides an adequate description of our data. If the data has some feature not observed in the plot, then further modeling is required. If we have many covariate values X_i in our data, we may simulate one new observation $Y_{i,p}$ for each X_i, then plot the new observations in a profile plot to compare to the original data. If the new plots look like our original data profile plots, then the model fits the data.

5.4.1.2   Weight Loss Data

Figures 5.3(a) and (b) illustrate a multivariate inference plot and a multivariate prediction plot for the Weight Loss data. The starting population mean ranges from 184 to 196 pounds illustrating the uncertainty in its value. However, the time trend in each sample seems similar and in every sampled profile for the multivariate inference plot there is a weight loss over the 7 weeks of the study. Even if we were to sample thousands of profiles we would not see one in 5.3(a) that shows a weight gain. This indicates that we can be highly confident that weight loss occurred over the course of the study. Thus this plot supplements the formal statistical inference of the contrast between the weight at week 8 and the weight at week 1.

Both figures 5.3(a) and (b) suffer from the same display issues we saw in chapter 2 in figure 2.15. To properly visualize 5.3(b) will require a tall narrow plot. But at first perusal, this figure looks a lot like 2.15 and our model does not appear to fit the data too badly.

5.5   Problems

1. Write out the entire X_i matrix for the two matrices at the end of section 5.3.1 assuming 4 time points and $t_i = (1, 2, 3, 4)'$, $BAge_i = 45$, for two different subjects, one each from the treatment and the control groups.

2. Use the Small Mice results from subsection 5.3.4, in particular tables 5.2 and 5.3. Calculate the second and third column of table 5.4.

3. Explain how the multivariate normal model 5.3 is a special case of the multivariate regression model $Y_i = X_i\alpha + \epsilon_i$.

4. To create an example where the df changes with a modest change to the data set, fit a balanced data set with the random intercept covariance model and have a single covariate that is time-fixed. Check the degrees of freedom for the covariate estimate. Now alter the covariate so that it is not time-fixed, by changing the value of the covariate at only one time point for exactly one subject. Does the df remain the same for that covariate?

5. Consider the compound symmetry model (6.6), and consider adult weight, measured weekly. If helpful for the following problem, you may restrict yourself to members of one gender or to a particular nationality with which you are most familiar.

   (a) Give your guess as to the population mean weight α_1, the population variability τ^2, the within person correlation ρ. The product $\tau^2\rho$ is the variance of a single person's weight measurements around their average weight. What would you guess an individual person's weight variance to be?

   (b) What happens to your guesses for α_1, τ^2 and ρ if you switch from weekly measurements to monthly measurements? To annual measurements? Give both a qualitative statement and as best possible give a quantitative statement.

6. Repeat problem 5 for

   (a) Adult weekly height measurements.
   (b) Adult weekly blood pressure measures (systolic and diastolic).
   (c) Adult heart rate.
   (d) Student quarter or semester (pick one) GPAs at your University.

7. Consider the previous two problems 5–6, and order the different measurements (blood pressure, heart rate, weight, height) in terms of the intraclass correlation ρ. Assume daily measurements. Repeat for weekly, monthly, and annual measurements.

6
Tools and Concepts

Give us the tools and we will finish the job.

– Winston Churchill

The essence of mathematics is not to make simple things complicated, but to make complicated things simple.

– S. Gudder

Overview

- Likelihood ratio tests
- Model selection
  - Covariance model selection
  - Fixed effects model selection
  - AIC and BIC
  - Problems with forward selection and backward elimination
- Maximum likelihood and restricted maximum likelihood
- Assuming normality
- Computational issues
  - Lack of convergence
  - Maximizing a likelihood
  - Function maximization

- Ridges, saddlepoints, gradients, and Hessians
- Collinearity

- Back-transforming a transformed response
  - Estimates and predictions after transformation
  - The log/exponential transformation

- Some design considerations
  - Estimating the population mean
  - Cost and efficiency in longitudinal data
  - Two independent samples versus paired comparisons

In this chapter, we discuss a number of tools useful in developing and fitting various models for longitudinal data.

6.1   Likelihood Ratio Tests

Likelihood ratio tests (LRTs) can be used to compare two models when one model is a special case of the other. The alternative model allows additional parameters to vary, whereas the null model fixes those additional parameters at known values. The test statistic is 2 times the difference of the log maximized likelihoods for each model. When fitting a model, most software will report the maximized value of the log likelihood, or possibly ± 2 times the maximized value. The larger alternative model always will have the larger log likelihood l_A, where A indicates the alternative, whereas the null model has log maximized likelihood $l_0 < l_A$. The important question is how much larger. To consider this, we use the test statistic

$$\Lambda = 2(l_A - l_0).$$

To test H_0 versus H_A, we compare Λ to a chi-square with df equal to the number of additional parameters in the alternative model. When Λ (capital lambda) is larger than expected as indicated by the chi-square distribution, we reject H_0 in favor of H_A.

When testing covariance models, we keep the fixed effects the same for both models. When testing two fixed effects models, we keep the covariance models the same in both models. We do not usually compare models with different fixed effects and different covariance models. If we were to reject a null hypothesis, it might be due to the difference in fixed effects or due to the difference in covariance models. We usually prefer a simpler interpretation, and so we separately test fixed effects and covariance models. When we reject a hypothesis about the fixed effects, we wish to do so only for reasons related to the fixed effects, and not accidentally due to ancillary reasons like a significantly better covariance model in the alternative hypothesis.

When testing two nested covariance models, we usually use REML likelihoods, although ML likelihoods work almost as well. If a single parameter distinguishes the null and alternative models, we might use a t-test of the form estimate divided by its standard error to test the alternative, comparing the test statistic to maybe a standard normal. Most of the time, we do not have this situation. More commonly, we use LRTs to compare two nested covariance models. For example, we may compare parameterized covariance models to the unstructured covariance model (when we can fit it!) or to the independence covariance model. When we have a set of not necessarily nested covariance models, section 6.2 discusses approaches for comparing covariance models.

6.1.1   Issues with Likelihood Ratio Tests for Covariance Parameters

Likelihood ratio tests for covariance models are unfortunately problematic. We can calculate the test statistic easily enough. The problem is that there are many circumstances where the distribution of the LRT is known to not be chi-square but, for example, a mixture of chi-squares. In many other circumstances, the chi-square distribution is a poor approximation to the distribution of the test statistic under the null distribution. In spite of these problems, I still use the chi-square distribution with df as the difference in parameters as a comparison distribution. Several circumstances make this an acceptable compromise. First, scientifically, we are usually more interested in the fixed effects. The covariance parameter estimates often do not have a large effect on standard errors or estimates of the fixed effects, and two similar covariance models will often produce similar inferences for the fixed effects. Second, other approaches to covariance model choice discussed in section 6.2 use cutoffs for the LRT that are simple functions of the df, so we are not doing something so different from other current approaches to covariance model specification. Third, one must still use reasoning to select a covariance model. For example, between two covariance models of similar log likelihood, one might select the one easiest to describe to the intended audience and Occam's razor applied to covariance models recommends using more parsimonious covariance models over more complex models if other measures of the quality of the fit are similar.

6.1.2   Comparing Nested Random Effects Models

We mention an important special case. When testing the random intercept against the independence model, the test statistic is distributed as a 50-50 mixture of two chi-squares, one with 0 and one with 1 df. If we reject the null hypothesis using the chi-square with 1 df, then we are being conservative. Testing H_0, the random intercept, versus H_A, the random intercept and slope model, the likelihood ratio test is a 50-50 mixture of two chi-squares with 1 and 2 df. Again, rejecting the null using a chi-square with 2 df is conservative. In general, we may wish to test a null random effects model with r random effects against an alternative model that has $r + 1$ random effects. The two models must be nested inside each other. The null distribution of the likelihood ratio test statistic is a 50-50 mixture of two chi-squares with r and $r + 1$ df. We may reject the null hypothesis using the chi-square with $r + 1$ df and be conservative.

6.1.3 Fixed Effects

Likelihood ratio tests can be used to test hypotheses about fixed effects or sets of fixed effects. The test statistic is again 2 times the difference in log likelihoods under H_A minus that under H_0, $\Lambda = 2(l_A - l_0)$. Compare Λ asymptotically to a χ^2 with df equal to the number of fixed effects parameters fixed to known values under the null hypothesis. The ML likelihood ratio test behaves properly for fixed effects; the problems that occur with testing covariance models are not a problem for fixed effects. We must use ML likelihood to do the calculations. REML cannot be used to test fixed effects using likelihood ratio tests; among other problems, the difference in REML likelihoods can be either positive or negative and thus cannot be distributed as a chi-square statistic.

6.2 Model Selection

Model selection involves the choice of an appropriate model from among a set of candidate models. When specific scientific issues are involved, the issues often dictate choice of model or portions of the model. However, often we are still left with many potential models. Model selection is used when there is no particular clear choice among many different models. For fixed effects in an observational study, we would often like to include important demographic characteristics as covariates in predicting the outcome; Occam's razor suggests that we should not include characteristics that have no effect on the outcome. Model selection tools might well be used to guide choice of demographic characteristics. In contrast, in a randomized two-arm treatment study, the key covariate is the treatment effect; because of randomization, we expect subjects to be similar on demographic variables in both treatment arms. The main interest is in the treatment effect, and we would require that treatment be a covariate in the primary analysis. Model selection would not be used in this situation. Secondary analyses might well focus on the effects of demographic variables, nutritional variables, and other types of covariates; in these secondary analyses model selection for the fixed effects might well be of value.

Model selection tools are a useful set of techniques for screening through the many different covariance models. In chapter 8, we discuss a number of different covariance models. Fixed effects models and covariance models are rather different; they are usually treated differently in modeling, and we discuss model selection for the covariance parameters and for the fixed effects separately. The covariance models for our data are often of secondary interest; we do want to pick the best covariance model for our data but if we make a modest error in covariance model specification it is not as costly as an error in the fixed effects specification.

6.2.1   Model Selection for the Covariance Model

When faced with a new data set, we want to determine a best covariance model for use in fitting the fixed effects, and we try out a large number of covariance models. Our goal is typically to pick a single best or most useable model for use in further analyses of fixed effects. While we do our covariance model selection, we will pick a single set of covariates and try out many different covariance models.

One approach to model selection is to use likelihood ratio tests (section 6.1) comparing various models to each other or to the unstructured covariance model or to the model of independent observations. Using likelihood ratio tests, we can compare nested covariance models to each other. However, a number of covariance models are not nested within each other and have the same or similar numbers of parameters. How do we compare non-nested models?

A number of criterion-based approaches to model selection have been developed. Criterion-based model selection approaches compare adjusted log likelihoods penalized for the number of parameters in the covariance model. The penalty for model m increases with the number of covariance parameters q_m. Models with more covariance parameters should fit better, meaning they should naturally have a higher log likelihood, than models with fewer parameters. The penalty function levels the playing field compared to what would happen if we compared models using raw log likelihood. The model with the best score on the criterion is selected as "best."

The two most popular covariance model selection criteria are AIC and BIC, short for *an information criterion* or *Akaike information criterion* and *Bayes information criterion*, respectively. AIC for a given model m is defined as

$$AIC(m) = -2 \log \text{likelihood}(m) + 2q_m$$

and BIC is nearly identical, except that instead of the 2 multiplying the number of covariance parameters q_m, the penalty is $\log(N)$

$$BIC(m) = -2 \log \text{likelihood}(m) + \log(N)q_m.$$

Model selection for covariance parameters is typically done with REML log likelihoods; as such, the log likelihoods in the formulas for AIC and BIC should be REML likelihoods, q_m is the number of covariance parameters, and the fixed effects model must be the same in all models.

Model selection proceeds similarly for both criteria. Each covariance model under consideration is fit to the data, and the models are ranked according to either AIC or BIC. The model with the smallest value of AIC or BIC is selected as best. As can be inferred from the differing penalty functions, AIC tends to select models with more covariance parameters and BIC selects models with somewhat fewer covariance parameters.

There are 4 equivalent variants of AIC and BIC. Our version of AIC and BIC are in *smaller is better* form. Some programs multiply our definitions by -1 and have a "larger is better" version, and some programs may multiply AIC and BIC by $\pm.5$ in the definition. It is important to confirm which version is being used. There is no particular advantage to any of the versions. Our version usually keeps most of the numbers positive, and both AIC and BIC are on the -2 log likelihood scale, so that any intuition about differences on the -2 log likelihood scale can be used to measure differences in AIC or BIC.

In covariance model selection, it is important that all models have the same fixed effects. Otherwise model testing or comparison procedures would be comparing both different fixed effects and different covariance parameters at the same time. If one badly mis-specifies the fixed effects, the wrong covariance model may be selected. Key in this regard is to include fixed effects that are functions of time to describe the population mean over time. Very significant treatment, grouping or continuous covariates should also be included in the fixed effects when selecting the covariance model. Unfortunately, this problem is circular; some covariance model must be assumed to identify important covariates. In practice the problem is not so difficult. First we use graphical methods to identify the functional form for changes over time to put in the fixed effects, and the most important covariates are usually known or can be guessed at before the analysis even begins. For example, in studies of depression, gender is often the most important covariate in the analysis; we automatically include gender as a covariate from the beginning. Covariates that are not significant, or are barely significant usually do not have much impact on the covariance model selection. In contrast, omitting a strong linear time covariate may greatly alter the decision about the best covariance model.

6.2.1.1 Small Mice Data

BIC and AIC are used to select a covariance model in sections 8.3.1 for the mice data and 8.3.2 for the Pain data after the discussion of a number of covariance models. We look at a subset of the Small Mice data results here. Table 6.1 presents covariance model selection choices for the log REML

Cov Model	# Parms	-2 Log REML	AIC	BIC	LRT vs. IND	LRT vs. UN
UN	28	964.7	1020.7	1038.6	198.41	–
AR(1)	2	1035.3	1039.3	1040.5	127.83	70.6
RIAS	4	1038.6	1046.6	1049.2	134.5	73.9
CS	2	1105.3	1109.3	1110.6	57.8	140.6
IND	1	1163.1	1165.1	1165.7	–	198.41

Table 6.1. Results from fitting various covariance models to the Small Mice data set. Fixed effects include the intercept, linear, and quadratic time terms. Models are ordered by BIC from best to worst.

likelihood, AIC, BIC, and tests against the UN and IND models. Results for 4 covariance models are given, the unstructured (UN), autoregressive (AR(1)), compound symmetry (CS), and independent (IND) models.

The UN model is the best model on all measures. It has the lowest AIC by a large amount. It has the lowest BIC, although only by a little bit over AR(1). All three models with correlation among the observations are significantly better than the IND model; this is usual. In this set of results, the four other models are significantly worse than the UN model. In section 8.3.1, we will identify a model better that the UN model.

6.2.2 Comparison of AIC and BIC

Suppose we have two nested covariance models 1 and 2 with parameters θ_1 and θ_2, respectively, of lengths $q_1 < q_2$. In simple terms, nested means that if we fix $q_2 - q_1$ parameters of θ_2 in model 2 to zero or some other known values, we have model 1. The difference in -2 log likelihoods is distributed approximately as a chi-square with $q_2 - q_1$ df. Using AIC is equivalent to using $2(q_2 - q_1)$ as the cutoff for this chi-square test. When $q = 1$, this has a type-one error rate of $a = .16$. The error rate decreases as $q_2 - q_1$ increases; for $q_2 - q_1 = 7$, the error rate is $a = .05$, and for $q_2 - q_1 = 16$, the error rate is $a = .01$. This error rate does not depend on the sample size; as the sample size goes to infinity, under the null hypothesis that model 1 is correct, we continue to erroneously pick model 2 as the correct model with probability equal to the type-one error rate.

Decision theory suggests that, as sample size increases, we should use a decreasing type-one error rate. BIC does this for us automatically, by penalizing the model according to both the number of parameters q_m and the number of observations N. The log function does grow slowly in the number of observations, and so the penalty does not grow quickly in N. Because of its larger penalty on the number of parameters, BIC tends to pick smaller models than AIC does and so BIC more than AIC tends to pick the null model when the null model is in fact correct.

Conversely, for smaller data sets, one should make additional assumptions when modeling data; as the sample size increases, one should relax the assumptions and allow the data to determine modeling assumptions. In specifying a covariance model, additional assumptions means choosing a more parsimonious covariance model, one with fewer parameters. As the sample size n increases, one should allow for more complex covariance models with more unknown parameters. BIC goes against this advice, at least in comparison to AIC. Simulation studies have suggested that BIC tends to select a model with too few parameters. On the other hand, having too many extra parameters in the covariance model can interfere with inferences.

6.2.3 The Probability of a Covariance Model

Bayesian methods are finding greater and greater use in practical statistical analyses. When comparing two non-nested models, classical methods do not allow for a formal p-value test. Bayesian testing can compare nested or non-nested models and can compare two or more models at the same time. A Bayesian test of two models 1 against 2 is a probability statement that model 1 is the correct model given that either 1 or 2 is correct. BIC can be used to approximate the probability that a given covariance model among a set of covariance models is correct. Let BIC(1) and BIC(2) be the BICs for two models. The probability that model 1 is the correct model is

$$P(1|Y) \approx \frac{\exp[-.5\text{BIC}(1)]}{\exp[-.5\text{BIC}(1)] + \exp[-.5\text{BIC}(2)]}$$
$$= \frac{1}{1 + \exp\{-.5[\text{BIC}(2) - \text{BIC}(1)]\}}$$

and the probability that model 2 is correct is $P(2|Y) = 1 - P(1|Y)$. What is important in this calculation is the difference of BICs. If the difference of $\text{BIC}(2) - \text{BIC}(1) = 0$, then the two models are equally likely, $P(1|Y) = P(2|Y) = .5$. If the difference is $\text{BIC}(2) - \text{BIC}(1) = 2$, then $P(1|Y) = 1/[1 + \exp(-1)] = .73$. If $\text{BIC}(2) - \text{BIC}(1) = 6$, $P(1|Y) = .95$ and if the difference is greater than 12, the probability of model 1 is less .0025. At larger differences in BIC, the probability in favor of one model or the other becomes overwhelming. Given K models, the probability that model k is the correct model is

$$P(k|Y) \approx \frac{\exp[-.5\text{BIC}(k)]}{\sum_{j=1}^{K} \exp[-.5\text{BIC}(j)]}.$$

Suppose that three models are under consideration with $\text{BIC}_1 = -1.2$, $\text{BIC}_2 = -3.6$ and $\text{BIC}_3 = -5$ for models 1, 2, and 3. Then

$$P(1|Y) \approx \frac{\exp(.6)}{\exp(.6) + \exp(1.8) + \exp(2.5)} = .091$$

while

$$P(2|Y) \approx \frac{\exp(1.8)}{\exp(.6) + \exp(1.8) + \exp(2.5)} = .302$$

and

$$P(3|Y) = 1 - P(1|Y) - P(2|Y) \approx \frac{\exp(2.5)}{\exp(.6) + \exp(1.8) + \exp(2.5)} = .607.$$

We do not have overwhelming evidence in this example in favor of any one model, although model 3 is somewhat preferred over model 2, and both are somewhat preferred to model 1. The Bayesian approach treats all three models symmetrically, and models need not be nested in each other for it to work. The formulas for $P(m|Y)$ assume that the fixed effects are the same for the three models; other formulas are required for selecting fixed effects.

In table 6.1, the UN model has BIC = 1038.6, and the AR(1) model has BIC = 1040.5. If we restrict attention to just these two best models, then

$$P(\text{UN}|Y) \approx \frac{1}{1 + \exp[.5 \times (1038.6 - 1040.5)]} \approx .72$$

and the UN model is only slightly better than the AR(1) model. If we also consider the CS model, this probability does not change because BIC = 1110.6 for CS and an extra term of $\exp(-36)$ in the denominator does not noticeably change the final result.

6.2.4 Model Selection for Fixed Effects

The huge variety of different circumstances involving fixed effects makes giving advice for model selection for fixed effects more complex than for covariance models. The first step in choosing the fixed effects is to determine if there is a strong time trend, as in the mice data. If there is, we specify fixed effects to adequately describe the time trend. This time trend must be included in all further analyses including covariance model specification.

Now suppose we wish to wade through a large number of covariates and to include those that are predictive of the response. The traditional model selection methods of forward selection and backward elimination can be used with longitudinal models. Suppose we are contemplating a finite set of potential additional covariates. Forward selection works by first specifying a base model. Each potential covariate is added to the base model in turn. The covariate with the most significant t or F statistic is added to the base model and this new base plus covariate model becomes the new base model. The remaining covariates are added in to this model one at a time and the most significant is included in the model. The process continues until no remaining covariate is significant enough to include in the model. A significance level is set as a minimum level required to allow a new covariate into the model.

Backward selection works similarly, but starts with a model with all of the candidate covariates as predictors. The least significant covariate is dropped and the model refit. The least significant covariate in the new model is dropped, unless its significance level is above some minimum level.

Many variations of forward selection and backward elimination have been suggested, included algorithms that switch between forward and backward steps in various orders.

6.2.4.1   Problems with Forward and Backward Selection

Forward selection has issues with whether one wishes to consider including interactions. After two covariates are in, should we consider their interaction? In practice, covariates with very significant main effects are the ones most likely to have interactions. Non-significant main effects are relatively unlikely to have strong interactions, although never say never. In backwards selection, if one considers all interactions, the model may become so large as to be unfittable or uninterpretable in practice. If one does manage to include two-way interactions in the first model, then one must not delete main effects if they have interactions included in the current model. Similarly with quadratic terms; linear terms must be included if the quadratic terms are included.

Another problem with both forward and backwards selection is that there can be so many potential covariates that either process becomes unwieldy in the absence of a pre-programmed routine to do the selection. Of course, if the analysis is pre-programmed, it is easy to ignore the intermediate result steps, where there can be information about relationships among covariates and whether something starts out significant but later drops or increases in significance level.

6.2.4.2   Some Suggestions

Because of the problem of interactions and an inability to fit the full model as the first step of backward elimination, I usually prefer forward selection type algorithms over backward elimination, particularly for longitudinal data analyses. As I step through the models, I keep an eye open for whether interactions should be included or not. A simplified forward stepwise algorithm that I have had some success with is as follows. Fit the first step of the forward stepwise algorithm. Take all variables above some significance cutoff and put them into a single, final model. If the single final model shows basically the same story as the individual models, then the methodology is a success. If a covariate drops loses its significance, then it is recommended to figure out why. Is there another covariate correlated with the non-significant variable that explains similar variability in the response? Or is there an interaction with that covariate that may be of interest?

Many data sets have a randomized treatment variable. The first analysis will often include the randomized treatment variable only, plus functions of time if needed, and possibly interactions between treatment and time. In secondary analyses, we explore demographic or other classes of covariates for significance, and these analyses may use model selection to explore many covariate effects. Unless randomization is poor, or if the data set is small, including main effects of covariates are unlikely to affect the estimate of the treatment variable. A third analysis may ask whether the treatment differs according to demographic characteristics. In the first step of a forward selection algorithm, one would include both the main effect and the interaction between the demographic variable and the treatment effect.

In studies of growth (weight or height from birth) or decline (diet studies or Alzheimer sufferers' mental ability), the key issue for a treatment is whether the treatment speeds or slows the rate of growth or decline. The important treatment effect is whether the coefficient of time is different between the treatment and control groups. If we wish to assess demographic variable impacts on the treatment effect, we are now talking about three-way interactions between treatment, time, and the demographic variable.

All possible subsets is another possible approach. If there are K variables, there are 2^K possible models assuming no relationships among the variables and no interactions. AIC and BIC can be used to select among these models. The formulas need to change from the versions that we used for covariance model selection. First of all, we need to use maximum likelihood, rather than REML estimation. Second, we should keep the covariance model the same for all models, while the covariates may vary. Third, the number of parameters for model m is now $K_m + q_m$, the number of fixed effects K_m plus the number of covariance parameters q_m. The penalty function, either 2 or $\log N$ remains the same. Thus AIC for model m is defined as

$$AIC(m) = -2 \text{ ML log likelihood}(m) + 2(K_m + q_m)$$

and BIC is

$$BIC(m) = -2 \text{ ML log likelihood}(m) + \log(N)(K_m + q_m).$$

The $BIC(m)$ for fixed effects can be used with the Bayes formulas of section 6.2.3 to give an approximate calculation of the probability of one fixed effects model versus another.

6.2.4.3   More Problems with Stepwise Procedures

As they do with regression models, forward selection and backward elimination in longitudinal models suffer from several additional serious problems. These algorithms explore a very limited set of models, so they frequently do not come close to finding an optimal model. Because of the repeated testing involved, forward selection and backward elimination often find spurious effects, declaring significance where no effect really exists. Stated p-values are often over stated and confidence intervals are overly narrow.

6.3   Maximum Likelihood and Restricted Maximum Likelihood

This section is somewhat technical, and parts require more advanced mathematics and statistics than the overwhelming majority of the book. This is (1) for thoroughness, (2) to inform more advanced readers, and (3) to expose readers who will eventually take more advanced mathematical statistics courses to these formulas for the first time.

There are many approaches to statistical inference. We briefly discuss two methods for statistical inference in longitudinal models: one is *maximum likelihood* (ML) and the other is *residual or restricted maximum likelihood* (REML).

Maximum likelihood is a traditional estimation method. It requires developing a sampling model $f(Y_i|\alpha, \theta)$ for the data given the parameters. In normal linear regression $y_i = x_i'\alpha + \delta_i$, with n independent subjects and $\delta_i \sim N(0, \sigma^2)$, the y_i's are distributed $N(x_i'\alpha, \sigma^2)$ and the sampling density of $Y = (y_1, \ldots, y_n)'$ is

$$f(Y|\alpha, \sigma^2) = \prod_{i=1}^{n} \frac{1}{(2\pi\sigma^2)^{1/2}} \exp\left[\frac{(y_i - x_i'\alpha)^2}{2\sigma^2}\right].$$

This is the density of an n-dimensional bell-curve. The sampling density is a function of data Y and parameters (α, σ^2). As a sampling density, we think of it as a function of Y given or *conditional on* the parameters. After we observe data Y, we fix Y and consider the sampling density strictly as a function of the unknown parameters. Although this is the same function $f(Y|\alpha, \sigma^2)$, we give it a new name, the *likelihood—)*, and we write it as

$$L(\alpha, \sigma^2|Y) = f(Y|\alpha, \sigma^2).$$

Maximum likelihood finds estimates $(\hat{\alpha}, \hat{\sigma}^2)$ of unknown parameters (α, σ^2) that maximize the likelihood function $L(\alpha, \sigma^2|Y)$

$$L(\hat{\alpha}, \hat{\sigma}^2|Y) \geq L(\alpha, \sigma^2|Y).$$

The hat on top of an unknown parameter denotes a maximum likelihood estimate for that parameter. In linear regression, the ML estimate for α is the familiar *least squares estimate*

$$\hat{\alpha} = (X'X)^{-1}X'Y,$$

but the ML estimate for the variance

$$\hat{\sigma}^2 = \frac{1}{n}\sum_{i=1}^{n}(y_i - \hat{y}_i)^2.$$

is slightly different from the more usual $\tilde{\sigma}^2$ estimate given in section 5.3.3 that uses $n - K$ in the denominator. The difference between the two estimates is modest unless $n - K$ is small.

In normal longitudinal models, the ML estimate of α is the weighted least squares estimate $\hat{\alpha}$ already discussed in section 5.3.3. The maximum likelihood estimate $\hat{\theta}$ of θ is rarely available in closed form and a number of algorithms for calculating $\hat{\theta}$ have been implemented in software. We discuss computational issues at a general level in section 6.5.

The likelihood function for multivariate regression models is

$$
\begin{aligned}
f(Y|\alpha,\theta) &= L(\theta,\alpha|Y) \\
&= \prod_{i=1}^{n} \frac{1}{(2\pi)^{n_i/2}|\Sigma_i(\theta)|^{1/2}} \\
&\quad \exp\left[.5(Y_i - X_i'\alpha)'\Sigma_i^{-1}(\theta)(Y_i - X_i'\alpha)\right].
\end{aligned} \tag{6.1}
$$

This expression involves the *determinant* $|\Sigma_i(\theta)|$ and the inverse $\Sigma_i^{-1}(\theta)$ of the covariance matrix $\Sigma_i(\theta)$. For all that, finding the maximum likelihood estimate for α for known θ is given by $\hat{\alpha}(\theta)$ in equation (5.13).

Maximum likelihood algorithms generally proceed by substituting $\hat{\alpha}(\theta)$ into (6.1) and maximizing the log of $L(\hat{\alpha}(\theta),\theta|Y)$ as a function of θ. Maximum likelihood estimates $\hat{\alpha}$ and $\hat{\theta}$ can have problems when used with data sets with small n and N and large K. In linear regression, where $n = N$, statisticians use the estimate $\tilde{\sigma}^2$ rather than the ML estimate $\hat{\sigma}^2$ for the variance parameter.

6.3.1 Residual Maximum Likelihood

In longitudinal models, we do not typically have estimates of θ available that are unbiased. However, there exists a method of inference called *residual* (or *restricted*) *maximum likelihood* (REML) that gives $\tilde{\sigma}^2$ as an estimate in linear regression and generalizes to longitudinal models. Generally we will use REML estimates throughout the book, but there are enough circumstances where we use ML that we usually mention which form of estimation is being used in any one example.

The REML estimate $\tilde{\theta}$ maximizes a slightly different likelihood $L(\theta|Y)$ rather than $L(\hat{\alpha}(\theta),\theta|Y)$. What happens is that we *integrate* $L(Y|\alpha,\theta)$ with respect to α. We *marginalize* the likelihood $L(Y|\alpha,\theta)$ with respect to α. Marginalization of the likelihood is done because it can be done easily and maximizing the marginalized likelihood produces better estimates of θ.

The REML likelihood function for longitudinal models is

$$
\begin{aligned}
L(\theta|Y) &= \int f(Y|\alpha,\theta)d\alpha \\
&= (2\pi)^{K/2}\left|\sum_{i=1}^{n} X_i'\Sigma_i^{-1}(\theta)X_i\right|^{1/2} \prod_{i=1}^{n}\left(\frac{1}{(2\pi)^{n_i/2}|\Sigma_i(\theta)|^{1/2}}\right. \\
&\quad \left. \exp\left\{-\frac{1}{2}[Y_i - X_i'\hat{\alpha}(\theta)]'\Sigma_i^{-1}(\theta)[Y_i - X_i'\hat{\alpha}(\theta)]\right\}\right).
\end{aligned} \tag{6.2}
$$

This is a function of θ only. The value of θ that maximizes (6.2) is $\tilde{\theta}$. It is in some ways similar to formula (6.1), but formula (6.2) has the weighted least squares estimate $\hat{\alpha}(\theta)$ substituted in place of the unknown parameter α. A second difference is that there is a determinant in the numerator; this is the determinant of the variance of $\hat{\alpha}(\theta)$. There are a few extra $(2\pi)^{1/2}$ thrown in for good measure, but multiplicative constants do not matter in maximizing a likelihood, nor do they change the inferences.

Happily, the algorithms that produce ML estimates also produce REML estimates with minor modification, so the same software can produce either estimate.

The REML estimate of α is then $\tilde{\alpha} = \hat{\alpha}(\tilde{\theta})$. The REML standard error of $\tilde{\alpha}$ is

$$\text{Var}(\tilde{\alpha}) = \left[\sum_{i=1}^{n} X_i' \Sigma_i(\tilde{\theta})^{-1} X_i \right]^{-1}$$

REML fitted and predicted values and standard errors follow the same formulas as those for ML, merely substituting $\tilde{\theta}$ for $\hat{\theta}$.

6.4   Assuming Normality

The normality assumption (5.7) is used throughout this book except in chapter 11. Technically, the normal assumption is strictly appropriate only when Y_i is distributed multivariate normal around $X_i\alpha$ and with covariance matrix $\Sigma_i(\theta)$ for some values of α and θ. The normal assumption can be hard to check if n_i is variable and $X_i\alpha$ is different for each case. In practice, we hope that the Y_i have a symmetric and unimodal distribution around $X_i\alpha$.

The Y_{ij}s or ϵ_{ij}s in many data sets appear to be from unimodal, symmetric, or nearly symmetric distributions of which the normal is certainly one possible candidate distribution. When researchers became interested in longitudinal data, the push was made to extend linear regression models to longitudinal data and the familiar normal assumption was carried along as well. We do need to make some assumption to make progress, and if we make the normal distribution assumption, we are able to do the calculations for longitudinal linear regression models, as we have been able to for regular linear regression for many decades.

The central limit theorem is often mentioned as a justification for assuming normal residual errors. Early linear regression software began with the normal assumption and tradition has taken over as a partial justification for assuming normality. Limited research has shown that errors such as the ϵ_{ij}s are usually not normal but often visibly long-tailed, and the normality assumption is in the process of falling into disrepute. Long-tailed data means that the data has more observations farther from the postulated mean than the normal distribution would predict. That is, typical data has outliers! The t distribution has long tails and is occasionally used in place of the normal to model regression and longitudinal data.

Another reason for assuming the normal distribution as an error distribution is that it returns least squares estimates. There are theoretical justifications for least squares estimates. In particular, an important one is an asymptotic justification that as the sample size n gets large enough, least squares estimates get closer and closer to the right answer; that is, $\hat{\alpha}$ gets closer and closer to α, even if the covariance model $\Sigma_i(\theta)$ is incorrectly specified. We often do not think the normality assumption affects our conclusions greatly. If we could demonstrate that the normal assumption mattered substantially and that our data was non-normal, then we would not make the normal assumption. Rather, we would come up with a distributional assumption that provides a better description of the particular data set that we are modeling. Because there are currently few options, and, because the options are problematic and require substantial additional resources to produce a single analysis, we do not further consider alternatives to the normal distribution. Chapter 14 gives some references.

6.5 Computational Issues

I used one software package to fit model 5.3 to the Big Mice data. It gives an error message, but it does provide estimates of Σ although without standard errors. Inspection of the reported estimates indicates that the estimates are incorrect; the variance of the first observations is reported to be $\hat{\sigma}_{11} = 246482$ or a standard deviation of almost 500! Looking at any of figures 2.1–2.3 shows that the range of weights at day zero run from greater than 0 to less than 200; the standard deviation should easily be less than the range divided by 4 or about $50 = (200 - 0)/4$. A better covariance model we will meet in chapter 8 estimates $\sigma_{11} = 240.78$. The standard deviation is then around $\hat{\sigma}_{11}^{1/2} = 15.5$, considerably more reasonable than 500.

With other data sets and models one may get various error messages – the software is unable to provide parameter estimates at all; estimates have not converged; the Hessian is not positive definite; the number of iterations has reached a maximum; the list of problems is extensive.

Several problems occur with software in trying to fit models to longitudinal data. These usually boil down to a *lack of convergence* of the algorithm. Lack of convergence is a fancy way of saying that the software could not find good estimates and that the software knows that it did not find good estimates. It is incumbent upon the user to do basic sanity checking of statistical output to make sure that reported estimates are sensible, as sometimes the software does not know that it did not find good estimates!

It is easy to check that an estimated variance is plausible. Take the square root of the variance σ^2, multiply by 4, and that should approximately be the range of the responses at a given time. If our covariates explain a huge amount of the variation in our data, 4σ may even be considerably less than the range. As a rule of thumb, the standard deviation should not be larger than one quarter the range of the data.

6.5.1 Maximizing a Likelihood

In ML inference, we are attempting to find the parameter values that maximize a likelihood function $L(\alpha, \theta | Y)$. As previously mentioned, the likelihood is a function of the fixed effects parameters α and the covariance parameters θ with the data Y held fixed. If we know the variance parameters θ, then the fixed effects parameter estimates $\hat{\alpha}(\theta)$ can be calculated in closed form using the weighted least squares formula (5.13). This simplifies the problem, as we can write our likelihood function as $L(\hat{\alpha}(\theta), \theta | Y)$. We have reduced the dimension of the problem from needing to estimate K unknown fixed effects and q unknown variance parameters to merely needing to estimate q unknown variance parameters. The function $L(\hat{\alpha}(\theta), \theta | Y)$ we now have to maximize is more complicated, but the dimension q is smaller than $q+K$, and this is usually a helpful trade-off. The function $L(\hat{\alpha}(\theta), \theta | Y)$ is similar to the REML likelihood that we need to maximize to find REML estimates, and our discussion on computational issues applies to both ML and REML equally.

Most function maximization algorithms are *iterative*. Given starting values $\theta^{(0)}$, they attempt to find a set $\theta^{(1)}$ of better values using a simple algorithm. The algorithm is then re-applied starting at $\theta^{(1)}$ to give $\theta^{(2)}$. At iteration l, we have estimates $\hat{\alpha}^{(l)} = \hat{\alpha}(\theta^{(l)})$, and $\theta^{(l)}$. The superscripts in parentheses (l) indicate the iteration number. The algorithm continues until a *stopping rule* is triggered. Stopping rules usually have several different ways to stop the algorithm. A simple reason is that a maximum number L_0 of iterations has been achieved. A second reason is that all candidate values $\theta^{(l+1)}$ for the next iteration have a decreased likelihood

$$L(\hat{\alpha}(\theta^{(l)}), \theta^{(l)} | Y) > L(\hat{\alpha}(\theta^{(l+1)}), \theta^{(l+1)} | Y)$$

rather than an increased likelihood. The idea here is that there are situations when the algorithm can identify that improvement in the likelihood is possible, but it cannot identify an actual new point θ^{l+1} where the likelihood is higher. The algorithm will stop. A third reason is that the algorithm discovers that there are no immediate directions in θ space in which the likelihood increases. This last reason may mean that the latest estimates $\hat{\alpha}(\theta^l)$ and θ^l are the maximum likelihood estimates we seek. If we can then estimate standard errors of the estimates, then the algorithm stops and claims convergence. Sometimes we may not be able to estimate standard errors of the estimates, and this often means that we have not found maximum likelihood estimates and that the algorithm has failed.

6.5.2   A Function Maximization Analogy

How do computational algorithms work? We give qualitative details here. A reasonable analogy for function maximization is the problem of finding the highest point of a mountainous terrain surrounded by a flat meadow while in a dense fog. Wherever we start, we cannot look over the entire piece of terrain; we can only see the terrain within an arm's length of where we currently stand. If we are near the highest peak, then many algorithms will get us to the top of the highest mountain. If we are in the meadow, and the meadow is flat, then we may not even be able to tell where the mountains are. This is an example of requiring good *starting values*. Where we start often determines whether the algorithm succeeds in finding the top of the mountain.

Most algorithms attempt to estimate the slope of the land at the point where we currently stand. They estimate the direction of steepest slope, and the distance to the peak, and that is the next location $\theta^{(l+1*)}$ that they try. Often that is not actually a good location, so it will test values along the line connecting $\theta^{(l)}$ and $\theta^{(l+1*)}$ to identify a good new location $\theta^{(l+1)}$. This general type of algorithm works well much of the time. But we can make some general statements about where it fails.

If we are standing in a flat meadow, then we cannot estimate a slope. The meadow looks flat in all directions, so we cannot determine where the mountains are. All candidate next steps that we propose will be equally good, or equally bad depending on our personal optimism. The algorithm stops because it cannot find a good next step. The algorithm has not converged, it does not necessarily even get started. This may be what happened in the introductory example given at the beginning to this section.

Another meadow problem occurs when the meadow rises up to a cliff overlooking the ocean. We may accidentally get started off toward the ocean, and only then discover that we really have not gotten anywhere useful, as we really needed to go in the other direction to get to the top of the mountain.

A serious problem occurs when there are many mountaintops. Our algorithm will often get us to the top of one, but it may not be the highest mountaintop. Almost the only practical solution to this is to try many wildly divergent starting values and then to compare the various estimates one gets. Otherwise it is quite difficult to identify that we are in this situation.

6.5.2.1   Ridges

Another set of problems involve a knife-edge-like mountain ridge. If we can follow the ridge as it climbs, we can reach the mountaintop. But suppose that we can tell we are on the ridge, but we cannot find the exact direction along the ridge where the ridge climbs – every where we look, the land falls away from the ridge. This would happen if the knife-edge curves slightly as it rises. So all new attempted points $\theta^{(l+1)}$ will be worse than the current point, and we are stuck. Cannot climb up the ridge, and can only find new points of terrain below the current spot, yet the software may know that it is on the knife-edge.

A somewhat more rounded ridge may let us find the direction where we can climb up the ridge, but the algorithm may only be able to take short steps, for example if the ridge wanders a little bit left and right as we climb it. Then it can take many many steps to get to the top of the mountain, even if the ridge leads to the top; this can cause our algorithm to use too many iterations and to quit because of this. A rounded ridge may lead us to the mountaintop if we just let the algorithm run long enough; many algorithms print out the current maximized log likelihood – if it is increasing slowly but steadily and for a long time, we may well be on a ridge. The algorithm may quit because the number of iterations is greater than the allowed maximum. Increasing the allowed maximum number of iterations may let us find the maximum likelihood estimates.

6.5.2.2   Saddlepoints

Occasionally, the algorithm can find itself in the flat portion of a saddle-shaped region. The algorithm may have followed a ridge up to get to the saddle, and another ridge leads away, but the algorithm cannot find the direction of the new ridge. Indeed, the algorithm may think it is at the top. However, when the algorithm calculates the curvature of the mountaintop, it discovers that the terrain does not curve downwards in all directions. The algorithm may declare that the "Hessian matrix is non-positive definite," which is a fancy way of saying that it does not think it is at a mountaintop but that it does not have a direction to go to climb up the mountain further.

6.5.3   Gradient and Hessian

Function maximization evaluates the log likelihood $\log L(\hat{\alpha}(\theta), \theta|Y)$ at the current guess $\theta^{(l)}$ to give $\log L(\hat{\alpha}(\theta^{(l)}), \theta^{(l)}|Y)$. We use the log rather than the likelihood for computational, statistical, and mathematical convenience. Algebraically, the math is often easier on the log scale; statistically, on the log scale some of the quantities that we calculate to maximize the function have statistical interpretations; and computationally, the numbers do not get so large or small that we have numerical overflow problems.

Locally we approximate the terrain's shape by a quadratic. A quadratic function is easy to find the maximum of; we pretend we have a quadratic, and estimate the maximum of the quadratic; that is our next step in the iteration. The shape of the terrain of a quadratic is determined by two functions, the *gradient* and the *Hessian*. The steepness of the slope is called the gradient $G(\theta)$. The gradient is the *first derivative* or slope of the log likelihood function. It is a vector of length q. The Hessian $H(\theta^{(l)})$ is the second derivative, or curvature, of the log likelihood function. In general, it is a q by q matrix when the likelihood function has q unknown parameters. Given the Hessian and the gradient, we can calculate $\alpha^{(l+1)}, \theta^{(l+1)}$. With a single parameter $\theta^{(l)}$, the next step is $\theta^{(l+1)} = \theta^l - H^{-1}(\theta^{(l)})G(\theta^{(l)})$.

When we get to the top of the mountain, the gradient is zero, because the mountaintop is flat in the sense that there is no upward trend in any direction at the top. The Hessian will be negative at the top. The negative inverse of the Hessian matrix is the estimated variance for the parameter estimates. A relatively gentle mountaintop means a small Hessian (small curvature) and large variances for the parameters. A sharply peaked mountaintop means a very large Hessian (large curvature) and very small variances. A mountaintop that is long and narrow means a mix of larger and smaller standard errors. Nonsensical parameter estimates often indicate that we are in a meadow. The Hessian in a meadow is often very very small (no curvature). If our starting values are too far from the mountains, it will lead to silly parameter estimates; often the estimated standard errors are also nonsensical and very large as well.

6.5.4   Collinearity

In linear regression, collinearity of two predictors causes numerical problems in computation of least squares estimates. Collinearity occurs if two or more predictors are highly correlated. A mountain ridge in the likelihood is essentially akin to collinearity. It indicates that two parameter estimates are very highly correlated. A slight change in the value of one estimate will lead to a substantial change in the other. In general in linear regression, we do not necessarily need to include two predictors that are collinear; we often delete one covariate. This works if we want to predict our outcome variable and do not care how we do it. When we are interested in root causes, variable deletion is not appropriate. Either or both variables could potentially be a cause of the outcome. Variable deletion may well lead to erroneous conclusions by deleting a causal predictor. A solution that can solve the numerical problem without sacrificing the scientific problem is to define two new covariates; one is the average and the second is the difference of the two collinear covariates. Of course, taking an average is appropriate assuming that the two covariates trade off in a one-to-one fashion, that is, that a change in one estimate causes a one-to-one change in the other covariate. Suppose in a regression model that one covariate is baseline height and the other covariate is height a year later. In predicting an outcome, presumably the two heights trade off in a one-to-one way. If we measure one height in inches and the other in yards, then the two coefficients are likely to trade off in the ratio of 36 to one rather than one-to-one.

In longitudinal models, if the iterative function maximization algorithm moves up the ridge slowly, one possible solution is to redefine the covariance parameters that are correlated. Often there are not the same issues as for collinear covariates; we can eliminate one parameter as not being a useful addition to the model.

6.5.5   Dealing with Lack of Convergence

The larger q is, that is, the more covariance parameters we have, the harder the function maximization algorithm has to work. Going back to the mountainous terrain analogy, the more covariance parameters, the farther from the mountaintop we are, and often, the farther into the meadow we are. The problem with the introductory example in this section is that there were too many parameters in the covariance model, and the Big Mice data set does not have enough data to estimate all those covariance parameters.

As a general statement, if we cannot get the software to create estimates for our data set, simplify the covariance model; reduce the number of parameters in the covariance model until we can fit the model to the data. Then go back and, if possible, add a few parameters at each model fitting until we get closer to the model that we need or want. Covariates that are functions of time and have a very significant effect in explaining the response must be included in the model. Until these needed time covariates (i.e., time, time squared) are in the model, one can have real difficulties in fitting even simple data sets. The Small Mice data can be difficult to fit without time in the fixed effects model. With time as a predictor, we can fit the unstructured covariance model to the data. Without it, my software does not converge. Admittedly this is a silly model, but it would better to have the software fit the model and then have data plots and diagnostics tell us that it does not fit well, rather than us having to guess at the problem because the software does not converge. Extra covariates that do not explain much of the response are not helpful in getting algorithms to converge and can be omitted until we reach a covariance model that can be fit to the data. Too many extra useless covariates can also cause problems in their own right.

Collinear covariates cause numerical problems with computations in longitudinal models the same way that they can for linear regression. The solutions are similar; redefine the covariates to eliminate the collinearity or delete one of the covariates.

The units of time can cause additional problems for our algorithms in longitudinal data. Suppose we collect data every year, and the correlation between consecutive yearly observations is, say, .9. If we measure time in days, then the correlation parameter will be $.9^{1/365} = .9997$ which is the correlation between observations taken on consecutive days. Software may have trouble estimating a correlation this close to 1. Better is to measure time in years, so that the natural correlation is near .9 and not near 1. Similarly, if time is a covariate in our analysis, it is best to measure time in modest units so that the range of time is in single digits, and does not vary too greatly over the data set. If observations are measured yearly, measure time in years, not days. If observations are taken every second, don't use milliseconds for the units, but use seconds.

6.5.6 Discussion

Our discussion here has been specifically for normal linear models for longitudinal data. The discussion should, for the most part, apply also to hierarchical models for hierarchical data sets, and general linear multivariate models for multivariate data as well. When we get into discrete response data in chapter 11 or nonlinear normal models for longitudinal data, additional problems arise. A key issue is that the fixed effects parameters cannot be estimated in closed form even if the variance parameters are known; this means that the estimation algorithms require an additional layer of numerical computations with concomitant increase in the numbers of possible problems. We will not discuss Bayesian methods here, but Bayesian computational algorithms are quite competitive with maximum likelihood type algorithms for complex models in terms of quality of estimates, convenience of inference, convergence problems and the possibilities of simple error checking for lack of convergence. For complex data sets and models, Bayesian methods should be seriously considered.

6.6 Back-Transforming a Transformed Response

In section 2.3.2, we recognized that the Pain data needed a transformation to reduce skewness, and we saw that a log transformation removed most of the skewness. We do our analysis on the log-transformed data instead of the original data. We make estimates of mean response for a given set of covariates, and we can make predictions also using our current technology. However, these estimates and predictions would be on the transformed scale. The process of *back-transformation* undoes the original transformation. It allows us to have estimates and inferences on the original scale. This is helpful, as judgment about the practical effect of differences is usually much easier on the original scale rather than on the transformed scale.

For this section, let W_{ij} be the original, or raw, observation for subject i at time t_{ij}. Let $Y_{ij} = g(W_{ij})$ be the transformed observation. We restrict transformations $g(\cdot)$ to be *monotone*; for all W_{ij} and W_{ik}: if $W_{ij} < W_{ik}$

then we want $Y_{ij} < Y_{ik}$ also. Monotone could also mean that if $W_{ij} < W_{ik}$ then $Y_{ij} > Y_{ik}$, reversing the inequality for all W_{ij} and W_{ik}. Reversing the inequality on the transformed scale has the effect of inverting larger and smaller; this is usually inconvenient for communication purposes and does not expand our statistical options. If a response is better the higher it is, it is convenient that the transformation also follows the same interpretation. If the transformation reverses the worse/better direction, it can be fixed by taking the negative of the transformation. Then the worse to better direction remains the same as the original scale.

Examples of transformations are the log: $Y_{ij} = \log W_{ij}$, and the square root: $Y_{ij} = W_{ij}^{1/2}$. We would then analyze the Y_{ij} in our models. These two transformations are both members of the power family of transformations where the zero-th power corresponds to the *log* transformation

$$Y^{(\lambda)} = \left\{ \begin{array}{ll} Y^\lambda & \lambda = 0 \\ \log Y & \lambda \neq 0 \end{array} \right. .$$

Power transformations are most commonly used on positive or non-negative data to fix a right tail that is longer than a short left tail. The null transformation is the 1 power, $Y_{ij} = W_{ij}^1$. The farther the transformation is from 1, the stronger it is. The inverse square root power is stronger than the log transformation because $-1/2$ is farther away from 1 than zero is and the log transformation is stronger than the square root transformation. The stronger the transformation, the more it adjusts the skewness. It is rare to take power transformations where the power is larger than 1; these correspond to increasing the length of the right tail and shortening the left tail.

The base on the log transformation whether 2, 10 or base e does not matter scientifically, as different bases on the log are equivalent to multiplying the Y_{ij} by a constant; $\log_2 W_{ij} = \log W_{ij} / \log(2)$. Log base 2 may be easier to use in practice, as it is easier to calculate powers of 2 than powers of e allowing the analyst to quickly back-transform results. If the data do not cover a large dynamic range, i.e., several powers of 10, then log base 10 is equally difficult to invert without a calculator.

In regression analyses, we often transform data; usually we restrict ourselves to transformations that are easily interpretable, such as $\log Y_{ij}$ or square root, $Y_{ij}^{1/2}$ and occasionally inverses, Y_{ij}^{-1}. General recommendations learned from linear regression also apply to longitudinal data. For the Pain data, we took logs; for non-negative responses, the log transformation is easily the most common and useful. If the data are non-negative, i.e., it has zeros, a convenient fix is to add a small constant to the data before taking the log transform. A first default value to add is the smallest positive value in the data set. If the response is a count, we often add 1 before taking a log. We rarely take other less interpretable transformations.

Discretization, turning a continuous response into a discrete or categorical variable, is virtually never recommended.

6.6.1   Estimates and Predictions After Transformation

Suppose we have estimated a linear function $x'\hat{\alpha}$ of α with standard error $\mathrm{SE}(x'\hat{\alpha})$ using REML or ML. We assume that the x is a possible covariate x_{ij} for some, possibly hypothetical, subject at time t. Then $x'\hat{\alpha}$ estimates $x'\alpha$, and

$$(x'\hat{\alpha} - 2\mathrm{SE}(x'\hat{\alpha}), x'\hat{\alpha} + 2\mathrm{SE}(x'\hat{\alpha})) \qquad (6.3)$$

is a 95% confidence interval for $x'\alpha$. The interpretation of $x'\alpha$ is a mean at a given time on the transformed $Y = g(W)$ scale for a subject at a given time t with covariates x. Under the normality assumption, the mean is also a *median*, so the above confidence interval is also an interval for the median.

The inverse transformation is $W = g^{-1}(Y)$. For the log transformation, $W = \exp(Y)$ and for the square root transformation, $W = Y^2$. Back on the original, untransformed W scale, $g^{-1}(x'\hat{\alpha})$ is an estimate of the median of the observations at covariate value x and time t, but is no longer an estimate of a mean of those observations. Similarly, we can apply $g^{-1}(\cdot)$ to the endpoints of the confidence interval (6.3) create a confidence interval for the median $g^{-1}(x'\alpha)$

$$(g^{-1}[x'\hat{\alpha} - 2\mathrm{SE}(x'\hat{\alpha})], g^{-1}[x'\hat{\alpha} + 2\mathrm{SE}(x'\hat{\alpha})])$$

This is called *back-transformation*.

We can also back-transform a prediction. Again, for an observation with covariate vector x at time t, and standard error of prediction $\mathrm{SEpred}(x'\hat{\alpha}$ from (5.15). The back-transformed interval is

$$(g^{-1}[x'\hat{\alpha} - 2\mathrm{SEpred}(x'\hat{\alpha})], g^{-1}[x'\hat{\alpha} + 2\mathrm{SEpred}(x'\hat{\alpha})]).$$

Interval (6.6.1) is a 95% prediction interval for a new observation measured on the original W scale at covariate value x at time t.

6.6.2   The Log/Exponential Transformation

If the transformation g is the log transform, inference is slightly easier back on the original scale than for other transformations. For individual coefficients, $\exp(\hat{\alpha}_j)$ is the ratio of the median estimates $\exp(x^{*'}\hat{\alpha})$ to $\exp(x'\hat{\alpha})$ where x^* is equal to x except that the jth covariate has been increased by 1 unit. To convert to a percentage, $[\exp(\hat{\alpha}_j) - 1]100\%$ is the percentage increase in the median for a unit increase in x_{ijk}. A 95% confidence interval on $[\exp(\hat{\alpha}_j) - 1]100\%$ is constructed by transforming the endpoints of the confidence interval for $\hat{\alpha}_j$ in this same way.

We can also transform back and get an estimate and interval for the mean on the W scale. The mean $E(W|x)$ of the responses on the original scale for covariate vector x at time t is a function of both the mean and the variance on the log scale

$$E(W|x) = \exp[x'\alpha + \text{Var}(Y)/2].$$

We need to estimate this; because there is uncertainty in $\hat{\alpha}$, that will inflate the mean estimate $E(W|x)$ somewhat. Our estimate of the mean on the original scale becomes

$$E(W|x) = \exp[x'\hat{\alpha} + \text{SEpred}^2(x'\hat{\alpha})/2].$$

6.6.3   General Back-Transformation

In general, back-transformation is somewhat messy. Suppose first we wish to transform a point estimate $\hat{\mu} = x'\hat{\alpha}$ of $\mu = x'\alpha$ and associated standard error back on the original scale. We require that x be a covariate vector of a potential observation and not a contrast or other linear combination of the fixed effects. Let $h(W) = Y$ be the inverse transformation of $W = g(Y)$. The *delta method* can be used to estimate the standard error back on the original scale. The delta method requires taking the first derivative of the $h(\mu)$ function in calculating the required standard error. For the power family, the derivative of $h(\mu)$ with respect to μ is

$$\frac{d(c\mu^\lambda)}{d\mu} = c\lambda\mu^{\lambda-1}$$

for $\lambda \neq 0$ and where c is an arbitrary non-zero constant. For the inverse of the log transformation

$$\frac{d\exp(c\mu)}{d\mu} = c\exp(c\mu).$$

The delta method says that the standard error of $h(\hat{\mu})$ is

$$\text{SE}[h(\hat{\mu})] = \text{SE}(\hat{\mu}) \left|\frac{dh(\hat{\mu})}{d\hat{\mu}}\right|.$$

For the power transformation family and $\lambda \neq 0$

$$\text{SE}(c\hat{\mu}^\lambda) = c\,\text{SE}(\hat{\mu})\lambda\hat{\mu}^{\lambda-1}$$

and for the log transformation

$$\text{SE}[\exp(c\hat{\mu})] = c\,\text{SE}(\hat{\mu})\exp(c\hat{\mu}).$$

Usually we do not use this standard error to create a confidence interval on the back-transformed scale because normal theory does not apply as readily as it does on the transformed scale. We would create a confidence interval on the original scale and then back-transform the endpoints as described earlier.

Let x_1 and x_2 be two possible covariate vectors. We wish to compare the average population responses $\mu_1 = x_1'\alpha$ and $\mu_2 = x_2'\alpha$ at the two vectors on the original scale. The back-transformed estimated difference is

$$h(\hat{\mu}_1) - h(\hat{\mu}_2) = h(x_1'\hat{\alpha}) - h(x_2'\hat{\alpha}) \tag{6.4}$$

and the squared standard error of this difference is

$$
\begin{aligned}
\mathrm{SE}^2[h(\hat{\mu}_1) - h(\hat{\mu}_2)] \;=\; & \left[\frac{dh(\hat{\mu}_1)}{d\hat{\mu}_1}\right]^2 \mathrm{SE}^2(\hat{\mu}_1) + \left[\frac{dh(\hat{\mu}_2)}{d\hat{\mu}_2}\right]^2 \mathrm{SE}^2(\hat{\mu}_2) \\
& -2\frac{dh(\hat{\mu}_1)}{d\hat{\mu}_1}\frac{dh(\hat{\mu}_2)}{d\hat{\mu}_2}\mathrm{Cov}(\hat{\mu}_1,\hat{\mu}_2)
\end{aligned}
$$

This can be used to compare the difference between any two estimates on the original scale. Often the two estimates we wish to compare differ only by a single unit in the jth covariate. In this case, this formula may be approximated to good effect by

$$\mathrm{SE}^2[h(\hat{\mu}_1) - h(\hat{\mu}_2)] = \left[\frac{dh(\hat{\mu}_1)}{d\hat{\mu}_1}\right]^2 \mathrm{SE}^2(\hat{\alpha}_j). \tag{6.5}$$

Equations (6.4) and (6.5) can be used to determine the effect of a single covariate on the outcome on the back-transformed scale. The effect size and standard error depend on the level $h(\hat{\mu}_1)$ and the derivative $dh(\mu_1)/d\mu_1$ evaluated at $\hat{\mu}_1$ and these will vary depending on covariates x_1.

6.6.3.1 Pain Data

We illustrate back-transformation for two Pain data analyses. One uses the log base 2 transformation and the other uses the slightly stronger negative inverse square root $-W_{ij}^{-1/2}$ transformation. We chose the $-W_{ij}^{-1/2}$ because in figure 2.7 we saw some remaining skewness in the data after the log transformation. Both analyses use the unstructured covariance matrix and both have two covariates: the intercept and an indicator for being a distracter. Results are given in table 6.2.

The point estimates on the transformed scales are very different for the two transformations; these numbers are not comparable, for example 4.57 versus $-.212$. When we transform back, the transformed estimates are measured in seconds and are now comparable, 23.8 for the log versus 22.2 for the other. In the back-transformed scale, the results from the log analysis have a longer right tail than the $-W_{ij}^{-1/2}$ transformation. Estimates of the lower end of the confidence interval for attenders are almost identical, but as the estimates get higher and higher, as for distracters and for the upper end of the confidence interval, the log analysis back-transformed estimates are further out than the $-W_{ij}^{-1/2}$ analysis.

Transform						Original units		
	Est	SE	t,p	Low	Upp	Est	Low	Upp
$\log W_{ij}$								
A	4.57	.16	28.2	4.25	4.90	23.8	19.0	29.8
D	5.08	.16	31.4	4.75	5.40	33.8	27.0	42.2
D-A	.51	.23	2.2, .031	.05	.96	1.4	1.0	1.9
$-W_{ij}^{-1/2}$								
A	-.212	.010	-21.6	-.232	-.192	22.2	18.6	27.0
D	-.183	.010	-18.6	-.203	-.164	29.8	24.3	37.4
D-A	.029	.014	2.1, .042	.001	.057			

Table 6.2. Analysis of the Pain data analysis with the unstructured covariance model and two different response transformations. Fixed effects are an intercept and indicator for coping style. The top analysis uses the $\log_2$ transform, and the bottom analysis uses the negative inverse square root transformation $Y_{ij} = -1W_{ij}^{-1/2}$. Columns are the analysis name, the effect being estimated, the t-statistic and p-value using 62 df as given by the software, lower and upper end points for a 95% CI, the back-transformed estimate, and the back-transformed confidence interval endpoints.

6.7   Some Design Considerations

We discuss basic design issues in the context of a balanced data set with compound symmetry covariance model with constant population mean over time. This model is simple enough that the computations can be done in closed form. For several longitudinal designs, we compare the variance of the grand mean and we consider the costs of the designs. Next we elaborate our model to have two population means, and we compare the two-sample design to the paired comparison design for estimating the difference between the means of the two populations.

This section uses more algebra than is typical for the main portion of the text.

6.7.1   Estimating the Population Mean

Our design consists of a sample of size n with J observations per subject. The fixed effect consists of a constant mean α_1 over time.

$$Y_{ij} = \alpha_1 + \epsilon_{ij}. \tag{6.6}$$

The ϵ_{ij} have variance τ^2 and covariance $\tau^2\rho$ between two distinct time points for the same subject.

Suppose that we estimate the population mean μ by $\bar{Y}$, the average of all the observations

$$\bar{Y} = \frac{\sum_{ij} Y_{ij}}{nJ}.$$

This can be written as the average of the n individual subject means

$$\bar{Y}_i = \frac{\sum_{j=1}^{J} Y_{ij}}{J}$$

so that

$$\bar{Y} = \frac{\sum_i \bar{Y}_i}{n}.$$

Observations within a subject are correlated, but subjects are independent. The sample grand mean $\bar{Y}$, being an average of the n individual subject means $\bar{Y}_i$ has the same expected value as the individual $\bar{Y}_i$ and has variance $1/n$ times the variance of a single $\bar{Y}_i$. We plug in model (6.6) into the above formula for $\bar{Y}_i$ and simplify.

$$
\begin{aligned}
\bar{Y}_i &= \frac{\sum_{j=1}^{J} Y_{ij}}{J} \\
&= \frac{1}{J} \sum_{j=1}^{J} (\alpha_1 + \epsilon_{ij}) \\
&= \alpha_1 + \frac{1}{J} \sum_{j=1}^{J} \epsilon_{ij}.
\end{aligned}
$$

The expected value of subject i's average $\bar{Y}_i$ is equal to α_1, the quantity that we want to estimate, plus the expected value of the ϵ_{ij}s. The ϵ_{ij}s have expectation zero, so

$$\mathrm{E}[\bar{Y}_i] = \alpha_1.$$

The variance takes a little more work.

$$
\begin{aligned}
\mathrm{Var}[\bar{Y}_i] &= \frac{1}{J^2} \mathrm{Var}(\sum_{j=1}^{J} \epsilon_{ij}) \\
&= \frac{1}{J^2} \left[\sum_j \mathrm{Var}(\epsilon_{ij}) + 2 \sum_{k<j} \mathrm{Cov}(\epsilon_{ij}, \epsilon_{ik}) \right] \\
&= \frac{1}{J^2} \left(\sum_j \tau^2 + 2 \sum_{k<j} \tau^2 \rho \right).
\end{aligned}
$$

Now expand $\tau^2 = \tau^2 \rho + \tau^2(1-\rho)$ in the first sum, and collect J^2 of the $\tau^2 \rho$ terms and J of the $\tau^2(1-\rho)$ terms in the sums, such that

$$\mathrm{Var}[\bar{Y}_i] = \tau^2 \rho + \frac{\tau^2(1-\rho)}{J}.$$

If there were no J in the denominator of the second term, then this would equal τ^2, the variance of a single observation. Rearranging terms gives

$$\mathrm{Var}[\bar{Y}_i] = \frac{\tau^2}{J}[1 + (J-1)\rho].$$

If we had J independent observations, then ρ is zero, yielding a variance of τ^2/J. Because there is correlation ρ between observations, we have the added $(J-1)\rho$ term inside the square brackets. This is the penalty for having correlated observations when attempting to estimate an average.

The grand mean is then distributed as

$$\bar{Y} \sim N\left(\alpha_1, \frac{\tau^2}{Jn}[1 + (J-1)\rho]\right).$$

If $J = 1$, we are back in the situation of the simple random sample $y_{i1} \sim N(\alpha_1, \tau^2)$ and $\mathrm{Var}(\bar{y}) = \tau^2/n$. On the other hand, suppose J is very large, so that $\mathrm{Var}(\bar{Y}) = \tau^2[1 + (J-1)\rho]/Jn \approx \tau^2\rho/n$. In the extreme, if additional longitudinal measurements are free, we decrease the variance of our estimate $\bar{Y}$ by availing ourselves of as many observations J on subjects as we can.

6.7.2   Cost and Efficiency in Longitudinal Data

Suppose that recruiting a new person costs C_1 dollars and measuring each observation, including the first, costs $C_2\$$. The total cost of a study with n people and J longitudinal measures is

$$nC_1 + nJC_2.$$

Consider two designs, the simple random sample (SRS), with $2n$ people but with $J = 1$, and the longitudinal design (LD) with n subjects and $J = 2$ observations per person. The costs are $2n(C_1 + C_2)$ and $n(C_1 + 2C_2)$ respectively; the LD study is always cheaper than the SRS study.

Now let us compare the $2n$ subject SRS study to a LD with n subjects and 4 observations. The respective variances of $\bar{Y}$ are $\tau^2/(2n)$ and $\tau^2(1 + 3\rho)/4n$. The simple random sample has smaller variance if $(1 + 3\rho)/2 > 1$, or solving for ρ, if $\rho > 1/3$. Otherwise the longitudinal design has smaller variance. The simple random sample is cheaper if $C_1 < 3C_2$ otherwise the longitudinal design is cheaper. The designs and conditions under which each is cheapest or gives the minimum variance estimate of α_1 is given in table 6.3. Table 6.3 also gives results for the general SRS with sample size mn for unspecified $m \geq 1$ and for a longitudinal design with n observations and $J \geq 1$ repetitions.

Precision is the inverse of variance. By paying for a bigger sample size, we buy increased precision of our estimates. One way to jointly evaluate both the cost and variance is to think of the cost per precision or, in other words,

	Design	
	SRS	LD
SS	$2n$	n
Repeats	1	4
Cost	$2n(C_1 + C_2)$	$n(C_1 + 4C_2)$
Var of $\bar{y}$	$\tau^2/(2n)$	$\tau^2(1 + 3\rho)/(4n)$
Var best	$\rho > 1/3$	$\rho < 1/3$
Cost best	$C_2/(C_1 + C_2) > 1/3$	$C_2/(C_1 + C_2) < 1/3$
Cost×Var	$(C_1 + C_2)\tau^2$	$(C_1 + 4C_2)\tau^2(1 + 3\rho)/4$
SS	mn	n
Repeats	1	J
Cost	$mn(C_1 + C_2)$	$n(C_1 + JC_2)$
Var of $\bar{y}$	$\tau^2/(mn)$	$\tau^2[1 + (J-1)\rho]/Jn$
Var best	$\rho > \frac{J-m}{m(J-1)}$	$\rho < \frac{J-m}{m(J-1)}$
Cost best	$\frac{m-1}{J-1} < \frac{C_2}{C_1+C_2}$	$\frac{m-1}{J-1} > \frac{C_2}{C_1+C_2}$
Cost×Var	$(C_1 + C_2)\tau^2$	$(C_1/J + C_2)\tau^2[1 + (J-1)\rho]$

Table 6.3. Summary of two designs. Top portion is for a SRS of size $2n$ against a longitudinal design with n subjects and 4 longitudinal measures. Bottom is for a general SRS versus a general balanced LD design.

cost×variance. The design with the lower cost×variance is then best. The product of cost times variance is included in table 6.3 as well.

In longitudinal studies, C_2 is typically much less than C_1. In some circumstances, the cost to acquire the $(n+1)$st subject can be larger than even the C_1 cost to acquire each of the first n subjects. Suppose for one example that 50 attention deficit hyperactivity disorder (ADHD) children come into our clinic for treatment each year. In 1 year, we can recruit $n = 50$ children. The cost of the 51st child and beyond is infinite, or we must recruit into a second year, which incurs extra costs, or recruit at another clinic, which minimally requires involving another doctor and another bureaucracy in our study. Alternately, we may take longitudinal measures and stay with a sample size of 50.

For a second example, consider a blood pressure study. Blood pressure measurements are well-known to be highly variable. Using the random intercept model, this means that τ^2 is large and ρ is small. Also we know that taking multiple blood pressure measurements is inexpensive, particularly if the subjects are already coming to the clinic for other reasons. To quantify these qualitative statements, let us suppose that $\rho \approx .5$ and that $C_1 \gg C_2 \approx 0$. A SRS of size mn will produce a $\mathrm{Var}(\bar{Y}) = (\tau^2)/mn$. On the other hand, averaging many longitudinal observations within a person costs almost nothing, and we can reduce $\mathrm{Var}(\bar{Y})$ for the LD study to be $\tau^2\rho/n = \tau^2/(2n)$ by taking J large. In this example, the longitudinal design requires a sample size and cost that is half that of a simple random sample.

For a third example, suppose we are studying the cost of housing. Repeatedly measuring the cost of a person's rent/mortgage payment on a weekly or even monthly basis is unlikely to provide worthwhile additional information about rent or mortgage payments, except in the relatively rare event that the person moves, refinances their mortgage, or the rent increases. Now τ^2 is small and ρ is extremely high, perhaps nearly 1. Collecting longitudinal measurements would not be of great value.

6.7.3   Two Independent Samples versus Paired Comparisons

Consider trying to compare two means μ_1 and μ_2. Typically, we can use a two independent samples design, where we take n independent univariate samples from a population with mean α_1, and another n univariate independent samples from a second population with mean $\alpha_1 + \alpha_2$. Another popular design is the paired design, where we take two measurements Y_{i1} and Y_{i2} on each person $i = 1, \ldots, n$. We change the conditions between the two measurements so that the mean of Y_{i1} is α_1, and $\alpha_1 + \alpha_2$ is the mean of Y_{i2}.

We again use the compound symmetry model to model our data. For the two independent samples, let $\bar{Y}_1$ and $\bar{Y}_2$ be the means of the two samples. With the sampling variance τ^2 we have

$$\mathrm{Var}(\bar{Y}_2 - \bar{Y}_1) = \frac{2\tau^2}{n}.$$

For the paired design, we sample n people. However, we take differences

$$d_i \equiv Y_{i2} - Y_{i1} = \alpha_2 - \alpha_1 + \epsilon_{i2} - \epsilon_{i1}$$

and the variance of d_i is

$$\mathrm{Var}(d_i) = \mathrm{Var}(\epsilon_{i2} - \epsilon_{i1}) = 2\tau^2 - 2\tau^2\rho = 2\tau^2(1 - \rho).$$

Define the mean difference

$$\bar{d} = \frac{\sum_{i=1}^{n} d_i}{n},$$

and the variance of $\bar{d}$ as

$$\mathrm{Var}(\bar{d}) = \frac{\mathrm{Var}(d_i)}{n} = \frac{2\tau^2(1 - \rho)}{n}.$$

The paired design always has smaller variance than the two-sample design. Further, because the cost C_2 of a second observation is cheaper than the cost $C_1 + C_2$, of a new person plus their first observation, the paired design costs less money as well.

The longitudinal design is more flexible than the two-sample design. In the LD, we may take several baseline measurements, all with mean α_1. Then we intervene with the treatment, and take several additional measurements with mean $\alpha_1 + \alpha_2$. This allows us to further reduce the variance of the estimate of α_2. It also allows us to track the course of the improvement, that is, we can see if subjects decline prior to treatment, and if after treatment they have a trend over time as well.

There are many two-group problems where we cannot do a paired sample design. If we wanted to compare males and females or blacks and whites, a paired design is not possible. Still, with a LD, we can take multiple observations per subject to reduce variance of our estimates and find out about time trends, both not possible in the SRS design.

6.8 Problems

1. Compare the simple random sample (SRS) design with $2n$ subjects to the longitudinal design with $J = 2$ and n subjects. Assume $C_1 = 2C_2$ and assume the compound symmetry covariance model. We wish to estimate the population grand mean α_1. Under what conditions is the SRS a more efficient (i.e., smaller variance) design? When is the SRS a cheaper design? When is the SRS smaller in terms of cost$\times$variance?

2. Repeat problem 1 with $J = 3$ through 10. Create a table of results.

3. Continued from problem 2. For each $J \in (1, 2, \ldots, 10)$, and with the SRS having sample size mn, $m > 1$, at what relative cost $k > 1$ where $C_1 = kC_2$, do the SRS and LD designs have equal costs?

4. From problem 3, use the k that was calculated for each J. For which Js does the longitudinal design have smaller variance of $\bar{y}$?

5. Often in back-transforming an estimate, we want an estimate for the mean on the original $W = g^{-1}(Y)$ scale, rather than a median. One way to do this is to sample a number of new observations Y_l^* for $l = 1, \ldots, L$ on the transformed scale, transform them back to $W_l^* = g^{-1}(Y_l^*)$, then average the observations to estimate the mean on the original scale. Let $x'_{ij}\hat{\alpha}$ and $\mathrm{SEpred}(x'_{ij}\hat{\alpha})$ be the mean estimate and standard error of prediction on the transformed scale. Sample L normal random variables z_l, $l = 1, \ldots, L$ with mean zero and variance one. Calculate

$$Y_l^* = \mathrm{SEpred}(x'_{ij}\hat{\alpha})z_l + x'_{ij}\hat{\alpha}$$

and transform back to $W_l^* = g^{-1}(Y_l^*)$ and average the resulting predictions. The average is an estimate of the predicted mean on the original scale. This is called the *population averaged* response on the original scale. It does not necessarily follow the back-transformed path of any individual subject. The standard deviation of the W_l^* is a standard deviation of prediction on the original scale, and a $(1 - a)100\%$ prediction interval can be constructed by finding the

$(a/2)100\%$ observation from either end of the sample and having those observations be the end points of the prediction interval.

6. Use the methods of the previous problem to estimate the mean for distracters and for attenders for the Pediatric Pain data back on the original scale.

7. Suppose that we take 3 measurements before treatment and 3 measurements after treatment, and that the covariance matrix is compound symmetry. Subject means are α_1 prior to treatment, and are $\alpha_1 + \alpha_2$ after treatment. Assuming n subjects and balanced data, we might estimate the pre-treatment population mean by the average of the pre-treatment responses, and similarly for the post-treatment population mean. What is the variance of the difference of the post-treatment average response minus the pre-treatment average response? How many subjects would you need to have in the two independent samples design to get the same variance for estimating α_2?

7
Specifying Covariates

Prediction is very difficult, especially if it's about the future.

– Niels Bohr

The great tragedy of science – the slaying of a beautiful hypothesis by an ugly fact.

– Thomas Huxley

Overview

In this chapter, we discuss specification of the covariates used to model the population mean response over time.

1. We discuss how to specify the population mean

   (a) when there is no time trend;
   (b) when the time trend can be modeled simply by a polynomial, and
   (c) when it must be modeled by a more complex nonlinear function of time.

2. We discuss covariates, which can be

   (a) continuous or discrete and
   (b) time-fixed or time-varying;

3. We discuss

(a) Two-way and three-way interactions where time is involved in the interaction;

(b) How to specify time in our analyses.

4. We discuss the right way to adjust for baseline in an analysis.

We discuss how to specify covariates in our model for the mean response $E[Y_i] = \mu_i = X_i\alpha$. Appropriately specifying the X_i matrix allows the parameters α to describe the differences in population means between females and males, treated and untreated, older and younger, and how those differences change over time. The parameter estimates will tell us whether being higher on a covariate leads to a higher or lower response and whether that difference changes over time. Appropriate choice of covariates allows us to test the hypotheses of interest. We will graphically illustrate our models in theoretical versions of inference plots.

Setting up the covariate matrix for longitudinal data models is a more complex task than in linear regression. Covariates for longitudinal data models can be all the usual types: continuous, discrete, multinomial, ordered multinomial that are familiar from linear regression. Table 7.1 gives some examples. The issue in linear regression is how the population mean changes as the covariates change. In longitudinal models, the population mean is a function of time. If the population mean is flat over time, then the issue is usually like that in linear regression: describing how the population mean changes with covariates. However, much more can occur, and key concerns include how the average response changes over time and how covariates affect the time trend. To make life more interesting, some longitudinal models have time-varying covariates, and statisticians have only begun to develop ways to model how time-varying covariates may affect the mean response.

7.1   Time-fixed Covariates

In this section we discuss including time-fixed covariates in the covariate matrix. Time-fixed covariates do not change through the duration of the study. Some covariates, for example gender or race, do not change during a subject's life. Variables such as age, height, or weight will change only slightly during a short study. The changes will be small or not particularly relevant to the outcome, and we treat age, height or weight as time-fixed covariates. We collect information on time-fixed variables at baseline but typically do not re-measure them during followup visits to free up resources for collecting other more useful information. Even with variables that change over time, we often use only the baseline measurement as our covariate to keep interpretation simple.

Covariate type	Examples Name	Description
Continuous	Age	Age in years
	Dosage	Amount of drug given
Dichotomous	Gender	Female or male
	Treatment	Treatment or control
Multinomial	Race	Asian, Black, Hispanic, White
Ordered multinomial	Economic status	Q: How well off are you? Allowed answers are (1) Trouble buying food; (2) Able to eat and afford a home; (3) Able to buy some luxuries; (4) Well off
	HIV status	(1) Asymptomatic; (2) Symptomatic; (3) Advanced to AIDS
	Education	(1) Less than high school (2) High school grad (3) Some college (4) College grad

Table 7.1. Types of covariates, with examples.

In setting up the covariate matrix X_i for time-fixed covariates, columns X_{ik} are constant over time. Stated another way, each $x_{ijk} = x_{ilk}$ for any times t_{ij} and t_{il}. The x_{ijk} values change from subject to subject, but not from time to time within a subject. If all covariates are time-fixed, then all rows are the same $x'_{ij} = x'_{il}$. Then $x'_{ij}\alpha = x'_{il}\alpha$ is constant and the population mean $X_i\alpha$ is constant over time. That each subject has a constant population mean over time does not mean that all subjects have the same population mean. Two subjects with different covariates will usually have different means.

In the next subsections, we first discuss categorical time-fixed covariates and then continuous time-fixed covariates.

7.1.1 Categorical Covariates

Examples of categorical covariates include gender, race/ethnicity, acculturation, education, employment status, relationship status, and residential stability. Categorical covariates can be coded in our analysis using indicator variables, also known as grouping, 0-1 or dummy variables. The procedure is basically the same as in linear regression except that each subject is observed at n_i times and so has n_i values of the categorical covariate, one for each time t_{ij}. The categorical covariate typically takes values from 1 up to g if there are g groups and we need $g-1$ indicator variables to identify to which of the g groups the subject belongs. The g groups usually have names; when reporting results, it is important to use the names rather than the numerical coding. One group, usually the first or last in some ordering of the categories is treated differently, and is indicated in the covariate matrix by setting all $g - 1$ 0-1 variables equal to zero. The remaining $g - 1$ groups are identified by setting exactly one of the $g - 1$ 0-1 variables equal to one at all times and the rest of the variables are set equal to zero at all times.

A categorical covariate with exactly two categories is also called a binary or dichotomous covariate. With more than two categories, categorical covariates are also called multinomial, polytomous and polychotomous covariates.

7.1.2   Parameterization

We describe in detail how to work with a two category covariate. Subjects belong to one of two categories, labeled 1 and 2. We let $G_i = 1$ when subject i is a member of group 2, and otherwise $G_i = 0$, G_i is the indicator that subject i is a member of group 2. To create a column of the X_i matrix, we multiply G_i times $\mathbf{1}$ a vector of ones of the appropriate length. If the subject is in group 2, then $G_i\mathbf{1}$ is an n_i vector of ones, otherwise $G_i\mathbf{1}$ is an n_i vector of zeros. Suppose we have only an intercept and the grouping variable as covariates in our analysis. The first column X_{i1} of X_i is the intercept, a column of ones for all subjects. The second column X_{i2} is set to $G_i\mathbf{1}$. When subject i is in group 1, we have

$$X_i = \left.\left(\begin{array}{cc} 1 & 0 \\ \vdots & \vdots \\ 1 & 0 \end{array}\right)\right\} \text{ length is } n_i \qquad (7.1)$$

and when subject i is in group 2, we have

$$X_i = \left.\left(\begin{array}{cc} 1 & 1 \\ \vdots & \vdots \\ 1 & 1 \end{array}\right)\right\} \text{ length is } n_i \qquad (7.2)$$

For each subject, the appropriate X_i matrix either (7.1) or (7.2) has n_i rows; the number of rows may vary from subject to subject. We will not continue to label the number of rows. The number of columns is the same for all subjects.

The fixed effects parameter α has length 2,

$$\alpha = \left(\begin{array}{c} \alpha_1 \\ \alpha_2 \end{array}\right).$$

The parameter α_1 is the intercept and is the mean response of the subjects in group 1. The parameter α_2 is the difference in mean response between

group 2 and group 1. English can be imprecise; "the difference ... between"
does not clearly specify which group mean is subtracted from which. It is
necessary to be clear about whether group 1 or 2 has the higher mean.
Saying that α_2 is the group 2 mean minus the group 1 mean is more careful
language. If $\alpha_2 > 0$, responses from subjects in group 2 will be higher on
average than those in group 1. If $G_i = 0$ then

$$E[Y_i|\alpha] = \mathbf{1}\alpha_1,$$

else when $G_i = 1$

$$E[Y_i|\alpha] = \mathbf{1}(\alpha_1 + \alpha_2).$$

Figure 7.1(a) illustrates the two population means, for group 2 (dashed
line) with a higher population mean than group 1 (solid line). In particular,
$\alpha_1 = 2$, and $\alpha_2 = 1.1$, and $\alpha_1 + \alpha_2 = 3.1$.

   Suppose now there are three groups labeled 1, 2, and 3. We need two
grouping variables, G_i and H_i. Set both $G_i = H_i = 0$ if subject i is in
group 1; set $G_i = 1$, $H_i = 0$ if i is in group 2; and set $G_i = 0$, $H_i = 1$ if
subject i is in group 3. We include two columns $G_i\mathbf{1}$ and $H_i\mathbf{1}$ to the X_i
matrix. There are three different types of subjects and three corresponding
X_i matrices. For subjects in group 1,

$$X_i = \begin{pmatrix} 1 & 0 & 0 \\ \vdots & \vdots & \vdots \\ 1 & 0 & 0 \end{pmatrix}.$$

For group 2 subjects,

$$X_i = \begin{pmatrix} 1 & 1 & 0 \\ \vdots & \vdots & \vdots \\ 1 & 1 & 0 \end{pmatrix},$$

and when subject i is in group 3,

$$X_i = \begin{pmatrix} 1 & 0 & 1 \\ \vdots & \vdots & \vdots \\ 1 & 0 & 1 \end{pmatrix}.$$

   The *regression parameter* is $\alpha = (\alpha_1, \alpha_2, \alpha_3)'$. Parameter α_1 is the intercept, and represents the mean response of subjects in group 1 at each
time. Parameter α_2 is the difference in mean response for group 2 minus
group 1, and α_3 is the average difference in response of group 3 minus
group 1. Figure 7.1(b) illustrates the three group situation, with group 3
(dotted line) with the highest average, then group 2 (dashed line) in the
middle, and group 1 (solid line) on the bottom. The figure corresponds to
$\alpha_1 = 2$, $\alpha_2 = 1.1$, and $\alpha_3 = 2.3$; in vector notation $\alpha = (2, 1.1, 2.3)'$, and
the means of the three groups are $2 + 0 \times 1.1 + 0 \times 2.3 = 2$ for group 1,

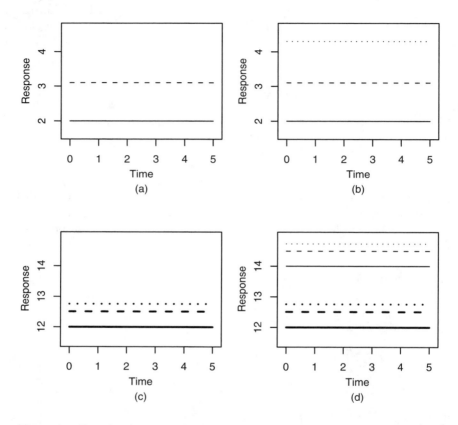

Figure 7.1. Population means for four time-fixed examples. (a) Two groups with group 2 (dashed line) above group 1 (solid line) and $\alpha = (2, 1.1)'$. (b) Three groups. First two groups are as in (a), and group 3 (dotted line) is above group 2, $\alpha = (2, 1.1, 2.1)'$. (c) Population means with a single continuous baseline covariate taking on values of 0, 2, 3 (respectively, solid, dashed, and dotted lines) for three different subjects and a coefficient $\alpha = (12, .25)'$. (d) Like plot (c) with an additional second grouping variable G_i with coefficient $\alpha_3 = 2$. Six lines are plotted: those for $G_i = 0$ have a thicker line, dash or dot type, whereas those for $G_i = 1$ have thinner lines.

$2 + 1 \times 1.1 + 0 \times 2.3 = 3.1$ for group 2, and $2 + 0 \times 1.1 + 1 \times 2.3 = 4.3$ for group 3.

Other parameterizations of grouping variables are of course possible. For two groups, some people like to code the difference in group variable as $G_i = 1$ if group 2, and $G_i = -1$ if group 1. This is called the *parameter effects coding*. The mean at time t_j for subject i is

$$\mu_{ij} = \alpha_1 + G_i \alpha_2$$

so for a subject in group 1,

$$\mu_{ij} = \alpha_1 - \alpha_2$$

else if group 2,

$$\mu_{ij} = \alpha_1 + \alpha_2.$$

With this parameterization, the meaning of the intercept is the average of the two group means. This may or may not be an interesting quantity. The parameter α_2 is $1/2$ of the difference group 2 minus group 1 means. This parameterization may seem awkward to use, however, there are computational advantages to this parameterization, and computer packages may invisibly use it internally, even though the user specifies a different parameterization.

Another parameterization deletes the intercept. Let G_i be 0 or 1 for group 1 or 2. Now let the rows of X_i be $x_{ij} = (G_i, 1 - G_i)'$. Now α_1 is the mean of group 1, whereas α_2 is the mean of group 2. This parameterization is called the *cell means coding*. This parameterization is perfectly sensible; to use it, first we tell our statistics package to delete the intercept. Second, we will also have to make sure that the package knows to use both G_i and $1 - G_i$ as covariates.

For $g > 2$ groups, there is an obvious version of each of these alternative parameterizations; for the parameter effects coding there are $g - 1$ covariates with values -1 and 1. For the cell means, we have g covariates each of which is the population mean for one group. In the end, all of these parameterizations are the same model, they provide exactly the same fit to the data and they provide the same possibilities for inference. Because of the way statistical computations are presented, some parameterizations may be more convenient to use than others, as when particular tests or confidence intervals are of interest.

7.1.2.1   Pediatric Pain

Consider a model for log base 2 response time with two groups for the two coping styles. We ignore the treatment effects at this time, and we assume an unstructured covariance model. We set X_{i1} to be the intercept and X_{i2}, the second column of X_i to be an indicator for distracter, so that α_2 is the distracters mean minus the attenders mean. The parameters were

Log base 2 scale

Parameter	Est	SE	t, p	95% CI
Attender	4.57	.16		$(4.25, 4.89)$
Distracter	5.08	.16		$(4.76, 5.39)$
Dist - Att	.51	.23	$t = 2.2$	$(.06, .95)$
			$p = .031$	

Original scale

Parameter	Est	95% CI
Attender	23.8	$(19.1, 29.6)$
Distracter	33.8	$(27.1, 42.1)$
Dist/Att	1.42	$(1.04, 1.94)$

Table 7.2. Results for the log base 2 Pain data with coping style as covariate using the unstructured covariance matrix. Bottom results convert the point estimate and 95% confidence interval back to the original seconds scale.

estimated using REML to be $\hat{\alpha}_1 = 4.57$, while the distracter population mean estimate was $\hat{\alpha}_1 + \hat{\alpha}_2 = 5.08$; the distracter mean was .51 higher than the attender mean. Table 7.2 gives results.

Some comments about table 7.2 are in order. First, the analysis has been run on the log base 2 seconds scale. Second, we have rounded results to 3 or sometimes 2 digits. P-values usually are not of interest beyond one significant figure, so the .031 could be rounded to .03. If we expected to perform further calculations from these results, then many more significant digits would be kept; one should not round first and compute second. Third, a t-statistic and p-value are only reported for the difference, and not for the estimated mean baseline pain response, for which there is no interest in testing that the population means on the $\log_2$ scale are equal to zero or not. Fourth, the contrast is moved into the table along side the parameter estimates, and not presented separately in another table even though that is often how computer output formats the contrast. Fifth, rather than make the reader of the table do the conversion by hand, the bottom half of the table converts results from the log scale back to the original seconds scale.

Results show that on average, distracters have longer pain tolerance than attenders and the difference is barely significant with a p-value of .03. On the original scale, we estimate a median value of 34 seconds for distracters versus 24 seconds for attenders and a point estimate of 1.4 for the ratio of times.

7.1.3 Continuous Covariates

Most time-fixed continuous covariates are baseline covariates, measured at or before t_{i1}, the time the first response measure is taken. Popular continuous covariates in many models are demographic variables such as age or income at study entry or medically relevant variables such as disease status, height, weight, or dosage. We consider how baseline covariates may affect our average response. Let B_{i1} be a single baseline covariate. The simplest and most common modeling assumption is to include a column $\mathbf{1}B_{i1}$ in the X_i matrix.

$$X_i = (\mathbf{1}, B_{i1}\mathbf{1}),$$

Or written out fully

$$X_i = \begin{pmatrix} 1 & B_{i1} \\ \vdots & \vdots \\ 1 & B_{i1} \end{pmatrix}.$$

This model says that each subject's average profile is flat, and that the level depends on B_{i1} in a linear fashion. Figure 7.1(c) illustrates with $\alpha = (12, .25)$ for three subjects whose values of B_{i1} are 0, 2, and 3. Because the coefficient of B_{i1} is positive, the subject with higher $B_{i1} = 3$ has a higher mean than the subject with $B_{i1} = 2$ or $B_{i1} = 0$.

Let us add a dichotomous grouping variable G_i to our model, and now $X_i = (\mathbf{1}, \mathbf{1}B_{i1}, G_i)$. Figure 7.1(d) illustrates the means over time for 6 subjects, three from group 1 and three from group 2. Both groups have B_{i1} values of 0, 2 and 3. The parameter vector is $\alpha = (12, .25, 2)'$. Group 1 has the darker lines, while group 2 lines are thinner. The baseline covariate and grouping variable do not *interact*; we have an *additive* model. Within each group, the relationship among the three lines is the same. In both groups, the subject with $B_{i1} = 2$ has a line that is $.5 = (2 - 0) \times .25$ above the line of the subject with $B_{i1} = 0$. If there was an interaction between the covariate B_{i1} and the grouping indicator, G_i, then the three lines would not be the same distances apart for the two groups.

If we have several baseline covariates, $B_{i1}, \ldots, B_{il}$, all enter in this same fashion.

$$X_i = (\mathbf{1}, B_{i1} \times \mathbf{1}, \ldots, B_{il} \times \mathbf{1}).$$

The population means will be flat for any set of covariate values, but with many covariates and patterns of covariate values, each subject could potentially have their own distinct population mean.

7.1.3.1 Cognitive Data

For this particular analysis of the Cognitive data, we take Raven's as the response, with data available on 529 subjects and a total of 2494 observations used in this particular analysis. Raven's ranges from 0 to 31 with a mean and sd at baseline of 17.3 and 2.6 respectively. We present an analysis with gender and a measure of socio-economic status (SES) as covariates. Gender is coded as female 0, male 1. SES ranges from a low of 28 to a high of 211 with a median of 79. It is constructed from an extensive baseline survey of

Parameter	Est	SE	t	p
Intercept	17.58	.28		
Gender (male)	.63	.15	4.25	$< .0001$
SES	.0061	.0031	1.99	.05

Table 7.3. Results for the Raven's with gender and SES as covariates using the autoregressive covariance model.

social and economic variables. Within a semester, times are random. The covariance model is an autoregressive covariance model.

   Table 7.3 presents results for the fixed effects in this analysis. We see that males score significantly higher than females on the Raven's. Socio-economic status has a positive coefficient and is marginally significant at $p = .05$. We can compare the effect of socio-economic status to that of gender by comparing the means of two female children, one at the 25th percentile SES of 65 and the other at the 75th percentile SES of 96. Then the fitted mean for the one child is $17.54 + .0063 \times 65 = 17.95$ and for the other is $17.54 + .0063 \times 96 = 18.14$. The difference between the two means is $(96 - 65) \times .0063 = .20$ is less than about one third of the gender effect. The difference in SES between two children would have to be approximately $100 \approx .63/.0063$ units for the difference due to SES to be equivalent to the gender effect.

7.2   Population Means Varying as a Function of Time

The previous section discussed time-fixed population means. More often than not when we collect longitudinal data, the population mean changes over time. Sometimes the change can be described by a simple parametric function of time and sometimes the trend requires a more complex model.

7.2.1   Polynomials in Time

Including time t_i and the intercept as columns of the X_i allows for a linear trend over time. If time is the only covariate, the covariate matrix is

$$X_i = (\mathbf{1}, t_i).$$

The intercept coefficient α_1 is the population mean at time $t_{ij} = 0$. The slope coefficient α_2 is the rate of change in the population mean as time increases. The expected mean of observation Y_{ij} at time t_{ij} is

$$E[Y_{ij}|\alpha] = \alpha_1 + t_{ij}\alpha_2.$$

Figure 7.2(a) shows three examples with time ranging from 0 to 4. The top dotted line with increasing population mean line has parameters $\alpha =$

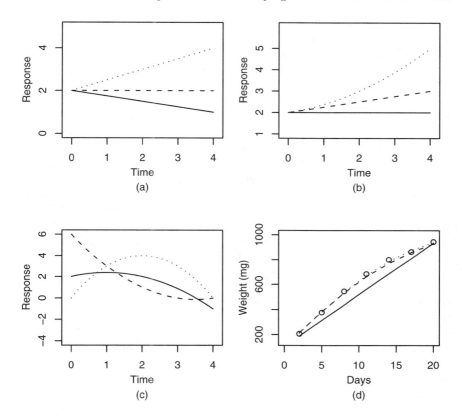

Figure 7.2. Population means. (a) Three linear time trends. (b) Flat, linear and quadratic time trends. (c) Three quadratic mean functions. (d) Small mice example showing a linear, quadratic, and cubic fit. The points plot the means of the data at each of the seven time points.

$(2, .5)'$, the flat, dashed line has $\alpha = (2, 0)'$, and the solid line illustrates a decreasing time trend with $\alpha = (2, -.25)$.

Modest curvature in the mean with increasing time can be accommodated with a quadratic time effect. Recall that t_i^2 is defined as the vector with elements t_{ij}^2. The coefficient matrix for subject i with a quadratic time trend is

$$X_i = \left(1, t_i, t_i^2\right).$$

The parameters $\alpha = (\alpha_1, \alpha_2, \alpha_3)'$ are the intercept α_1, slope α_2, and quadratic coefficient α_3, all of which combine to create the mean function. Figure 7.2(b) shows three population means. The solid line is a flat population mean with $\alpha = (2, 0, 0)'$, the dashed line shows a linear increasing population mean with $\alpha = (2, .25, 0)'$, and the dotted line has $\alpha = (2, .25, .125)'$ and shows a modestly increasing quadratic trend. The three means all start at the same point at $t = 0$. As time increases, the three curves separate, although initially the linear and quadratic trend are quite close. *Launch speed* is defined as the initial rate of growth of a curve. It is of interest particularly when the zero time is a well-defined event, such as birth, intervention date, or disease onset, and where subjects start at some beginning level as in figure 7.2(b) where the population mean at time zero α_1 is the same for all three curves. The launch speed is the coefficient of time t at $t = 0$ or α_2 which is the same for the dotted and dashed lines, both of which have a higher launch speed than the solid line, which has $\alpha_2 = 0$. Groups with higher launch speeds will tend to grow higher over the short term as compared with groups with lower launch speeds.

Figure 7.2(c) shows three stronger quadratic trends, two are *concave*, (solid, dotted curves) meaning that α_2, the coefficient of the quadratic term is negative and one *convex* or *concave up* (dashed curve) with α_2 positive. Weak quadratic curvature as in the solid curve may be used to approximate a mean function that starts level and then moves lower after some time. Similarly, the dashed line might approximate a mean function that drops precipitously at the beginning of the study and then levels off. For both of these functions, if observations continue much further, then the quadratic nature of the curve means that the curves will take off in a way that is possibly inappropriate. The dotted line shows a stronger quadratic trend where the mean starts low, increases to a peak and then drops. While this is possible, it is not common. We will occasionally see a strong quadratic trend in data or in residuals. The minimum if $\alpha_3 > 0$, or maximum if $\alpha_3 < 0$, of the quadratic occurs at the time $t = -.5\alpha_2/\alpha_3$. The three curves in figure 7.2(c) have their maxima or minima between $t = 0$ and $t = 4$

A quadratic time covariate included in the X_i matrix may satisfactorily accommodate modest curvature of the population mean. When the curvature is strong, or when the time range is large, a quadratic polynomial may not be satisfactory, and more complicated functions of time may be required. Higher order polynomials are not usually recommended although on occasion, data representing growth (slow initial rise followed by a more rapid rise followed by slower growth) may be modeled successfully as a cubic polynomial. Even those curves must have very little time in the initial and final slower growth phases for the cubic to fit passably.

7.2.1.1   Small Mice Data

We tried the linear, quadratic, and cubic polynomial models for fitting the Small Mice data at the end of section 5.3.3. The fits are plotted in figure 7.2(d). The dashed line is the quadratic fit. It is nearly identical to the dotted cubic fit. The linear fit does not seem to go through the data points, suggesting that perhaps our software did not find the best fit line to the data. I have often had trouble fitting a linear time trend to this data.

Unfortunately, the software did not give an error message identifying the fit as poor.

We can also try fitting polynomials to the Big Mice data. There are more data points, and the time frame is longer, with the addition of days 0 and 1. Now the cubic term is significant as are the quartic (day to the fourth power) and quintic (day to the fifth power), but not the sixth or seventh order terms. Still, a quintic is inconvenient to work with, and generally we do not recommend using high order polynomials to fit the mean function. Another approach might be useful.

7.2.2   Unstructured Mean

Sometimes the mean time trend is difficult to describe with a linear or low order polynomial in time. For balanced data with J time points, or balanced with missing data, an *unstructured mean* model may be useful. In this model, we have one parameter α_j that models the mean at each time t_j. We will need one column in the X_i matrix for every time point where we have observed data. The simplest X_i matrix omits the intercept and is an identity matrix of size J

$$X_i = \begin{pmatrix} 1 & 0 & \cdots & 0 \\ 0 & 1 & \ddots & \vdots \\ \vdots & \ddots & \ddots & 0 \\ 0 & \cdots & 0 & 1 \end{pmatrix}.$$

With this covariate matrix,

$$E[Y_i|\alpha] = X_i\alpha = \alpha$$

as the identity matrix times any vector returns that vector. Each parameter α_j is the population mean of observations at time $t_{ij} \equiv t_j$, assumed the same for all subjects.

7.2.2.1   Parameterizations of the Unstructured Mean Model

Other parameterizations of the unstructured mean are also useful. A second parameterization includes the intercept as the first column, but the second through Jth column are the same as the previous parameterization with a single 1 in the jth row of column j.

$$X_i = \begin{pmatrix} 1 & 0 & 0 & \cdots & 0 \\ 1 & 1 & 0 & & \vdots \\ 1 & 0 & 1 & \ddots & \vdots \\ \vdots & \vdots & \ddots & \ddots & 0 \\ 1 & 0 & \cdots & 0 & 1 \end{pmatrix}.$$

Week	n	Sample mean weight	Est mean weight	Diff. from week 1	Diff. from prior week
1	38	192.8	192.8	0.0	–
2	38	193.6	193.6	0.8	+0.8
3	38	190.9	190.9	-1.8	-2.6
4	38	188.0	188.0	-4.8	-3.0
5	33	185.8	186.6	-6.1	-1.3
6	31	190.8	188.1	-4.7	+1.5
7	25	188.4	186.9	-5.9	-1.2
8	24	190.3	185.3	-7.4	-1.6

Table 7.4. Raw data means and parameter estimates for different parameterizations for the Weight Loss data set. The first column gives the week, the second column gives the number of subjects with weights at that week, and the third column gives the sample mean weight. The fourth column gives the estimated population mean for each week using a random intercept and slope covariance model. The fifth column gives the differences jth week's estimated weight minus first week's estimated weight. The last column gives the estimated differences between consecutive weeks' means. Rounding causes some minor discrepancies in the table. Missing data for some subjects after week 4 causes the estimated mean weights to differ from the sample mean weight.

The first coefficient α_1 is the first time point population mean. Succeeding α_j are the difference of the mean at time j minus the mean at time 1. The population mean at time $j > 1$ is $\alpha_1 + \alpha_j$.

A third parameterization is

$$X_i = \begin{pmatrix} 1 & 0 & \cdots & & 0 \\ 1 & 1 & \ddots & & \vdots \\ \vdots & & \ddots & \ddots & 0 \\ 1 & \cdots & & 1 & 1 \end{pmatrix}.$$

The kth column X_{ik} starts with $k-1$ zeros followed by $J-k+1$ ones. Each row x'_{ij} has j ones followed by $J-j$ zeros. The first coefficient α_1 is again the mean for time $t_j = 1$, the same as in the other two parameterizations. The remaining α_j, $j > 1$ are now the population differences between consecutive observations in time. For example, α_2 is the difference in means at time 2 minus time 1, and α_7 is the difference in means between time 7 and time 6. The population mean at time j is

$$E[Y_{ij}|\alpha] = \sum_{k=1}^{j} \alpha_k.$$

7.2.2.2   Weight Loss Data

Table 7.4 gives estimates from the random intercept and slope covariance model for the Weight Loss data (version 1) where the population mean is modeled with an unstructured mean. There is a population mean parameter for each week. The second column gives the number of observations from that week and the sample mean weight for each week is given in the third column. We see that more observations are missing at later times than in the beginning and that average weights appear to be around 190 pounds. The sample averages do not suggest that much weight loss occurred over the course of the study. The fourth column gives the estimated population mean from our model, which was fit with an unstructured mean and a random intercept and slope covariance model. The estimated mean weights are identical to the sample means for the first 5 weeks where there was no missing data. The estimated mean weights are lower for weeks 6, 7, and 8 than the raw sample means where we had some missing observations. According to this model, the subjects whose observations were missing at the later times would have had lower average weights than the people whose data we already have. Taking raw data averages would be misleading about the extent of the weight loss. Figure 7.3(a) plots the weekly sample means and the weekly estimated mean weights over time.

The last two columns of table 7.4 present estimates from the two alternative unstructured mean parameterizations. The next to last column gives the estimated differences between week 1 and week j, positive numbers indicate that week 1 had a higher estimated mean weight. Parameter α_j for this parameterization is the difference $E[Y_{ij}|\alpha] - E[Y_{i1}|\alpha]$, except of course for α_1. We see an increase in average weight at week 2 but drops from week 1 at all later weeks and an apparent 7.4 pound loss over the full period, or slightly over 1 pound weight lost per week. The last column gives the consecutive weekly weight losses and the occasional gains. Parameter α_j for the last column is the difference $E[Y_{ij}|\alpha] - E[Y_{i(j-1)}|\alpha]$. There are weight gains from week 1 to week 2 and again at week 5 to week 6. We see fully 2.6 and 3.0 pounds weight lost from weeks 2 to 3 and 3 to 4. The weight lost in a later week is less than that in an earlier week.

We also have standard errors associated with all the estimates in table 7.4. For the estimated population means at each week, the standard errors were 4.2 pounds for the first two weeks and 4.1 pounds for the remaining 6 weeks. No obvious null hypothesis springs to mind regarding individual weekly means, but we can ask interesting scientific questions about the differences in weekly means. Table 7.5 gives standard errors, t-statistics, and associated two-sided p-values for the two alternate parameterizations. We see that all the means are significantly lower than at week 1 except for week 2, which is not significantly higher using this covariance model. All of the consecutive differences are significant, again except for the change from week 1 to week 2. The positive increase in weight from week 5 to week 6

Week	Diff from week 1				Diff from prior week			
	Est	SE	t	p	Est	SE	t	p
1	0.0	0	–	–				
2	0.8	.4	1.9	.07	+0.8	.4	1.9	.07
3	-1.8	.5	-3.9	.0001	-2.6	.4	-6.2	.0000
4	-4.8	.6	-8.7	.0000	-3.0	.4	-7.1	.0000
5	-6.1	.7	-9.3	.0000	-1.3	.4	-3.0	.003
6	-4.7	.8	-6.1	.0000	+1.5	.5	3.2	.002
7	-5.9	.9	-6.5	.0000	-1.2	.5	-2.4	.02
8	-7.4	1.0	-7.3	.0000	-1.6	.5	-2.9	.005

Table 7.5. Estimated differences between week 1 and week j, associated standard error, t-statistic and two-sided p-value (columns 2–5); estimated consecutive week differences, standard error, t-statistic, and p-value (columns 6 – 9). P-values of .0000 mean that $p < .0001$.

is also significant. This increase in weight is so surprising that rather than merely reporting it, I would investigate further to try to find an explanation.

A complex hypothesis can be tested using the means. We can ask whether the weekly means follow a linear trend. A linear trend would be very convenient, and would be much easier to report and to explain than the unstructured pattern we see in figure 7.3(a). The unstructured mean has 8 parameters, whereas the linear time trend $E[Y_{ij}|\alpha] = \alpha_1 + \alpha_2 t_j$ has two parameters. The difference is 6 degrees of freedom, and we can test this by fitting both models using maximum likelihood. Minus two times the difference in log likelihoods is $1479.4 - 1389.2 = 90.2$ is very significant and we reject the null hypothesis that the means follow a linear trend after comparing 90.2 to the $\chi^2(6)$ distribution. We can also test this hypothesis by fitting a single model that includes an intercept and continuous week and also the discrete week covariate with 8 indicator covariates. This is over-parameterized, and software will set two of the discrete week coefficients to zero, leaving six parameters. We can test the null hypothesis that the six coefficients are equal to zero using REML or ML. The F-statistics are 18.6 and 19.2, respectively, all highly significant and again we reject the null hypothesis.

7.2.2.3   Other Patterns

Figures 7.3(b)-(d) show three different time trends that are not easily describable with low-order polynomials. In 7.3(b), we see a not uncommon pattern of a first measurement which is higher than all remaining measurements which appear to vary around a constant mean. This can happen if, for example, people are selected to be high on the response measure at the first time, but then, because there is little correlation from one measurement to the next, they revert to average at the remaining time points. This pattern might happen in a study of blood pressure where people may enter

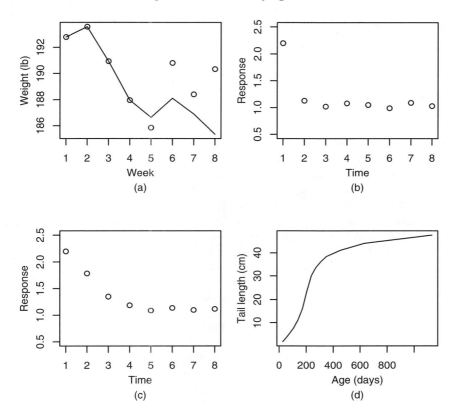

Figure 7.3. Population means (a) Weight Loss data; circles are sample means; profile in the same plot gives the estimated means using a random intercept and slope covariance model. (b) Nonlinear pattern: first week different. (c) Nonlinear pattern: decrease to asymptote. (d) Estimated Wallaby tail length from birth to 1100 days.

the study only if they have a high blood pressure reading at the first week. In intervention studies, the difficulty is deciding whether to attribute the drop to the intervention, given immediately after the first observation or to attribute it to the selection bias at the first time point, and this point cannot be resolved without a placebo group to compare the treatment group to. An unstructured mean will model this data but will use many more parameters than are necessary. Another way to model this pattern is not with a polynomial at all, nor with an unstructured mean, rather we would use an indicator of the first time point and a single constant mean for the remaining time points. The covariate matrix would be

$$
X_i = \begin{pmatrix} 1 & 0 \\ 0 & 1 \\ \vdots & \vdots \\ 0 & 1 \end{pmatrix}.
$$

The parameters $\alpha = (\alpha_1, \alpha_2)'$ are the mean at baseline and the mean after baseline, respectively.

In figure 7.3(c), we see the mean falling to a lower asymptote where it levels off. Low or high asymptotes are common in some subject areas. Growth of an animal will level off after it grows to adult size. A human adult's cognitive abilities will drop off and then stabilize at a very low level after the onset of Alzheimer's disease. Modeling these trends with a low-order polynomial is potentially possible, provided there are few or no observations at the level of the asymptote. If there are more than one or two observations then a low order polynomial will not fit. A nonlinear mean model may possibly give interpretable scientific results with fewer parameters than the unstructured mean. Alternately, an unstructured mean can be used.

Figure 7.3(d) illustrates a mean trend for the growth of Wallaby tails over time from birth up to around 1100 days of age. We see initial rapid growth followed by deceleration by day 400 to a slower rate of growth. If the data is balanced, or balanced with missing data, then it could be modeled with an unstructured mean. If the data is unbalanced, then a different approach is required, one that may also prove useful for data with a mean like in 7.3(c). We discuss *semiparametric* models for the mean in section 7.6.

No matter what parameterization one uses, more than one parameterization is likely to be of interest, requiring the user to either request contrasts, or requiring the user to run multiple analyses with different covariate matrices. The results of tables 7.4 and 7.5 could have been produced by fitting the model three times with different parameterizations each time. In this instance I fit the model only once, using an unstructured mean and asking for 14 contrasts to get the point estimates, standard errors, t-statistics and p-values presented in the two tables.

7.2.3   *Time-varying Group Membership*

Often treatment variables or other indicators vary over time. We may take a baseline measure and only then introduce a treatment, so that a treatment indicator variable will be time-varying; at baseline all subjects are in a single untreated group. After treatment is initiated, subjects are part of the treatment group or groups. In our Pain data, a counseling session was given prior to the fourth trial; three different types of counseling sessions were given, counseling to attend, to distract or a null counseling intervention

without advice. Treatment appears to require a polychotomous variable
with three states: A D or N, for the attend, distract or null counseling
treatments. But there is a fourth state. During trials 1 through 3, subjects
are in an untreated state; this is the fourth level of the treatment variable.
The treatment variable is time-varying and there are four different states,
with all subjects sharing the untreated state for trials 1, 2, and 3.

In the Cognitive study, children were given one of four school lunch
interventions: meat, milk, calorie, or control. The first three groups were
fed a school lunch supplemented with either meat, milk, or oil to create a
lunch with a given caloric level. The control group did not receive a lunch.
The baseline Raven's measurement was taken prior to the lunch program
onset. At baseline, all children belonged to the control group, and those in
the meat, milk, or calorie group switched into their respective groups after
the first time point.

Some treatments may be given in any order. Consider an allergy medicine
with a short-term effect, we can give either the medicine or a placebo during
any given allergic episode. Once the treatment has worn off and the allergy
episode runs its course, that particular episode is over. A single trial for a
subject consists of one allergy episode. When a new allergy episode begins,
we may again administer either medicine or placebo. Treatment will be
a time-varying variable as we will give a mix of medicine and placebo
treatments to each subject. Cross-over trials feature this sort of design.

7.2.3.1   Two Groups

In the case of a time-varying treatment, our treatment indicator variable
G_i will be a non-constant vector. That is, some element $G_{ij} = 1$, while
some other element $G_{il} = 0$. For example, if we have two observations with
subject i on placebo followed by three observations on treatment, then

$$G_i = \begin{pmatrix} 0 \\ 0 \\ 1 \\ 1 \\ 1 \end{pmatrix}.$$

The reverse experience, two treatment observations followed by three
placebo observations gives

$$G_i = \begin{pmatrix} 1 \\ 1 \\ 0 \\ 0 \\ 0 \end{pmatrix}.$$

A treatment, placebo, treatment, placebo, treatment series of observations
gives

$$G_i = \begin{pmatrix} 1 \\ 0 \\ 1 \\ 0 \\ 1 \end{pmatrix}.$$

We usually will have the X_i matrix include an intercept column and the
treatment indicator variable. As with the time-fixed grouping covariate of
the previous section, if there are two groups, we must introduce one indicator variable, if there are three groups, we must introduce two indicator
variables, and so on; for g groups, we introduce $g - 1$ indicator variables.
With the intercept, the coefficient α_1 of the intercept is the mean of the
group that is identified by setting all indicator variables to zero. The coefficient α_g of indicator g is the difference between the group g mean and
the group whose mean is the intercept. Without an intercept in the model,
we would need g indicator variables, and each coefficient is the estimated
mean of its corresponding group.

The X_i matrix for the two group problem modeled using an intercept
and one indicator G_i is

$$X_i = (\mathbf{1}, G_i)$$

and the parameters are $\alpha = (\alpha_1, \alpha_2)$. The intercept α_1 is the mean of the
population on placebo, and α_2 is the difference in the means of the observations on treatment minus the mean on placebo. If treatment/control can
vary from time to time, then many possible mean functions are possible
for different people. With 5 repeated measures, and two possible treatments, there are $2^5 = 32$ possible mean functions over time. Figure 7.4(a)
shows three possible population means, corresponding to the three choices
of G_i above. The dashed and dotted zig-zag G_i patterns have been offset
slightly for the second and third patterns so that the means are clearly
distinguishable. The parameter values are $\alpha = (2, 1)'$.

7.2.3.2   Three Groups

If we have three groups, we have two indicators G_i and H_i, with $G_{ij} =
H_{ij} = 0$ to identify a group 1 assignment at time t_{ij} for person i, $G_{ij} = 1$,
$H_{ij} = 0$ to indicate group 2, and $G_{ij} = 0$, $H_{ij} = 1$ for group 3. Our X_i
matrix is

$$X_i = (\mathbf{1}, G_i, H_i).$$

The parameter vector $\alpha = (\alpha_1, \alpha_2, \alpha_3)'$ with α_1 the population mean for
observations under group 1, $\alpha_1 + \alpha_2$ the population mean for group 2, and
$\alpha_1 + \alpha_3$ the population mean for group 3. As with the two group case, there

are many possible mean profiles, depending on the sequence of groups that a single subject cycles through. In theory, with 5 observations, there are 3^5 possible mean structures. Figure 7.4(b) illustrates means for $\alpha = (2, 1, 1.5)'$ for three possible X_i matrices

$$X_1 = \begin{pmatrix} 1 & 0 & 0 \\ 1 & 1 & 0 \\ 1 & 0 & 1 \\ 1 & 0 & 0 \\ 1 & 1 & 0 \end{pmatrix}$$

(solid line),

$$X_2 = \begin{pmatrix} 1 & 1 & 0 \\ 1 & 0 & 1 \\ 1 & 0 & 0 \\ 1 & 1 & 0 \\ 1 & 0 & 1 \end{pmatrix}$$

(dashed line) and

$$X_3 = \begin{pmatrix} 1 & 0 & 0 \\ 1 & 0 & 0 \\ 1 & 0 & 1 \\ 1 & 0 & 1 \\ 1 & 0 & 1 \end{pmatrix}$$

(dotted line). The first two mean structures cycle through the groups in order, starting with group 1 for subject $i = 1$, and starting with group 2 for subject $i = 2$. Subject $i = 3$ has two observations in group 1, then 3 more observations in group 3, with none at all in group 2.

In many of our figures in this chapter, time could be continuous or discrete. Measurements that are measured in continuous time units such as seconds or days are typically continuous. In contrast, when we measure time in units of trials, time is necessarily discrete. For the models of section 7.1 and the first part of this section, if we were able to take measurements at non-integer times, such as at time 1.4 or 3.6, the means that are drawn in figures 7.1, 7.2, and even 7.3 would be appropriate, because the group assignments or covariate values, being time-fixed, hold for non-integer and integer times equally. In contrast, with figure 7.4(a) and (b), one has to explicitly determine treatment or control for any observation, and this has only been done at times 1, 2, 3, 4, and 5, and is not defined at non-integer times. The line segments and particularly the diagonal line segments connecting integer times in figure 7.4(a) are not possible means of observations under the model with time-varying treatment, they merely serve to connect the response means at the feasible integer trial times.

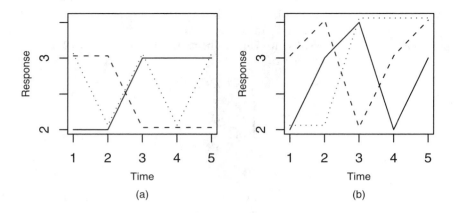

Figure 7.4. (a) Time-varying groups, two possible groups, $\alpha = (2,1)'$. Three subjects. (b) Time-varying treatment, three groups, three mean functions, $\alpha = c(2,1,1.5)$.

7.2.3.3   A Psychology Experiment

Some psychology experiments consist of many tens of trials. Each trial is performed under one of several stimulus conditions. Often the trials form a full factorial arrangement of experimental conditions. For example, if the stimulus has 4 different characteristics each at two levels, then there are $2^4 = 16$ different stimuli, and subjects will typically do 16 trials. If trials are quick, they may well do a small integer number of replications of the entire experiment; if trials take some time, then they may do what is called a fractional factorial experiment, doing 4 or even 8 trials.

For a simple example, suppose we are measuring reaction times to an audible response and we give a warning tone to participants prior to each stimulus. We vary the warning tone in several ways; it might be (1) close or far away in time from the stimulus; (2) loud or soft; (3) annoying or pleasant; and (4) similar/dissimilar to the stimulus. If the subject does 16 trials, one each at every possible treatment combination, then the X matrix will have 16 rows. A main effects model with no interactions would require 4 columns in the X matrix to describe the stimulus condition plus one column for the intercept. Each of the four columns describing the stimulus condition will have 8 ones and 8 zeros identifying the trials that had the high/low, loud/soft, annoying/pleasant, and similar/dissimilar stimulus characteristics.

7.2.3.4   Pediatric Pain Data

The Pediatric Pain data has two groups, coping style (CS) attender (A) or distracter (D), and as discussed earlier, really four treatments (TMT), the pre-treatment or baseline (TMT = B) observations and then TMT =

attend, distract or none (A, D or N). The CS assignments are time-fixed, and the columns of X_i involving CS were discussed in the previous section. No treatment was applied at the first three observations, then TMT is randomly assigned and given before the fourth trial. Treatment is therefore time-varying. To add treatment to our model, we will have to add three additional indicator variables to indicate which treatment was assigned. Let us suppose that TMT does not interact with CS – we will allow an interaction later. This leads to X_i matrices that are 4×5 for subjects with all four observations. We illustrate X_i matrices for three subjects. Detailed discussion of the the last three columns follows the three X_i matrices. Our first example has CS = A, TMT = A, denoted AA,

$$X_{AA} = \begin{pmatrix} 1 & 0 & 0 & 0 & 0 \\ 1 & 0 & 0 & 0 & 0 \\ 1 & 0 & 0 & 0 & 0 \\ 1 & 0 & 1 & 0 & 0 \end{pmatrix}.$$

Next for a subject with CS = D, and TMT = D denoted DD,

$$X_{DD} = \begin{pmatrix} 1 & 1 & 0 & 0 & 0 \\ 1 & 1 & 0 & 0 & 0 \\ 1 & 1 & 0 & 0 & 0 \\ 1 & 1 & 0 & 1 & 0 \end{pmatrix}.$$

Finally for a DN subject,

$$X_{DN} = \begin{pmatrix} 1 & 1 & 0 & 0 & 0 \\ 1 & 1 & 0 & 0 & 0 \\ 1 & 1 & 0 & 0 & 0 \\ 1 & 1 & 0 & 0 & 1 \end{pmatrix}.$$

The first column is the intercept $\mathbf{1}$ in all subjects. The second column is the CS grouping indicator, which is time-fixed; the first subject is an attender and the second and third are distracters. The third through fifth columns are the treatment indicators. We choose to have the baseline or pre-treatment condition correspond to all three indicators set to zero. Since the counseling treatment was given just before the fourth trial, these indicators are always zero for the first three time periods. Only one treatment is delivered before trial four to each subject, and so the last row of the last three columns will have but a single 1. A one in the last row of the third column indicates the A treatment, a one in the last row of the fourth column indicates the D treatment and a one in the last row of the fifth column indicates the N treatment.

Table 7.6 gives results from fitting this model to the data using REML output and the compound symmetry covariance model. Software does not usually report all of these results in this way; one has to assemble the results into a single compact table. As before, the analysis has been done on the log base 2 seconds scale and we have transformed results back to the seconds scale. The seconds scale results are given in the last three columns of the table. The usual regression output is given in the first five rows (not counting the column labels at the top) in the rows labeled "Intercept" through "TMT N" and in the first five columns. In the parameterization we use this time, the intercept is the mean response for attenders at baseline (AB). Because we have no interest in testing the null hypothesis that the mean attender response is equal to zero, we delete the t-test and p-value for this row from the table, even though software routinely (and uselessly!) reports those values. The parameter labeled "CS D-A" estimates the distracter minus attender mean baseline difference. We see that the estimate is not quite significant with a p-value of .06; the standard error is slightly over 50% of the parameter estimate. This differs slightly from the result of the analysis reported in table 7.2. There we did not include the treatment effects, and the baseline CS effect was significant with a p-value of .03. The difference between a p-value of .03 and .06 is not large. Sampling variability between different data sets, or changes in assumptions for the same data set, as between tables 7.2 and 7.6, will routinely lead to changes in p-values of this magnitude.

The next three rows, labeled "TMT A," "TMT D," and "TMT N" report estimates of the treatment effects. These parameters are the differences between the post-treatment and pre-treatment population means. The individual treatment effects range from borderline significant at $p = .03$ for the distract treatment, to borderline not significant with $p = .06$ for treatment N, to not significant for treatment A. Testing the joint significance of the three treatment effects shows borderline statistical significance with an F-statistic of 3.05 on three degrees of freedom and a p-value of .036. Inspecting the three estimates, we notice that the TMT D estimate is of the opposite sign of the TMT A and TMT N estimates. The difference between TMT D and TMT A effects is $.32 - -.29 = .61$ and is large compared to the standard deviation of the original numbers, and might well be very significant. We can *guesstimate*, i.e., guess at, or estimate, the standard error of the difference under the approximation that the two estimates are independent, by calculating

$$SE(\text{diff})^2 \approx SE(\text{est}_1)^2 + SE(\text{est}_2)^2.$$

where diff is the difference in estimates, and est_1 and est_2 are the two estimates in question and $SE(\cdot)$ is the reported standard error of the estimate. This approximation is exact when the estimates are independent. The correct formula needs to substract off twice the covariance of the two estimates

$$SE(\text{diff})^2 = SE(\text{est}_1)^2 + SE(\text{est}_2)^2 - 2 \times \text{Cov}(\text{est}_1, \text{est}_2).$$

The covariance among parameter estimates is not usually standard output from software, although usually some special keyword or option will produce it. Without the covariance, the approximation is the best we can do in

	Log base two seconds				Seconds		
Parameter	Est	SE	t	p	Estimate	Lower	Upper
Intercept	4.54	.17			23.3	18.4	29.5
CS D-A	.45	.24	-1.9	.06	1.37	.98	1.91
TMT A	-.12	.15	-.8	.57	.92	.75	1.13
TMT D	.32	.15	2.2	.03	1.25	1.02	1.53
TMT N	-.29	.15	-1.9	.06	.82	.66	1.01
Treatment contrasts							
TMT D-A	.44	.21	2.1	.04	1.36	1.02	1.82
TMT D-N	.61	.21	2.9	.005	1.53	1.14	2.05
TMT A-N	.17	.21	.8	.4	1.13	.84	1.51
CS× TMT means							
AB	4.54	.17			23.3	18.4	29.5
DB	5.00	.17			32.0	25.2	40.5
AA	4.42	.22			21.4	15.8	29.0
AD	4.86	.22			29.1	21.5	39.4
AN	4.25	.22			19.0	14.0	25.8
DA	4.88	.22			29.4	21.7	39.8
DD	5.32	.22			39.9	29.6	53.9
DN	4.71	.22			26.1	19.3	35.4

Table 7.6. REML output for Pain data analysis using CS and TMT in an additive model with compound symmetry covariance model. The first section gives parameter estimates. The second section gives contrasts among the treatment conditions, and the third and fourth sections give estimated log base two seconds means for each of the different combinations of CS and TMT. The first five columns give parameter names, estimates, and standard errors and, when appropriate, t-statistics and p-values. The last three columns transform the log base two seconds scale estimates, and endpoints of the 95% confidence intervals back to the seconds scale.

the absence of the covariance and we calculate $(.15^2 + .15^2)^{1/2} \approx .21$ which is slightly over $1/3$ of .61 so we strongly suspect the difference between the D and N treatments is significant.

This sort of approximate calculation is important when faced with software output that omits interesting contrasts. Lack of the exact calculation often occurs during intermediate stages of an analysis; usually we can request either the covariance or the contrast estimate and its standard error, and we will request it, possibly when we fit the next model. The approximation works well if the two estimates are almost independent and the approximation produces a conservative estimate of significance if the covariance between the two estimates is negative. Table 7.7 gives the correlations among the parameter estimates from the first five rows of table 7.6. Only one correlation is large, the correlation between the intercept and the CS D-A effect. The correlation between the TMT A and TMT D estimates is only .01 and our approximation turns out to be quite accurate.

	Int.	CS D-A	TMT A	TMT D	TMT N
Int.	1	-.70	-.07	-.08	-.08
CS D-A	-.70	1	.0001	-.005	-.004
TMT A	-.07	.0001	1	.01	.01
TMT D	-.08	-.005	.01	1	.01
TMT N	-.08	-.004	.01	.01	1

Table 7.7. Correlations among parameter estimates for the Pediatric Pain data analysis with coping style and additive treatment effect, REML estimation, and a compound symmetry covariance model.

We asked our software to calculate the contrasts and their significance and this is given in the next three lines of table 7.6. We see that our estimate and approximate standard error are correct to the number of significant figures reported, and that the D treatment is barely significantly different from the A treatment with $p = .04$ and is significantly different from the N treatment with $p = .005$. For the effects in rows 2 through 8, we can test whether the effect is significant by checking for 1 in the confidence intervals in the last two columns. We see that for TMT D-A and TMT D-N, the confidence intervals do not include 1, and the distract treatment is significantly better than the A and N treatments. The D treatment appears to lengthen the immersion time by an estimated $(1.25 - 1)100\% = 25\%$, while the attend and null treatment decrease the time in the water by 8% and 18%, respectively.

The last 8 rows give estimated population times under the eight different conditions, CS equal to A or D, and TMT equal to baseline B, or A, D or N. The last three columns transform the results back onto the seconds scale and are much easier to interpret than the log base 2 seconds scale. The last two columns provide 95% confidence intervals for the effects and for the means, and the third to last column is the point estimate.

7.2.3.5   Raven's Data

At baseline, that is, at all times less than zero, all subjects are in the control group. At time zero the school lunch programs are started. For subjects in the meat, milk or calorie groups, a school lunch is delivered to students in attendance provided school is in session. Cognitive measures such as the Raven's were assessed at up to five times, called *rounds*, for each subject. The round 1 average assessment time was -1.6 months; the average time of assessment for round 2 through 5 were at months 1.1, 5.6, 13.4, and 21.6. Times for individual subjects have a standard deviation of approximately .8 months around the average round times. We fit a model with gender and treatment as covariates and an autoregressive covariance model that accounts for the variable times of assessments.

Figure 7.5(b) plots fitted values, plotted at the average times for each round, with line segments connecting the dots within the different treat-

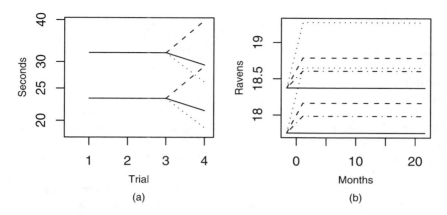

Figure 7.5. (a) Pediatric Pain fitted values for each coping style and treatment for all four trials. The fixed effects are CS and TMT with no interaction, and the covariance model is compound symmetry. At trial 4, the line types for the three treatments are: dotted, null; solid, attend; and dashed, distract. (b) Raven's results from model with gender and treatment. gender is time-fixed, treatment starts at $t = 0$. The treatment group line types are: solid, control; dash-dot, milk; dashed, calorie; and dotted, meat.

ment group and genders. At the average baseline time -1.6, all subjects are in the control group, so the fitted value is the same for all boys and also for all girls. Then after $t > 0$, subjects are in one of the four groups. The control group continues along flat and is the lowest of the four treatment groups, while the milk, calorie, and meat group are respectively, the dash-dot, the dashed, and the dotted lines. The bottom lines are the girls, and the top lines are the boys. The meat group effect estimate is so large that it swamps the difference between boys and girls. Girls given the meat lunch score higher than the boys in either of the control or milk group. The treatment group effect is highly significant, with an $F = 10.8$ on 3 degrees of freedom in the numerator and a p-value less than .0001.

This result looks very exciting for the researchers, however, there is a serious problem with our analysis. This model assumes that students will score at a constant level on the Raven for $21.6 - -1.6 = 23.2$ months, a period when they will have had almost 2 full years of schooling. We probably would like to consider the possibility that the time trend is not flat but increases linearly over time.

7.2.4 Group Interactions: Pediatric Pain

In the Pediatric Pain data set, the researchers' original hypothesis was that the attend treatment would do best for the attenders while the distract treatment would be better for distracters. This suggests an interaction between treatment and coping style. This full model has 8 covariates: 2 levels of CS times 4 levels of treatment for 8 total means. Results are given in table 7.8. The intercept is the mean of the distracters at baseline, CS A-D is the difference attender baseline minus distracter baseline. The next six parameters are one particular parameterization of the treatment effects. The parameter labeled TMT D A is the effect of the attend treatment on distracters, and similarly, TMT D D and TMT D N are, respectively, the effects of the distract and null treatment on distracters. The next three parameters labeled for example CS× TMT A-D A (or A-D D or A-D N) are the differences of the effects of the A (or D or N) treatment on attenders minus the effects on distracters. These three parameters are not interesting individually except that by testing the set of them being equal to zero we do get a test for whether the treatment effect is different in the attenders and distracters. The interaction CS× TMT is borderline not significant ($F = 2.44$, df $= 3$, $p = .07$).

We need to extend the analysis by producing a full set of interesting contrasts. My preferred parameterization combines the intercept and baseline coping style effect on the first two lines of table 7.8 with 6 parameters that measure the 2 coping styles $\times$ 3 treatments $= 6$ effects directly. The second section of table 7.8 gives these six treatment effect estimates. These are the changes in means due to treatment, the after minus before effect due to the 3 treatments first on attenders and then on distracters. Three of these effects are repeated from earlier in the table. None of the treatments has a significant effect for the attenders. For distracters, we have a strongly significant positive effect of the distract treatment and a strongly negative effect of the null treatment, and a negative, but not significant attend treatment. Tested jointly, the 6 treatment effects together are significantly different from zero ($F = 2.77$, df $= 6$, $p = .02$), that is, there is at least one treatment whose effect is not zero. Inspection of the second section of table 7.8 suggests that there are no treatment effects for the attenders but that the distract treatment worked really well for the distracters. The attend and null treatment for distracters leave the distracters with means roughly the same as the attenders, means that are lower than the distracter means at baseline.

The covariate matrix for this second parameterization has covariate matrix

$$X_i = (\mathbf{1}, \mathrm{CS}_i \times \mathbf{1}, T1_i, T2_i, T3_i, T4_i, T5_i, T6_i),$$

where the first column is an intercept; CS_i is the coping style membership, this time we have 1 =A, 0 =D; the next six variables are the intervention indicators. These six are all 0 for the first three time points, because the first three time points are pre-intervention. Then the last row for the six variables has 5 zeros with a single one indicating which CS–TMT combination person i was. Variables $T1_i$, $T2_i$, and $T3_i$ are for CS $= 1$ attenders, while $T4_i$, $T5_i$, and $T6_i$ are for distractors. One of $T1_i$ and $T4_i$ are one at the fourth time point if TMT was attend; one of $T2_i$ and $T5_i$ are one at the fourth time point if TMT was distract; and one of $T3_i$ and $T6_i$ are one at the fourth time point if TMT was null.

We give three example covariate matrices. For CS=A, TMT=A, X_i is

$$X_{AA} = \begin{pmatrix} 1 & 1 & 0 & 0 & 0 & 0 & 0 & 0 \\ 1 & 1 & 0 & 0 & 0 & 0 & 0 & 0 \\ 1 & 1 & 0 & 0 & 0 & 0 & 0 & 0 \\ 1 & 1 & 1 & 0 & 0 & 0 & 0 & 0 \end{pmatrix},$$

for CS=D, TMT=D

$$X_{DD} = \begin{pmatrix} 1 & 0 & 0 & 0 & 0 & 0 & 0 & 0 \\ 1 & 0 & 0 & 0 & 0 & 0 & 0 & 0 \\ 1 & 0 & 0 & 0 & 0 & 0 & 0 & 0 \\ 1 & 0 & 0 & 0 & 0 & 0 & 1 & 0 \end{pmatrix},$$

and for CS=D, TMT= N

$$X_{DN} = \begin{pmatrix} 1 & 0 & 0 & 0 & 0 & 0 & 0 & 0 \\ 1 & 0 & 0 & 0 & 0 & 0 & 0 & 0 \\ 1 & 0 & 0 & 0 & 0 & 0 & 0 & 0 \\ 1 & 0 & 0 & 0 & 0 & 0 & 0 & 1 \end{pmatrix}.$$

These parameterizations give estimates for the different effects of the treatments on the different CS groups.

The third part of table 7.8 gives contrasts between the treatment effects of part two of the table. There are no differences among the attend treatments. For distracters, the distract treatment is significantly better than the attend or null treatments. These two parameter estimates are approximately 1. Transforming back to the seconds scale, roughly $2^{.93} \approx 2^{1.06} \approx 2$ which means that we expect that the distracters given the distract treatment will tolerate the cold water approximately twice as long as those given the attend or null treatments! Apparently the effect of the distract treatment on distracters explains the main effect that we saw in the earlier additive model.

A third useful parameterization is the cell means parameterization. The bottom two portions of table 7.8 give the estimated means on the log seconds and seconds scale for the attenders and distracters baseline and post intervention conditions. Figure 7.6 presents point estimates and 95% confidence intervals for the 8 different means on the seconds scale. All attender groups seem quite similar. The distracter baseline (estimate= 32, 95% CI $(25, 41)$) is higher than the attender groups with estimates in the range 21–25 seconds. The distracters given attend or null interventions are right back with the attender groups, but the distracters given the distract treatment are figuratively off the chart with a point estimate of 48 seconds and a 95% confidence interval ranging from 34 to 68 seconds.

Parameter	Log seconds Est	SE	t	p	Seconds Est	Lower	Upper
Intercept	5.01	.17			32	25	41
CS A-D	-.48	.24	-2.0	.05	.7	.5	1.0
TMT D A	-.35	.21	-1.7	.10	.8	.6	1.1
TMT D D	.58	.20	2.9	.006	1.5	1.1	2.0
TMT D N	-.48	.22	-2.2	.03	.7	.5	1.0
TMT A-D A	.46	.30	1.5	.13	1.4	.9	2.1
TMT A-D D	-.54	.29	-1.8	.07	.7	.5	1.0
TMT A-D N	.37	.30	1.2	.22	1.3	.9	2.0
Treatment effects							
TMT A A	.11	.21	.5	.61	1.1	.8	1.4
TMT A D	.04	.21	.2	.85	1.0	.8	1.4
TMT A N	-.11	.21	-.5	.61	.9	.7	1.2
TMT D A	-.35	.21	-1.7	.10	.8	.6	1.1
TMT D D	.58	.20	2.9	.006	1.5	1.1	2.0
TMT D N	-.48	.22	-2.2	.03	.7	.5	1.0
Treatment contrasts							
TMT A D-A	-.07	.30	-.2	.82	1.0	.6	1.4
TMT A D-N	.15	.30	.5	.62	1.1	.7	1.7
TMT A A-N	.22	.30	.7	.47	1.2	.8	1.8
TMT D D-A	.93	.29	3.2	.002	1.9	1.3	2.8
TMT D D-N	1.06	.29	3.6	.001	2.1	1.4	3.1
TMT D A-N	.13	.30	.4	.67	1.1	.7	1.7
CS× TMT means							
AB	4.53	.17			23	18	29
DB	5.01	.17			32	25	41
AA	4.64	.26			25	17	36
AD	4.57	.26			24	17	34
AN	4.42	.26			21	15	31
DA	4.66	.26			25	18	36
DD	5.59	.25			48	34	68
DN	4.53	.26			23	16	33

Table 7.8. REML output of the Pain data analysis for the full model with a CS by TMT interaction with compound symmetry covariance model. The first section gives parameter estimates. The second section gives the treatment effects, and the third section gives contrasts among the treatments. The fourth and fifth sections give estimated log second means for each of the different combinations of CS and TMT. The first five columns give parameter names, estimates, and standard errors and, when appropriate, t-statistics and p-values. The last three columns back-transform the estimates and the lower and upper endpoints of 95% confidence intervals of parameters to report results on the more interpretable seconds scale.

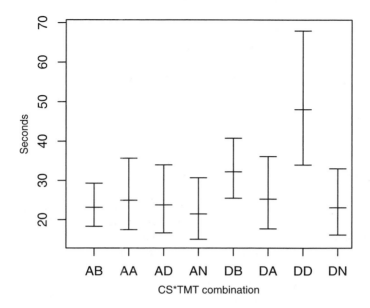

Figure 7.6. Pediatric Pain fitted values (middle horizontal line) and 95% confidence interval (lower and upper horizontal lines) for the 8 CS× TMT groups from the interaction model. Attender means are on the left, distracters on the right, first baseline (B), then the estimates after the attend, distract, and null treatments. The resulting estimates and confidence intervals have been transformed to the seconds scale.

The cell means parameterization are the population means for each of the eight groups, the two coping style groups D and A before intervention, and the 6 CS by TMT groups after treatment. The X_i form for the cell means parameterization for an individual with CS=D, TMT=A is

$$X_{DA} = \begin{pmatrix} 0 & 1 & 0 & 0 & 0 & 0 & 0 & 0 \\ 0 & 1 & 0 & 0 & 0 & 0 & 0 & 0 \\ 0 & 1 & 0 & 0 & 0 & 0 & 0 & 0 \\ 0 & 0 & 1 & 0 & 0 & 0 & 0 & 0 \end{pmatrix},$$

for CS=A, TMT=D

$$X_{AD} = \begin{pmatrix} 1 & 0 & 0 & 0 & 0 & 0 & 0 & 0 \\ 1 & 0 & 0 & 0 & 0 & 0 & 0 & 0 \\ 1 & 0 & 0 & 0 & 0 & 0 & 0 & 0 \\ 0 & 0 & 0 & 0 & 0 & 0 & 1 & 0 \end{pmatrix},$$

and for CS=D, TMT=N

$$X_{DN} = \begin{pmatrix} 0 & 1 & 0 & 0 & 0 & 0 & 0 & 0 \\ 0 & 1 & 0 & 0 & 0 & 0 & 0 & 0 \\ 0 & 1 & 0 & 0 & 0 & 0 & 0 & 0 \\ 0 & 0 & 0 & 0 & 0 & 0 & 0 & 1 \end{pmatrix}.$$

In this parameterization, there is only a single one in each row.

7.3 Groups and Time

In this section, we consider how to model the population mean when we expect a time effect and we have a time-fixed grouping variable. The models and model comparisons of interest are similar to the situation in linear regression with a continuous covariate and a grouping variable, but as usual with longitudinal data, there will be a twist.

We discuss the two group situation at length. Let $G_i = 1$ be the indicator of a subject in group 2 or $G_i = 0$ for a subject in group 1. The first model is the *additive model*, which has an intercept, the grouping variable $G_i 1$, and time as covariates in X_i

$$X_i = (1, G_i 1, t_i).$$

This model has three parameters $\alpha = (\alpha_1, \alpha_2, \alpha_3)'$. The population mean function is

$$E[Y_{ij}|\alpha] = \alpha_1 + G_i \alpha_2 + t_{ij} \alpha_3$$

The intercept α_1 is the average response of group 1 subjects at time $t_{ij} = 0$

$$E[Y_{ij}|t_{ij} = 0, G_i = 0, \alpha] = \alpha_1.$$

The intercept difference α_2 is not only the population difference of group 2 minus that of group 1 at time $t_{ij} = 0$, it is the difference in population means at any time

$$E[Y_{ij}|t_{ij}, G_i = 1, \alpha] - E[Y_{ij}|t_{ij}, G_i = 0, \alpha] =$$
$$(\alpha_1 + \alpha_2 \times 1 + \alpha_3 \times t_{ij}) - (\alpha_1 + \alpha_2 \times 0 + \alpha_3 \times t_{ij}) = \alpha_2.$$

The third parameter α_3 is the common time slope for both groups. Two example covariate matrices are

$$X_i = \begin{pmatrix} 1 & 0 & t_{i1} \\ 1 & 0 & t_{i2} \\ 1 & 0 & t_{i3} \\ 1 & 0 & t_{i4} \\ 1 & 0 & t_{i5} \end{pmatrix}$$

and

$$X_{i'} = \begin{pmatrix} 1 & 1 & t_{i1} \\ 1 & 1 & t_{i2} \\ 1 & 1 & t_{i3} \\ 1 & 1 & t_{i4} \\ 1 & 1 & t_{i5} \end{pmatrix}.$$

The first array is for group 1, the second for group 2. Both arrays assume five observations per subject like the Raven's data. The times may be different for different subjects as in the Raven's data. We often write this model in English as

$$\text{Response} = \text{Group} + \text{Time}, \tag{7.3}$$

The intercept is implicitly assumed. We would explicitly mention not having an intercept. Computer programs tend to use notation like formula (7.3), although usually without the plus sign.

There are two special cases of the additive model. We can test the hypothesis of no group effect $H_0 : \alpha_2 = 0$. If true, then both groups on average have the same mean response over time. And we can test that there is no time effect $H_0 : \alpha_3 = 0$, a hypothesis that leaves us with a time-fixed model.

A more complex model is the model that adds a group by time interaction to the additive model. The model is

$$\text{Response} = \text{Group} + \text{Time} + \text{Group} \times \text{Time},$$

or alternatively

$$X_i = (\mathbf{1}, G_i, t_i, G_i \times t_i)$$

and there are now 4 parameters $\alpha = (\alpha_1, \alpha_2, \alpha_3, \alpha_4)'$. The difference in means between groups 2 minus group 1 is $\alpha_2 + \alpha_4 t_{ij}$ which changes over time. Two example covariate matrices are

$$X_i = \begin{pmatrix} 1 & 0 & t_{i1} & 0 \\ 1 & 0 & t_{i2} & 0 \\ 1 & 0 & t_{i3} & 0 \\ 1 & 0 & t_{i4} & 0 \\ 1 & 0 & t_{i5} & 0 \end{pmatrix}$$

and

$$X_{i'} = \begin{pmatrix} 1 & 1 & t_{i1} & t_{i1} \\ 1 & 1 & t_{i2} & t_{i2} \\ 1 & 1 & t_{i3} & t_{i3} \\ 1 & 1 & t_{i4} & t_{i4} \\ 1 & 1 & t_{i5} & t_{i5} \end{pmatrix}.$$

The first matrix is for a subject in group 1, the second for one in group 2, and again both arrays assume five observations per subject. The parameter

α_1 is the intercept for group 1, $\alpha_1 + \alpha_2$ is the intercept for group 2, so that α_2 is the difference in group means at time $t = 0$. The parameter α_3 is the slope for group 1, and α_4 is the difference in slopes group 2 minus group 1, alternatively, $\alpha_3 + \alpha_4$ is the slope for group 2. There is one model we have not yet discussed, the model with an intercept, time and the time by group interaction

$$\text{Response} = \text{Time} + \text{Group} \times \text{Time},$$

or alternatively

$$X_i = (\mathbf{1}, t_i, G_i t_i)$$

This model is virtually never used in linear regression, but it is common in the longitudinal world. It says that the two groups begin at the same mean but that they grow at different rates. This model is sensible for treatment groups created by randomizing subjects into two groups at time $t = 0$. At $t = 0$, we take our first observation on subjects and the two groups should have the same mean. They then diverge on succeeding days under the influence of the treatment differences.

7.3.1   Raven's Data: Gender and Time Interaction

Figure 7.7 illustrates the four main models with the Raven's data and covariates time and gender. Raw data means for each gender and round are plotted in each figure; the boys' means are circles and the girls' means are crosses. Figure 7.7(a) plots the sole line from the model with no gender effect; 7.7(b) plots the parallel lines from the model with equal slopes, different intercepts; 7.7(c) plots the different slopes, equal intercepts model, and 7.7(d) plots the difference slopes and different intercepts model.

Inspecting the plots and even without knowing the sample standard errors of the mean of the raw data or the standard errors for the fitted lines, we suspect that the model with no gender effects is over-simplified. Comparing 7.7(b)–(d), the model with equal intercepts in (c) seems less than adequate. Because our two groups are defined by gender and not by randomization, we do not expect the two groups to have the same intercept. The equal slopes different intercepts model in (b) does not seem obviously inadequate, nor does the interaction model plotted in (d). The differences between these two models does not seem great, a slight rotation of the lines in (d) as compared to (b); the lines separate over time in (d).

Because the graphics do not distinguish between the two models, we turn to statistical tests to compare the additive model to the interaction model, and for illustration we do illustrate other tests that are useful in different circumstances. Table 7.9 presents results for the four models. It gives fitted values, standard errors, t-statistics, and p-values. As usual, we do not report inferences regarding the intercept, and inferences are rounded.

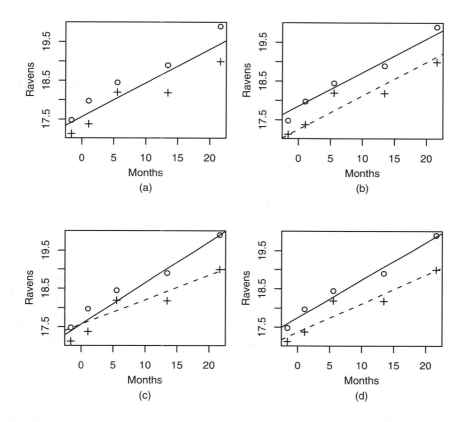

Figure 7.7. Raven's data. All figures plot the raw data means for boys (○) and girls (+) for each round at the average round time. The lines illustrate the fitted lines for each of four models: (a) equal slope and intercept for boys and girls; (b) equal slopes, different intercepts; (c) different slopes, equal intercepts; (d) different slopes and intercepts.

The t-tests allow us to compare each more complex model with the models nested inside with one fewer parameter. The time effect in model (a) is highly significant, showing that we do need to include a time effect in the model and placing our previous time-fixed models into disrepute. Looking at model (b), the gender effect is highly significant and looking at model (c), the time× gender effect is also highly significant. In the interaction model (d), neither time×gender nor gender is quite significant. Apparently either model (b) or (c) is adequate for this data, and the complexity of (d) is not needed. How do we choose between models (b) and (c)?

We turn to three pieces of information for guidance. First, we can compare the -2 log likelihoods of models (b) and (c). We do not have a formal hypothesis test available because the models are not nested. However the

	Parameter	Est	SE	t	p	# parms	-2 ll
(a)						2	12731.6
	Intercept	17.6	.1				
	Time	.087	.008	11.	<.0001		
(b)						3	12714.0
	Intercept	17.3	.1				
	Gender	.60	.14	4.2	<.0001		
	Time	.087	.008	11.5	<.0001		
(c)						3	12714.9
	Intercept	17.6	.1				
	Time	.064	.009	6.8	<.0001		
	Time× gender	.044	.011	4.1	<.0001		
(d)						4	12711.4
	Intercept	17.4	.1				
	Gender	.37	.20	1.9	.06		
	Time	.074	.011	6.8	<.0001		
	Time× gender	.024	.015	1.6	.11		

Table 7.9. Table of ML parameter estimates, standard errors, t-statistics, and p-values for the four models involving time and gender. The number of fixed effects parameters and the -2 log likelihood is given for each model. The log likelihood is calculated using ML, and the covariance model is the spatial autoregressive model. (a) Equal intercepts and slopes; (b) common slopes, different intercepts; (c) different slopes, common intercept; and (d) different intercepts and slopes. Time is measured in months, and the gender effect is coded 1 for boys and 0 for girls.

-2 log likelihood of model (b) with the intercept difference is slightly lower and therefore slightly better. We can use the approximate Bayes calculation that says that conditional on choosing between the two models, the probability of model (b) is

$$P(b|b \text{ or } c) = \frac{1}{1 + \exp(.5(12714.0 - 12714.9))} = .61.$$

This is exceedingly tenuous evidence in favor of model (b). More importantly, inspection of figure 7.7 suggests that the linear fits of model (b) go through the data points better than that of model (c). And thirdly and perhaps most importantly, if we expect a gender difference at later times we would expect to see a gender difference at earlier times also. Model (c) says that the two genders should be equal at the point where the school lunch treatment was initiated, but it seems unlikely that we would have happened to start randomization just at a point where boys and girls had been equal. Randomization should keep treatment groups similar at baseline, but randomization will not have any effect on boys' and girls' average responses. These three reasons suggest settling on the additive model with gender and time as covariates.

7.3.2   The Four Models Involving One Grouping Variable and Time

The interaction model (d) is similar to the linear regression model with two groups, one continuous covariate and a group by covariate interaction. In linear regression the models (a), (b), and (d) are usually the only ones of interest. In longitudinal data, when the covariate is a demographic variable or other unrandomized variable we usually consider these three models.

However, when we have a randomized treatment, then we expect the two groups to be similar at the beginning. If there is any difference between groups, it will be due to different time trends starting from the same intercept. We are mainly interested in the test between models (a) and (c), and we are not usually interested in model (b) or model (d) at all. If one of models (b) or (d) were to hold, then there is a difference at baseline between treatment and placebo, and that means that the randomization has failed in the sense that randomization did not deliver equivalent groups upon which to try out the treatments.

When randomization fails, we must do something else other than testing the interaction between group and time in model (c). Instead, we would do a difference of differences analysis. We would compare the increases in the two groups from baseline to the end of the study. Let $\alpha = (\alpha_1, \alpha_2, \alpha_3, \alpha_4)'$ be the parameters for model (d). Then α_1 is the intercept in the reference group, α_2 is the group effect, α_3 is the slope in the reference group, and α_4 is the difference in slopes between the two groups. Let $t_1 = 0$ be the time at baseline and let t_2 be the time of the end of the study. The end time minus beginning time population mean in the reference group is

$$\delta_1 = (\alpha_1 + \alpha_3 t_2) - (\alpha_1 + \alpha_3 \times 0) = \alpha_3 t_2,$$

and in the second group, the difference is

$$\delta_2 = [\alpha_1 + \alpha_2 + (\alpha_3 + \alpha_4)t_2] - [\alpha_1 + \alpha_2 + (\alpha_3 + \alpha_4) \times 0] = (\alpha_3 + \alpha_4)t_2.$$

The differences of differences analysis is therefore

$$\delta_2 - \delta_1 = (\alpha_3 + \alpha_4)t_2 - \alpha_3 t_2 = \alpha_4 t_2.$$

Because a hypothesis test of $\alpha_4 t_2$ gives the same result as the hypothesis test of α_4, we find that if the baseline groups are not equal, the scientific question we are interested in corresponds to a test of the significance of α_4 in model (d).

7.3.3   Three Groups

Suppose there are three groups; we use G_i and H_i as indicators of group 2 and 3 respectively. Then the equivalent of model (b) can be written

$$\text{Response} = \text{Group} + \text{Time},$$

or as a row of the covariate matrix, $x'_{ij} = (1, G_i, H_i, t_{ij})$. The mean function is $E[Y_{ij}|\alpha] = \alpha_1 + G_i\alpha_2 + H_i\alpha_3 + t_i\alpha_4$. If α_2 is zero, then groups 1 and 2 have the same population mean profile; if $\alpha_3 = 0$, then groups 1 and 3 have the same mean profiles. If $\alpha_2 = \alpha_3$, then groups 2 and 3 have the same population mean profile. There are three choose two equals three tests for the equality of two groups, and there is a 2 degree of freedom test for all three groups being equal.

Model (c) for three groups has $X_i = (\mathbf{1}, t_i, G_i \times t_i, H_i \times t_i)$ with one intercept and three slopes for each of the three groups. Finally the equivalent of model (d) has $X_i = (\mathbf{1}, G_i\mathbf{1}, H_i\mathbf{1}, t_i, G_i \times t_i, H_i \times t_i)$ and three intercepts and three slopes, one each for each group. If either α_5 or α_6 is different from zero, we have a group by time interaction.

7.3.4   Raven's and Treatment and Time

The Raven's data have a randomized treatment assignment with four treatments. We might reasonably expect the four treatment groups calorie, control, meat, and milk to be the same up to time $t = 0$, and then to grow differently. Before treatment onset, all subjects are effectively in the control group, and we expect them to all grow like the control group and so we label any observation with $t < 0$ as belonging to the control group. After treatment onset, we expect subjects in different groups to grow at different average rates according to their assigned treatment group. We fit models using REML and using the autoregressive covariance model.

For the hypothesis of no time by treatment interaction, the F test indicates that the treatment effect is highly significant, with an F statistic of 5.13 on 3 numerator degrees of freedom and a p-value of .0016. Table 7.10 gives results from fitting the interaction between time and treatment with no main effect of treatment. We present results for a typical parameterization, then estimates for the slopes of the four groups, and finally results for the pairwise differences among the treatment slopes.

The coefficient of time is the slope for the milk group, and it is significantly different from zero. The time$\times$calorie, time$\times$control, and time$\times$meat parameters are actually the differences in slopes between these groups and the milk reference group's time trend. Because the three parameter estimates are all positive, the time slopes are positive for all four groups. The highest slope belongs to the meat group, then control followed by calorie, and the milk group slope is the lowest of the four groups. If we try to add a treatment main effect into the model, it is not significant with an $F = 1.1$, and a p-value of .33.

Figure 7.8(a) plots the empirical data means for each group at each round including baseline. Figure 7.8(b) shows the fitted lines from the model of table 7.10. In both figures, the meat group is the solid line, the control group is the dashed line, the calorie is the dotted line, and the milk group is the lowest dash-dot line. The fitted lines smooth out the irregularities in

Parm	Est	Std. SE	t	p
Intercept	17.570	.101		
Time	.058	.012	4.9	$< .0001$
Time×calorie	.023	.015	1.6	.12
Time×control	.035	.015	2.3	.02
Time×meat	.058	.015	3.8	.0001
Time coefficients for each tmt				
Meat	.117	.012	9.5	$<.0001$
Control	.094	.012	7.9	$<.0001$
Calorie	.081	.012	6.8	$<.0001$
Milk	.058	.012	4.9	$<.0001$
Time by tmt pairwise comparisons				
Meat - control	.023	.016	1.5	.14
Meat - calorie	.035	.015	2.3	.02
Meat - milk	.058	.015	3.8	.0001
Control - calorie	.012	.015	.8	.42
Control - milk	.035	.015	2.3	.02
Calorie - milk	.023	.015	1.6	.12

Table 7.10. Table of parameter estimates, standard errors, t-statistics, and p-values for the model of time in months and treatment by time fit with REML and the autoregressive covariance model.

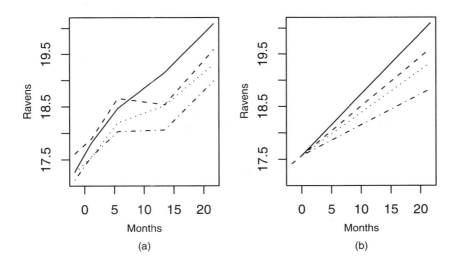

Figure 7.8. Raven's data. (a) The empirical summary plot for each treatment group at each round, connecting the dots within treatment group. (b) The fitted means. All four groups follow the same linear trend for $t < 0$. Lines are meat – solid; control – dashed; calorie – dotted; milk – dash-dot.

the observed data, but otherwise seem to generally imitate the empirical means. On the left, the control and milk group trends flatten between rounds 3 and 4, and generally the third round observation seems perhaps a little higher than might otherwise be expected from the other results displayed in the figures.

Table 7.10 indicates that the difference between the meat group slope and either of the calorie and milk group slopes is significant, and that the control slope is also significantly larger than the milk slope. It is somewhat alarming that the control group has a larger slope than two of the treatment groups and this is a problem for the researchers to puzzle over.

7.4 Defining Time

The definition of time is very important to the analysis of longitudinal data. In most studies, the choice of time is pretty clear, but poor choice of definition can make the analysis unwieldy or nearly impossible. In theory we can reproduce the same model no matter how we define time, but the necessary contortions can be minimized with a good definition. The first problem is to define the zero point. Is it the same calendar date for all participants or is it a subject-specific date? Examples of subject-specific dates are the time of first contact with study investigators, the first interview date or the onset of intervention in an intervention study.

As an example, suppose we are analyzing mental age in an intervention study of young children with autism. To explore children's mental growth, we may use (i) time of study entry, (ii) time of initial intervention, or (iii) birthdate as the zero point. Exactly which of these three choices is most appropriate might be determined by inspecting profile plots of the key response where each choice in turn is used to define time. We would be looking for the scale where the variance of the response at any given time is least. For normal children, we might well use birthdate as our zero point. For autistic children, this may not be as good a choice as intervention onset, as the variance of autistic children's mental age at a given physical age can be large. In contrast to these three potentially reasonable choices, a clearly poor choice would be to use mother's birthdate as the zero time. In a randomized intervention study, the control group may not have a well defined intervention onset date and we may would necessarily use first contact with the investigators or first interview date as our zero point.

7.4.1 Changing Units

We sometimes wish to change the units of time in our models. What happens, if, say, we change days to weeks, or months to years? The units should not affect our model. No matter how we define time, the theoretical and fitted values for Y_{ij} must be the same, so we can use this to figure out the relationship among coefficients for covariates on different scales. The main issue is when time enters as a covariate. Suppose that we have a fitted value of

$$E[Y_{ij}] = \alpha_1 + \alpha_2 \times \text{months}_{ij}.$$

Now we change to years

$$E[Y_{ij}] = \gamma_1 + \gamma_2 \times \text{years}_{ij}.$$

where we have intercept γ_1 and years coefficient γ_2. The $E[Y_{ij}]$ must be the same in either case, so

$$\alpha_1 + \alpha_2 \times \text{months}_{ij} = \gamma_1 + \gamma_2 \times \text{years}_{ij} \qquad (7.4)$$

At time $t = 0$, this equation reduces to $\alpha_1 = \gamma_1$, so the intercepts do not change when we change units. Substituting this into equation (7.4) and solving for γ_2, we get

$$\gamma_2 = \alpha_2 \times \frac{\text{months}_{ij}}{\text{years}_{ij}} = 12 \times \alpha_2.$$

The slope changes by a factor of 12 when we shift to years instead of months. The use of years versus time cannot change our inference, and the p-value and t-statistic for the test of hypothesis that $H_0 : \alpha_2 = 0$ must be the same as for testing $H_0 : \gamma_2 = 0$. Similarly, the standard error for the estimate of γ_2 differs from the standard error for the estimate of α_2 by the same factor of 12.

In general if we change the time units by a factor c, so that the new time is equal to c times the old units, then the new coefficient of time will be $1/c$ times the old coefficient, and the standard error will change by the same factor. The t-statistic and p value will remain the same.

7.4.2 Baseline Age and Time in Study

There are a set of three covariates related to time: time in study t_i with elements t_{ij}; baseline age, BA_i a scalar; and age Age_{ij}. These are related by the equation

$$\text{Age}_{ij} = \text{BA}_i + t_{ij}.$$

Any of the three covariates might be included in a longitudinal data analysis. Only two of them can be included in the model because the inclusion of the third causes collinearity in the covariate matrix. Collinearity in its simplest form occurs when some linear combination of the covariates adds up to be another covariate; in longitudinal models, we must have the same linear combination for all subjects and all times. Collinearity means that there are multiple values of the coefficients that give the same fitted values.

For example, if we have Age_{ij}, BA_i, and t_{ij} in our model, the linear combinations $2\text{Age}_{ij} + 1 \times \text{BA}_i + 1 \times t_{ij} = \text{Age}_{ij} + 2 \times \text{BA}_i + 2 \times t_{ij} = 3 \times \text{Age}_{ij}$ all give the same fitted values. The simplest solution to the problem of collinearity is to delete a relevant covariate.

The simplest solution for Age_{ij}, BA_i, and t_{ij} is to include BA_i and t_{ij} and delete Age_{ij}. Why would one want to include both BA_i and t_{ij} in the model? Consider a response which varies linearly with age, and suppose that the study involves a treatment that alters the natural course of the response over time. Then the coefficient of age and the coefficient of t_{ij} will not be the same. An example might be blood pressure, which we might expect to increase with age. An intervention that prevented blood pressure from rising at the usual rate with age would have a coefficient of t_{ij} that was less than the coefficient of BA_i.

Various hypotheses are of interest in the model with covariate matrix $X_i = (\mathbf{1}, \mathbf{1}\text{BA}_i, t_i)$. The test of $H_0 : \alpha_2 = 0$ checks for lack of a linear increase or decrease with baseline age, while $H_0 : \alpha_3 = 0$ checks for a lack of linear time trend in the response once entered in the study. A test of $H_0 : \alpha_2 = \alpha_3$ tests for no change after study entry in the mean trend as a function of age. An alternative of $H_A : \alpha_3 < \alpha_2$ is the test that study entry has caused a decrease in the rate of increase of the response with age. Assuming that higher response is bad, one does not need $\alpha_3 < 0$ for the trial to be a success.

7.4.2.1   Raven's Data

We fit the Raven's data with the AR(1) covariance model, REML, and covariates baseline Age_i, and time-since-intervention t_{ij}, with both baseline age and time measured in years. The Raven's data has children who are in the first form, equivalent to the first grade in the United States. Variable Age_i is perhaps a more interesting variable than might be expected by those used to Western schools. It has a mean of 7.6 years, standard deviation of 1.4, and ranges from 4.8 up to 15.2 years. Table 7.11 gives results for the model with baseline age and time-since-study-entry as predictors and with both measured in years. We see a modest but significant effect for Age_i, while time has a slope of 1 indicating that kids' Raven's score are going up approximately 1 point per year. In contrast, it takes 7 years difference in ages at baseline to create the same difference as 1 year in school. We see that school is substantially better ($t = -8.5$, $p < .0001$) than mere aging at improving children's Raven's score.

7.5   Three-Way Interactions

After the initial analysis of a randomized treatment or other important covariate, secondary analyses are undertaken to discover if and how the

Parameter	Est	SE	t	p
Intercept	16.40	.41		
Age_i	.15	.05	2.9	.004
Time (years)	1.05	.09	11.5	<.0001
Age_i - time	-.89	.11	-8.5	<.0001

Table 7.11. REML estimates for Raven's data with covariates Age_i at time 0 and time in years. Covariance model is autoregressive.

treatment effect differs in various sub-groups, perhaps defined by gender or race or other relevant grouping variables. When the important treatment effect is the time by treatment interaction, this leads naturally to three-way interactions of covariate by time by treatment. There are two ways to investigate the question of how the covariate affects the treatment by time interaction. One way is to split the data set into two separate data sets as defined by the covariate. One performs separate analyses on the two groups. The analyses are then compared to see if the time by treatment interaction is different in the two groups. This analysis is actually a very general interaction model that allows every parameter including regression coefficients and covariance parameters to vary between the two groups. This double analysis introduces a number of additional parameters that may not be necessary. Each separate analysis has reduced sample size; effects that had been significant in the pooled analysis may become non-significant due to the reduced sample size.

The second way to investigate the effect of the covariate on the treatment by time is to include the three-way interaction covariate by time by treatment. We would also include the main effect for the covariate as well as two-way interactions between the covariate and time. If we did not have a main effect for treatment in the original analysis, then there would not be a covariate×treatment interaction, otherwise there would be. We test the three-way interaction and declare that the treatment effect is different in the two groups if the three-way interaction is significant.

Whichever approach is used to test the three-way interaction, to understand how the effect differs by levels of the covariate requires a graphical display as the original parameters are usually difficult to interpret. Typically we plot the fitted values over time in the treatment by covariate groups. This can be supplemented by a similar plot of the raw data means in the different groups.

7.5.1 Raven's Example

We illustrate a three-way interaction between gender, time and morbidity level. Morbidity score for subject i is the average of individual morbidity scores based on usually 15 but occasionally fewer home visits. At each home visit, the child's health is checked for approximately 20 different

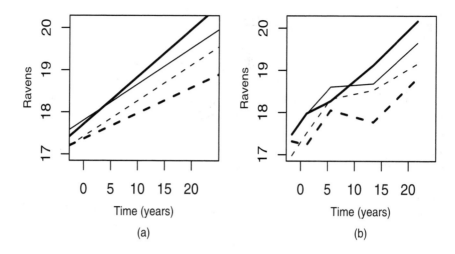

Figure 7.9. Raven's data. Plots illustrating the gender by morbidity by time three-
-way interaction. (a) Lines are fitted values from the REML fit with a random
intercept and slope covariance model. Solid lines are boys, dashed are girls, heavier width lines are the high morbidity groups, lighter is low morbidity. (b) Raw
data means at each round, plotted against mean round time of observation, same
line types.

symptoms or illnesses, each of which is identified as either a mild or severe
morbidity. The morbidity at a visit is rated as either 0, 1, or 2, denoting no
morbidities (0), mild but no severe morbidity (1), and presence of severe
morbidity (2), respectively. The average morbidity score is .75, and we
define a dichotomous morbidity level as 1 if the morbidity score is above
.75 and 0 if the score is at or below .75.

We fit a model with main effects, two-way interactions and the three-way
interaction of gender, morbidity level, and time using REML and a random
intercept and slope covariance model. The three-way interaction is barely
significant, with $p = .05$. Figure 7.9(a) shows fitted lines for each combination of gender and morbidity level. The solid lines are the fitted Raven's
scores for the boys, the dashed are for the girls. The heavier lines are for
high morbidity, and the lighter lines are for low morbidity. We see that the
boys start and end higher than the girls. However, the morbidity effects
are different in the different genders. In the boys, the high morbidity group
slope is higher than the low morbidity group. In the girls, the morbidity
effect is reversed, the high morbidity group has a lower slope than the low
morbidity group. In the boys, the high morbidity group actually does better than the low morbidity group. The girls low morbidity group appears to
be catching up with the low morbidity boys or at least staying parallel to
them, while the high morbidity boys are outstripping the high morbidity

girls. Figure 7.9(b) plots the mean Raven's score by round, gender, and morbidity level. The raw data is harder to interpret than the fitted lines, but the conclusions are the same, except that as in previous plots, we see several groups with a large increase at round 3 that does not appear to be maintained at round 4.

In modeling with interactions, we often build up to the three-way model in an exploratory fashion. We build up to the three-way interaction only if the two-way interaction terms are significant. If we had done this in modeling the Cognitive data, we would not have found a three-way interaction between gender, morbidity level, and time in the Raven's data, as the two-way interaction terms gender×morbidity, gender×time, and morbidity×time are not significant.

7.6   Step Functions, Bent Lines, and Splines

Longitudinal data collected over a long period of time often does not follow a linear trend, and a simple polynomial may also not suffice. Witness for example the Wallaby tail data whose estimated mean is plotted in figure 7.3(d). For balanced with missing data, we have the unstructured mean model, but that often has a large number of unknown parameters that need to be estimated. Even the Big Mice data set with unstructured mean can be challenging for software to fit, and many data sets are much more complicated than the Big Mice data.

A class of models called *splines* are useful for modeling smooth nonlinear mean functions. Splines are used to describe complicated curves when the true functional form is not known. The simplest spline model for the population mean is a sequence of flat steps. Each step approximates the mean response over a small interval of time. Figure 7.10(a) illustrates. The smooth curve is an unknown true mean response on the vertical axis as a function of time t on the horizontal axis. The flat lines are a *step function* that approximates the smooth curve. At the left-hand side of the figure for time between 0 and .2, the mean curve increases rapidly and we need a lot of steps to approximate the function. Even so, at the end points of each time interval, the step function can be far from the true mean function. The vertical line next to the curly brace indicates how far off the step function is at the right end of the third time interval. In contrast, at the right-hand side of the figure, the response curve varies quite slowly as a function of time, and a single time interval from $t = .6$ up to $t = 1$ with a single mean does a better job of approximating the slowly varying portion of the curve than all of the short little intervals at the left-hand side.

We might use a step function if we believe there are abrupt changes in our mean function. We would put the break points at the locations

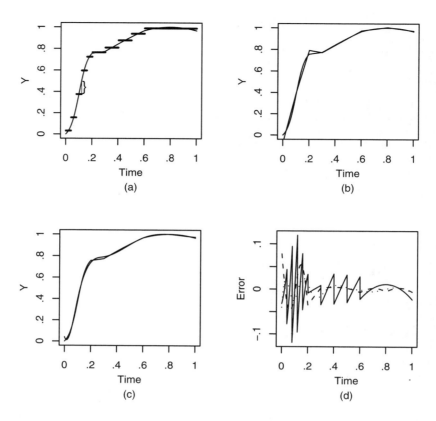

Figure 7.10. (a) Example of a step function with 10 steps (9 knots) approximating a smooth function. Vertical line and curly brace around $x = .12$, $y = .4$ shows the difference between the smooth function and the step function. (b) Bent line with 5 knots fit to same function. (c) Cubic spline fit with 4 knots. (d) Plots of the differences between the smooth function and the three fitted functions, solid line: step function error; dashed line: bent line error; dotted line: cubic spline.

where we expect a jump, but more commonly we use the jump function to approximate a continuous curve as in figure 7.10(a).

A slightly more complicated mean function is a *bent line*. A bent line is a continuous function that is linear on intervals of time but the slope is allowed to change from one interval to the next. Connected line segments were used to represent individual subjects profiles when plotting profiles in chapter 2. Connected line segments are the same thing as bent lines. We can also use bent lines to approximate the continuous curve in figure 7.10(b).

The step function requires one fixed effect parameter per step. The bent line requires two parameters for the first interval and one additional change in slope parameter for each succeeding interval. If the step function and the bent line have the same *change points* then the bent line will require one additional parameter. The change points are located at the jumps in the step function and at the slope changes for the bent line. The bent line does a much better job of fitting the curve in figure 7.10(b) than does the step function in figure 7.10(a). This particular bent line used only 7 parameters as compared to 10 parameters for the step function. Because the bent line generally fits better than a step function, we need many fewer change points and fewer parameters over all than if we use a step function to model the mean.

Bent line models are quite common in longitudinal data analysis models. They are used as a matter of principle when analyzing data where there is some expectation of an abrupt change in trend at a given time point. In educational studies on students over years, we may anticipate a change in students' performances at particular ages. For example, when students graduate from one school to the next, when they enter puberty, or if there is a well defined change in curriculum. In these instances, we might likely want to test whether the slope before the expected change point is different from the slope after the change point. We also may use the bent line model opportunistically to approximate a curve which operationally happens to look like a bent line when we plot the data.

Step functions are less common than bent lines, but I use them occasionally to explore noisy data where I believe the average response level is changing over time but where I do not know exactly what the general trend should be.

7.6.1 Step Function Parameterizations

The simplest X matrix for a step function is a series of non-overlapping indicator functions. Suppose we have observations at nine time points from $t = 1$ up to $t = 9$, and we have three step functions with steps at 2.5 and 5.5. We give a second parameterization with an intercept and a different interpretation for the α parameters.

time	V 1			V 2		
1	1	0	0	1	0	0
2	1	0	0	1	0	0
3	0	1	0	1	1	0
4	0	1	0	1	1	0
5	0	1	0	1	1	0
6	0	0	1	1	1	1
7	0	0	1	1	1	1
8	0	0	1	1	1	1
9	0	0	1	1	1	1

In both versions V 1 and V 2 of the covariate matrix, there are three parameters $\alpha = (\alpha_1, \alpha_2, \alpha_3)'$. In version V 1, the parameters are the population
mean responses in the intervals $(1, 2.5)$, $(2.5, 5.5)$, and $(5.5, 9)$. In version
V 2, α_1 is the mean over the first time interval, but α_2 is the difference
in population means interval 2 mean minus interval 1 mean. Similarly, α_3
is the difference interval 3 mean minus the interval 2 mean. These parameterizations are similar to the parameterizations for the unstructured
mean in subsection 7.2.2, except that a single parameter is the mean for
multiple time points. As with the unstructured mean, there are other possible parameterizations. The particular parameterization for the covariate
matrix is called a *basis* (plural *bases*) or a set of *basis functions*. We presented three bases in subsection 7.2.2 for the unstructured mean. The same
three parameterizations are possible for the step function; we presented two
parameterizations just now. Many bases are possible for any spline model.

7.6.2   Bent Line Parameterization

A bent line model with a single slope change point will have three covariates, an intercept, the slope up to the change point, and a covariate whose
coefficient is the change in slope at the change point. The usual parameterization for the bent line model is similar to version 2 of the step function.
We will have an intercept and a slope for two parameters. The third covariate is zero before the change point and equal to t minus the change point
time after the change point.

For a bent line model with two change points at times 2.5 and 5.5, the
covariate matrix looks like

$$
\begin{array}{cc}
\text{Time} & \text{Bent line} \\
\begin{array}{c}
1 \\ 2 \\ 3 \\ 4 \\ 5 \\ 6 \\ 7 \\ 8 \\ 9
\end{array}
&
\left(
\begin{array}{cccc}
1 & 1 & 0 & 0 \\
1 & 2 & 0 & 0 \\
1 & 3 & .5 & 0 \\
1 & 4 & 1.5 & 0 \\
1 & 5 & 2.5 & 0 \\
1 & 6 & 3.5 & .5 \\
1 & 7 & 4.5 & 1.5 \\
1 & 8 & 5.5 & 2.5 \\
1 & 9 & 6.5 & 3.5
\end{array}
\right)
\end{array}.
\tag{7.5}
$$

Now there are 4 parameters, $\alpha = (\alpha_1, \alpha_2, \alpha_3, \alpha_4)'$. Parameter α_1 is the
intercept and α_2 is the initial time slope. Parameters α_3 and α_4 are the
changes in the slope at times $t = 2.5$ and $t = 5.5$. The slope at $t = 3$ is
$\alpha_2 + \alpha_3$, and the slope after time 5.5 is $\alpha_2 + \alpha_3 + \alpha_4$.

7.6.3 Indicator Functions and Positive Parts

To mathematically write down the formula for the columns of the X matrix in a spline model, we introduce some notation. An *indicator function* is a function that takes on the value one when an *event* occurs and zero when the event does not occur. For example the indicator function

$$1\{t_{ij} \geq 2.5\} = \left\{ \begin{array}{ll} 0 & \text{if the time } t \text{ is less than } 2.5 \\ 1 & \text{if the time } t \text{ is } 2.5 \text{ or later} \end{array} \right.$$

describes the second column of version V 2 of the step function. The indicator of time being greater than 5.5

$$1\{t_{ij} \geq 5.5\}$$

is the third column of both V 1 and V 2. The second column of V 1

$$1\{2.5 < t_{ij} \leq 5.5\}$$

is a function that is 1 for the middle ranges of time and zero before time or at $t_{ij} = 2.5$ and zero again after time $t_{ij} = 5.5$.

The mathematical description of the covariates in matrix (7.5) for the bent line model is slightly more complicated. The second column of the X_i matrix is of course the time t_i. The third column is zero up to the change point, when it starts increasing linearly. We write

$$(t_{ij} - 2.5)_+ = \left\{ \begin{array}{ll} 0 & \text{if } t_{ij} - 2.5 \text{ is negative} \\ t_{ij} - 2.5 & \text{if } t_{ij} - 2.5 \text{ is positive} \end{array} \right. .$$

The subscript "+" indicates the *positive part* of the quantity in parentheses. If the quantity in parentheses is negative, it is replaced with zero, whereas if the quantity is positive, then it is left alone. The fourth column of (7.5) is $(t_{ij} - 5.5)_+$. The covariate $(t_{ij} - 5.5)_+$ is zero up to 5.5 then is positive after 5.5.

7.6.4 Knots

The break points at times $t = 2.5$ and $t = 5.5$ are formally called *knots*. On each interval between knots, and to the left of the first knot or to the right of the last knot, our function is flat (step function) or linear (bent line model). The bent line model is also called a *piece-wise linear* model. The number of parameters for the step function is equal to 1 plus the number of knots. The parameters for the bent line model is equal to 2 plus the number of knots, but for the bent line model, we often need many fewer knots than for the step function, and in practice the total number of parameters is fewer than for the step function model.

7.6.5   Higher Order Polynomial Splines

We can continue making our spline models more complex, using a piece-wise quadratic or piece-wise cubic model instead. The spline model has two parameters. One is the order of the piece-wise polynomial on each interval, and the second is the number of knots. If the spline is a kth order piece-wise polynomial, then for each knot there is a covariate that allows the coefficient of the kth order t_{ij}^k term to change. At each knot, there is a jump in the step function. For the bent line model there is a change in slope at each knot. The cubic spline model has an intercept, linear, quadratic, and cubic time term. At each knot, say at time t_{0k}, the cubic spline model also has a covariate of the form $(t_{ij} - t_{0k})_+^3$, the cube of the positive part of $(t_{ij} - t_{0k})$. This allows the coefficient of time cubed to change at each knot.

The number of parameters used in a spline model is equal to the degree of the polynomial plus the number of knots plus 1.

The most popular spline models are the bent line models and cubic splines. Bent lines are popular for their easy interpretability. Cubic splines because the curves that they generate are very flexible and smooth. Unlike the step function or bent line, the eye has great difficulty picking out the locations of the knots of a cubic spline, so the curve looks smoothly contin-uous without the jarring breaks that the bent line has. A further reason is that, as in this example, the cubic spline does a great job of approximating difficult curves. A cubic spline is illustrated in figure 7.10(c) approximating the complex curve. It is the most accurate of the three approximating func-tions that we show. Figure 7.10(d) shows the differences between the true curve and the various approximating functions. The difference between the true mean and the approximating function is plotted against time. The step function (solid line, 10 parameters) is the worst approximate function, the bent line (dotted, 7 parameters) is second worst and the cubic spline (dot-ted, somewhat hard to see) with 8 parameters is best. The three functions have 9, 5, and 4 knots respectively.

7.6.6   Placing Knots

Where do we place knots? Generally, we want to put them where the func-tion changes most rapidly. Flat portions of the function do not require any knots; linear portions do not require knots in the bent line or higher order polynomial splines. No knots are needed near quadratic portions for a cubic spline, but a quadratic trend is often difficult to distinguish from higher or-der trends. Kinks, jumps, gyrations, and sharp changes in slopes are places where more knots are needed. As an automatic procedure, one might place knots equally spaced throughout the range of time. Alternatively, after an initial guess at the knots, we might try removing some in areas where that

Parameter	Est	SE	t	p
Intercept	-1.86	.12		
Month	-.062	.005	-13.	<.0001
Slope18	.081	.008	9.9	<.0001
Slope36	-.020	.007	-3.1	.002
Season spring (3-6)	.16	.04	3.6	.0003
Season summer (7-10)	-.013	.045	-.30	.77
Gender (female)	.65	.13	4.9	<.0001
Slope before month 18	-.062	.005	-13.	<.0001
Slope between months 18, 36	.018	.005	3.8	.0002
Slope after month 36	-.002	.0029	-.68	.50
Season spring - summer	.17	.045	3.9	.0001

Table 7.12. BSI total. Results from fitting the adolescent data to a model with gender, season, and a bent line model with knots at 18 months and 36 months. REML estimation and an ARMA$(1, 1)$ covariance model were used. The ARMA$(1, 1)$ model is discussed in the next chapter.

covariate is not significant, and we should add knots in areas where the mean function does a poor job of fitting the data.

7.6.7 BSI Data

We illustrate a step function and a bent line in the same analysis. The BSI data is a psychological inventory composed of 53 items rated on a 0 to 4 integer scale assessing various symptoms. Each item represents a psychiatric symptom or a negative state of mind, and subjects indicate how much that particular symptom has troubled them during the past weeks. It has 9 sub-scales as well as the overall BSI scale. The score is the average rating on the items making up the scale or sub-scale. Because the resulting sample histogram is skewed and zero is possible, we add a small constant 1/53 to the scale and then take logs base 2. The sample studied in this example consists of adolescent children of HIV+ parents, and the sample observations are at a subset of the full data set. These observations are those where there is both parent and adolescent information.

The covariates in this analysis are gender and two versions of time: season, and time in months. Season was parameterized as a step function with three different periods. Winter indicates that the measurement was taken in November through February; spring is March through June, and summer is July through October. Similarly, a bent line model was developed for the mean response over time. The initial time $t_{i1} = 0$ is the initial interview time for the adolescent. Knots were placed at the end of month 18 and again at month 36. Results from fitting this model with REML and an ARMA$(1, 1)$ covariance model are given in table 7.12. The ARMA$(1, 1)$

covariance model is discussed in section 8.1.8. Linear combinations and contrasts of the parameters are given in the lower half of the table.

There are 7 fixed effects in this particular model, the intercept, 3 parameters for the slope over time, 2 parameters for season and one for gender. The initial slope up to month 18 is negative at $-.062 \pm .005$. The change in slope from before to after 18 months is large and positive at .081 ($t = 9.9$, $p < .0001$) and the resulting slope $.018 = -.062 + .081$ between months 18 and 36 is significantly ($t = 3.8$, $p = .0002$) positive. Due to rounding, the numbers do not quite add. Then there is another significant change and the final slope ($-.002$) after month 36 is not significantly different from zero. The units of the slopes are in terms of changes in log BSI per month. Very roughly, the log BSI decreases by $-.062 \times 18 \approx 1.1$ over the first 18 months, then increases by .3 over the next 18 months before continuing essentially flat for the remainder of the study. A decrease of 1 on the log base 2 scale means that the average score decreased roughly in half over the first 18 months, and this is followed by roughly a 25% increase over the next 18 months.

The season effect is that in spring, the average log BSI is significantly higher than at winter and summer, while summer and winter are not significantly different from each other. The coefficient of .16 means that BSI is approximately 11% higher in spring, where $11\% = (2^{.16} - 1)100\%$.

7.6.8   Big Mice Data

We fit a cubic spline with knots at days 3.5, 8.5, and 13.5 to the Big Mice data. The knots were chosen to be distributed throughout the range of days, but slightly toward the earlier times, as there appears to be less rapid change in the mean at the later times. The X_i matrix for each subject has 7 columns, the first four columns are the intercept, day, day squared, and day cubed, and the last three columns are the cubed positive portion of time minus the knot time, for example $(t_{ij} - 3.5)^3_+$. We used an ARH covariance model and ML estimation. The ARH covariance model is identified in the next chapter as the best covariance model for the mice data. It is a generalization of the autoregressive covariance model that allows the variances to change over time. It is discussed in section 8.2.

Table 7.13 presents the results from fitting the three knot cubic spline model to the mice data. The parameters are not particularly interpretable, but we notice that the last two spline coefficients are not significant and we wonder if we can perhaps remove one knot. Moving to the model with 2 knots at times 3.5 and 13.5 results in a model with a similar log likelihood where both knots are significant and we settle for the 2 knot model.

Label	Est	SE	t	p
Intercept	148.8	2.8	53.72	<.0001
Day	6.0	4.2	1.4	.16
Day^2	13.6	2.1	6.4	<.0001
Day^3	-1.23	.27	-4.6	<.0001
$(t - 3.5)^3_+$	1.05	.35	3.0	.003
$(t - 8.5)^3_+$	.15	.18	.8	.41
$(t - 13.5)^3_+$	.38	.23	1.7	.10

Table 7.13. Big Mice data summary of fit for the spline models using ML estimation.

7.6.9 Wallaby Tails

Splines are probably the right tool for modeling the Wallaby tail length data. The data set is quite large and unbalanced, and the mean pattern and pattern of variation is quite complex. Some of the models that I wanted to try could not be fit with my software and computer (long-term options include possibly changing software and probably computer and/or operating system and possibly writing my own software if fitting this data was important enough), so some compromises were made in the fitting. Initially, I tried a few covariance structures that can handle unstructured data, (see chapter 8). The AR model was strongly improved on by a random intercept and slope, and in turn a random quadratic (each subject has their own quadratic mean function over time) and then random cubic (each subject has their own cubic mean function) were continued improvements. At that point I was unable to make the correlation model more complex, and I settled for a random cubic model. I also restricted my data set to include data only up to 730 days, but that still left a complicated data set. REML was used in exploring the covariance model, and ML for exploring the mean model.

For the splines, I went straight to a cubic spline model with knots every 100 days. The knots at 300, 500, and 700 were not significant and were dropped from the model. I now checked for a time by gender interaction by seeing if separate splines for each gender were needed. The interaction was significant, with a twice the log likelihood improvement of over 50 on 7 degrees of freedom. In chapter 9, we will learn about some tools for checking the fit of random effects models. These include subject-specific fitted values $\hat{Y}_{ij}$ and subject-specific residuals $e_{ij} = Y_{ij} - \hat{Y}_{ij}$. The subject-specific fitted values will be the population fitted values plus estimates of the individual random effects. In the meantime, we take it for granted that we can get subject-specific fitted values and residuals which we plot in profile plots. See section 10.4 for more details.

Inspection of the residuals and fitted values for the model with 4 knots and the time by gender interaction suggested a strong need for additional knots in the first few hundred days. Wallaby data for 20 wallabies are plotted in figure 7.11(a). Figure 7.11(b) plots residuals for the 20 wallabies and the model with 4 knots and 7.11(c) plots the fitted values for those same 20 wallabies. Ideally we would like the residuals to not show any strong time trends, and to all be small. Unfortunately, the residuals gyrate wildly around zero in a non-random fashion before 200 days. Less obvious, but clear on close inspection, the fitted lines at times less than 200 days wiggle around more than do the raw data. Thus knots were added at days 50 and 150. These were highly significant, the improvement in the 2×maximized likelihood were approximately 500 likelihood units for the additional 2 degrees of freedom. Figure 7.11(d) plots the fitted values for 20 wallabies from the new model. The wiggles are much less at the earliest times and the fitted values appear to mimic the raw data much better. Inspection of the residual plot (not shown, sorry!) for the model with 6 knots shows that further improvement is still very possible. Additional knots will prove to be quite useful, and the covariance function warrants further improvement or exploration.

7.6.10   Other Basis Functions

There are several choices of parameterizations or basis functions for cubic splines. The obvious parameterization is the one we specified above. A particularly popular set of basis functions are the *B-splines*. This is the same cubic spline model as we described above, but it has different mathematical properties. When it comes time to do the computations, B-splines, for example, may allow software to handle more complex models than the simple parameterization that we specified. This is because the B-spline basis has columns of its X matrix that are not nearly as collinear as the simple cubic spline that we described above, where time squared, time cubed, and, for example, $(\text{time} - \text{knot-time})^3_+$ can be highly collinear.

7.7   Adjusting for Baseline; Generalizations of Paired Comparisons

In section 3.3, we discussed the simple paired comparison and recommended the improved *difference of differences* (DoD) design. Longitudinal data structures generalize both the paired comparisons and the difference of difference designs. Missing data often occurs in longitudinal data when not all subjects have the same number of baseline or followup observations. The paired *t*-test has no easy mechanism for incorporating unbalanced or unpaired data. In contrast, longitudinal data analysis allows the analyst to accommodate unbalanced designs and missing data. Longitudinal models are more flexible and can be adapted to use all of the observations available, thus estimates of treatment effects have the potential to be more efficient

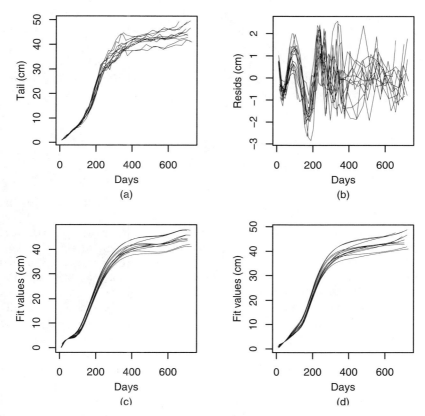

Figure 7.11. All figures show results for the Wallaby tail length data, with y axis in units of centimeters. The first 20 cases with the lowest animal id numbers are shown. (a) Raw data profile plot. (b) Residuals from model 1 with 4 knots. (c) Fitted values from model 1 with 4 knots. (d) Plot of the fitted values from the model with 6 knots.

than methods that require complete and balanced data in before treatment and after treatment pairs.

Often we read that an analysis is "adjusted for baseline." In the paired t-test design, this means that we have subtracted the baseline pre-treatment measurement from the post-treatment response, and similarly in the DoD design. A slightly more complicated analysis is the analysis of covariance (ANCOVA) model. Suppose that we have $g \geq 2$ groups, and a baseline measurement Y_{i1} and a post-treatment measurement Y_{i2}. The ANCOVA model is a regression model with response Y_{i2}. The covariates are a set of $g - 1$ indicators for the different groups and, typically, a single continuous covariate Y_{i1}. In the DoD design, the coefficient of Y_{i1} is fixed at -1; in the ANCOVA model, we estimate the coefficient of Y_{i1}.

What does it mean to adjust for baseline in a longitudinal analysis? There are several possible implied analyses. Without careful explanation by the data analysts, the reader of an analysis that was "adjusted for baseline" does not know which analysis was actually performed. We explain three of these analyses in turn, then explain the right way to adjust a longitudinal analysis for baseline. Let

$$ Y_i = \left(\begin{array}{c} Y_{i1} \\ Y_{i2} \end{array} \right) $$

be the response vector for subject i. The first observations Y_{i1} is the baseline observation that we wish to "adjust for."

One common analysis deletes the baseline observation from the response. Instead, Y_{i1} is included as a (time-fixed) covariate in the analysis to "adjust for baseline." This is one extension of the analysis of covariance model to the longitudinal framework. A second occasional analysis subtracts the baseline observation from all future observations and then analyzes the differences in longitudinal fashion. These "adjusted" approaches are especially common in difference of differences designs, for example with data where the treatment onset is immediately after baseline. Baseline observations are pre-intervention, and all remaining observations are during, or post-intervention. These "adjusted" approaches to longitudinal analysis are not recommended.

It is entirely possible to adjust for baseline but without subtracting off baseline from the remaining observations or deleting it from the response vector. The right way to accomplish this is that the baseline observation should be included in the Y_i vector of observations that are modeled as $Y_i = X_i \alpha + \epsilon$. Then we carefully fit a model to the data that estimates the mean $\hat{Y}_{i*}$ at a suitable end point t_{i*} and also estimates the baseline response $\hat{Y}_{i0}$ at $t_0 = 0$, both for the same set of covariates. The correct way to adjust for baseline is then to estimate $\Delta = \hat{Y}_{i*} - \hat{Y}_{i0}$. The adjustment is not done at the data stage, rather it is done after fitting the model, at the inference stage of the analysis. To compare two treatments, estimate the difference Δ for each group, and subtract them. If you randomly assign subjects to the two groups, the groups should have the same baseline estimate $\hat{Y}_{i0}$ and when you subtract the two Δs, these baseline differences cancel. However, if the two groups have different baseline $\hat{Y}_{i0}$s, then this difference of difference analysis adjusts appropriately for the baseline differences in the two groups.

Covariates X_i are chosen to model all of the Y_i vector including at baseline and all followup responses. Commonly, a time-fixed covariate that predicts the response, such as gender, say, also is predictive of the baseline observation. Removing the baseline observations from the Y_i vector removes the information in the baseline about covariates such as this.

7.8   Modeling Strategies

Selection of covariates in longitudinal models usually goes in several stages. Details must of course vary with the data set, and particularly the purpose for collecting the data and for analyzing the data set. Still, some useful generalities can be given. I try to develop a good model for the population mean over time before anything else. Without a good understanding of the temporal pattern of the response, it is difficult to start asking questions about what the effect of a particular covariate is, much less answering it. For example, in the Wallaby data, it is necessary to develop a reasonable spline model for tail length growth before moving on to answer the question about what the differences are between the two genders. When we incorrectly model the mean in our data, the error (called bias) is moved from the mean model into the variance model, inflating the estimates of variance. In the Wallaby data, a poor model for the tail length over time may cause us to incorrectly identify the covariance model as well. By putting as much of the variation in the data into the temporal mean model, we reduce the estimated variance of the observations. This in turn means that we have increased power to identify other effects in the data. If we attempt to identify covariate effects when the estimated variance is high, we have high standard errors, and will miss many subtle but potentially interesting covariate effects.

Having developed a good model for the mean over time, I now add in covariates that I know will be a priori important, for example, gender is almost always important in psychological studies such as the BSI data and it is added in even if it is not of direct interest in the analysis. In the Cognitive data, in addition to gender, we would include a covariate such as age or height at baseline. This variable would help to adjust for the very major differences among children at baseline where ages vary by over 10 years. In the BSI data or in the Cognitive data, we would not want differences in ages or gender between groups to affect the conclusions about treatments or other important covariates without adjusting for those covariates. Finally, I add in the covariates whose effects I am mostly interested in. For Cognitive this would be the treatment by time interaction effects.

In exploratory analyses with many potential covariates, I again start by modeling the response mean over time. Then next I introduce the covariates into the model. These may be added in as time-fixed effects. It is straightforward to do forward selection in longitudinal models, and that is what I often do. A variation on forward selection that skips many of the steps is to check each covariate for significance one at a time, as in the usual first step of forward selection. One selects all the variables whose significance level passes some arbitrary cutoff and adds all selected variables in a single step to the model, and that is the "final model."

Backward elimination in its general form is almost not possible in longitudinal models, particularly if one considers all the potential interactions with time. However, an acceptable compromise is sometimes to put all time-fixed main effects in the model, and then to backwards eliminate those covariates that are not significant; covariates that describe the time pattern should probably not be removed in this process. If there are not too many variables, putting them all into the model, and leaving them can be a reasonable compromise. A second compromise is to not delete variables one at time, but to remove the non-significant variables in a single step and stop as in the previous forward selection algorithm.

In all of these forward or backwards selection algorithms, we do not remove the covariates describing the temporal pattern, nor do we remove the originally identified important covariates such as gender (BSI or Cognitive data sets) or age at baseline (Cognitive data set).

7.9   Problems

1. Lengths and weights are usually considered to be ratio scales. However, would you consider human height or weight to truly be on a ratio scale? Discuss.

2. Give two alternative wordings that carefully and correctly describe the interpretation of α_2 in (7.1)–(7.2).

3. What happens to the interpretation of α_1 and α_2 if you reverse the coding of group 2 and group 1 in equations (7.1)–(7.2)?

4. Consider a fixed-effects model with a single three-group covariate.

    (a) Write out the X_i matrices for the parameter effects coding and the cell means coding.
    (b) What is the length of α? Does the parameterization matter when answering this question?
    (c) Give the interpretation for the elements of α for both parameterizations.
    (d) For subjects in each group, give the linear combinations of the parameters that give the population means.

5. For g groups, how many contrast statements are required to estimate all the pairwise differences in group means?

6. Suppose that you have two grouping covariates, with g_1 and g_2 groups for each covariate. If g_1 and g_2 are small, you can describe the effects of the grouping covariate by reporting means and standard deviations in a table or a plot for all possible covariate combinations, even if there is no interaction. Suppose that both g_1 and g_2 are fairly large (say 7–10), how could you summarize the effects of each covariate, again assuming no interaction.

7. In an analysis with race (White, Black, Hispanic, or Asian) as a covariate, you find that there is a very small number of one group.
   Consider the following procedures

   (a) Delete the members of that group from the analysis.
   (b) Pool the members of that group with the most similar group
       economically.
   (c) Pool the members of the group with the largest group.
   (d) Pool with the members of the next smallest group.
   (e) Estimate the effects of the four different groups, and pool
       the smallest group with the group with the closest parameter
       estimate.

   Compare and contrast these approaches. Discuss the effect on the
   estimate of the group that the smallest group is pooled with, and
   what is the effect on degrees of freedom for testing the significance of
   race as a covariate.

8. Suppose you have a study that follows subjects every 3 months for a
   decade. Subjects are asked how many cigarettes they smoke per day
   on average in the last 3 months. You may specify your own outcome
   measure if it is useful in answering this question.
       Consider using number of cigarettes smoked as a covariate in an
   analysis. Some people have always and will always smoke zero cigarettes and may be classified as non-smokers. Some people may be
   former smokers and some are current smokers. Over time, some smokers may bounce back and forth between zero (non-smoking smokers)
   and more than zero cigarettes smoked (smoking-smokers). It may be
   that in some analysis, the issue is whether you smoke or not, in other
   studies the issue is how much you smoke, and still in others, both
   issues are relevant.

   (a) The simplest covariate is a time-fixed grouping variable defined
       by baseline smoking status. Give at least three different smoking
       variable definitions, two with 2 groups and one with 3 groups.
   (b) Consider a study where some people do not smoke, some smoke,
       and some are off-again on-again smokers. Define a time-varying
       grouping variable for smoking. How many groups should there
       be?
   (c) In the survey questionnaire, do you ask subjects the same ques-
       tion(s) at baseline about smoking as at followups? How are they
       different? Should you ask more questions at baseline or fewer as
       opposed to at followups?
   (d) Consider a baseline smoking covariate and separately the time-
       varying smoking variable. Describe some important hypotheses
       that you might wish to test for. Assume that the smoking effects,

if they exist are additive level shifts; don't worry about time-varying effects.

(e) Consider drinking, illicit drug use, seat belt use, and risky sexual activity. Explain how using these variables as covariates in an analysis are similar or dissimilar to the smoking variable. Consider using baseline measures as time-fixed covariates, and also consider time-varying covariates.

9. Consider a model with a time-fixed continuous baseline covariate B_{i1} and a time-fixed grouping variable G_{i1}. What is the simplest form that an interaction could take? Sketch a plot like 7.1. What is the most complex form the interaction could take? Again sketch a plot of the means over time for the two groups.

10. For the quadratic mean model, with $E[Y_{ij}|\alpha, t_{ij}] = \alpha_1 + \alpha_2 t_{ij} + \alpha_3 t_{ij}^2$, show that the maximum or minimum time occurs at $t = -.5\alpha_2/\alpha_3$.

11. A particular residual plot for the mice data shows a strong quadratic trend.

(a) For the Small Mice data, draw a profile plot of the raw data. Inspection of this plot suggests that the individual mice all start at about the same initial point, but perhaps that they grow approximately linearly at different rates. This suspicion of a linear trend is not a certainty, but might be a starting point in our model building.

(b) Construct a set of empirical residuals for the Small Mice data. Fit a line to each mouse's data and construct the fitted values $\hat{Y}_{ij}$ for each observation. Construct the empirical residuals by subtracting the original data minus the fitted values $e_{ij} = Y_{ij} - \hat{Y}_{ij}$.

(c) Plot the empirical residuals in a profile plot. What structure do you see? What does this suggest about the population mean model for the Small Mice data? What covariate(s) do you need to add to the X_i matrix?

(d) Fit a model or models as needed to show that adding your suggested covariates are statistically important.

12. Consider the Weight Loss data set. Suppose you took sample means of the weight lost each week, and calculated the sample standard errors of those means. Would you expect those standard errors to be larger or smaller or about the same as the .4 or .5 standard errors associated with the seventh column of table 7.4? Do actually calculate the standard errors of the mean weekly differences. Then fit the random intercept and slope model to the Weight Loss data and see which approach gives smaller standard errors for the mean weekly

differences. How would you summarize the gain in efficiency, if any, due to using the statistical model?

13. Fit the unstructured mean model to the Weight Loss data using a random intercept and slope covariance model and again using the heterogeneous antedependence covariance model. Compare the point estimates for the three parameterizations for the two covariance models, and then compare the standard errors, t-statistics, and p-values for the differences reported in table 7.5. Are there any important differences in the point estimates? How about in the standard errors, t-statistics and p-values?

14. For the last three figures of figure 7.3, describe an X_i matrix for modeling those means.

15. The quadratics in figure 7.2(c) have (intercept, linear, quadratic) terms equal to $(2, .75, -.375)$, $(6, -3.5, .5)$ and $(0, 4, -1)$. (a) Identify which coefficients belong to which curve, (b) calculate the time of the minimum or maximum and (c) the height at that time. (d) Does launch speed appear to be an appropriate summary for these three curves? Why or why not?

16. Verify the fits to the Small Mice data presented in table 5.2. Plot the fits over time, with time running from $t = 2$ to $t = 20$. Plot the fitted values a second time, this time over a much wider time frame, extending the range of time out past the range of the data. Which curve, if any, does the best job of extrapolating how you might expect the mice to continue to grow?

17. There are two alternative versions of the Small Mice data. One has all the observations on every third day beginning on day 1, and the other has all the observations on every third day beginning on day 0.

    (a) Create these two data sets. How many subjects and how many observations for each of these two data sets? Give a short but appropriate identifying name to each data set; this will assist you in discussing the fits among the different models, otherwise your discussion will be difficult.
    (b) To each data set, fit polynomials of order 1, 2, 3, etc. as necessary until the highest order polynomial term is not significant. Show your results in a table similar to table 5.2. Use the ARH covariance model.
    (c) How do your results compare with the original Small Mice data fits?
    (d) Plot the three linear fits in a single plot. In your plot, plot the means at each time point using distinct symbols for the three different data sets.

(e) Repeat the plot but for the quadratic fits and again for the cubic fits.

(f) Discuss the differences between the fits implied by these figures and discuss which plots fit the data the best.

18. For the Small Mice data and for each of the other two Small Mice data sets of problem 17, fit the unstructured mean model. Use the ARH covariance model.

(a) For each of the three data sets: does the unstructured mean model fit better than the quadratic model? Is it better than the cubic model? What model would you suggest if you had to choose a single best model for all three data sets?

(b) Create a table of fitted values and standard errors that includes a pair of columns for the unstructured mean model. Do this for each of the other two data sets. Be sure to round appropriately.

19. Suppose that we expect to have a quadratic time trend, and that we have two groups. What would be the equivalent of models 1 through 4 as in table 7.9 in this case? Write out the X_i matrices. How many fixed effects parameters are there for each model, and what is the degrees of freedom for the hypothesis tests between the models?

20. (Continued) In an analysis with three groups and time, write out the degrees of freedom of the tests between the 4 models. Briefly interpret the tests, that is, in English what are the null and alternative hypotheses that we are testing? Which models and tests make sense in the context of a randomized treatment and which make sense in terms of a fixed covariate like race?

21. Consider a three-way interaction model as in section 7.5. Suppose that you fit the time by treatment interaction model separately in the two groups. Suppose you now wish to combine the two estimates of the time by treatment interaction. Sketch how to do a combined test of $(b_1 + b_2)/2$ where b_1 and b_2 are the estimates from the two analyses. Show that it has standard error of sqrt[Var(b_1)+Var(b_2)]/2.

22. In the Cognitive data, there are three additional response variables: arithmetic score (arithmetic), verbal meaning (vmeaning), and total digit span score (dstotal). The first two are considered measures of intelligence/education, whereas digit span is a test of memory. Pick one of the three measures to analyze. Pick also one or several covariates (ie time; time and gender; time and treatment; time, gender and treatment)

(a) Do basic data analysis for the analysis you intend. Examples of important initial data analysis procedures include, but are not limited to

i. Inspect the raw response values. What are the range and summary statistics?

ii. Plot histograms of the response and covariates.

iii. Plot histograms or kernel density estimates of the response stratified by round, by treatment and round×treatment.

iv. Make empirical summary plots for your response, and for your response for each level of any indicator variable that you will be analyzing. For a continuous covariate, you may split the covariate's range into two or three parts, and draw empirical summary plots for the subjects in each portion of the range.

v. Rather than continuous time, use round for this step. Explore the variances and correlations of your variable. Draw or calculate a (i) correlation matrix, (ii) covariance matrix, (iii) scatterplot matrix, and (iv) correlogram. What do you conclude? Which analysis do you find most useful? For the method that you find easiest to use and most useful, calculate or plot that method for your response separately for each treatment group.

vi. Draw a profile plot of the data. Because there are so many cases, you will likely want to plot only a subset of the data on any given profile plot.

vii. It is important to check the randomization. A key feature of the data after randomization is that the baseline levels of the response should be the same in each treatment group. Using just the round 1 response, do an analysis of variance (ANOVA) with your response. Are there differences among the groups?

For each step, write up a one- or two-sentence summary of the conclusions that you draw from that part of the analysis. Part of learning to analyze data analysis is to realize when there is not much to say and when there is something to say about each step of the process.

(b) Analyze your response as a function of gender and time.

(c) Analyze your response as a function of treatment and time. Be sure to check the model with treatment, time and treatment by time also to confirm that there is no significant baseline treatment effect.

(d) Fit your response as an appropriate function of treatment and time.

(e) Now explore the effects of including some covariates in the model.

    i. Include gender and age-at-time0 as covariates. Are gender or age-at-time0 significant? Do they change your conclusion about the interaction of treatment and time?

    ii. Include (separately) gender by time and age-at-time0 by time interactions in your model. (Be sure to keep the main effect covariate in the model.) Draw conclusions from the analyses. How does including these terms affect the treatment by time interaction?

    iii. Draw appropriate graphical conclusions (i.e., inference plots).

(f) Repeat the previous problem but for family socio-economic status (SES1, higher is better) and child's morbidity score (morbscore, higher is worse). The morbidity score is the average morbidity score across 15 visits where 0 means no morbidities, 1 means at most a mild morbidity at that visit, and 2 means a severe morbidity at that visit.

(g) Repeat the previous problem but for mother's baseline read-test and write-test scores.

(h) Combine the previous several problems into a single problem to analyze one of the responses, Raven's, digit span, arithmetic or verbal meaning. Write a two-page report on your complete analysis. Include necessary tables and plots in your appendix. The goal of your analysis is (a) to determine the effect of treatment on the outcome, and (b) to determine whether including any of the covariates socio-economic status, morbidity score, mother's read test, mother's write test, age at time 0, or gender does or does not alter your result. This last issue was a concern of the researchers. It usually does not happen; covariates usually don't explain away an interaction result such as our finding of a treatment effect for Raven's. Particularly this is true in the case of a randomized treatment. However, missing data can reduce the sample size so that the effect becomes non-significant. In observational studies, covariates are more likely to possibly explain away some significant result.

23. Do we need a round (time treated as a grouping variable) effect in the model for Raven's instead of the linear time trend? How about a round by treatment interaction? Is there any evidence of a quadratic trend in time?

24. In the model for Raven's where we explored the three-way interaction of time, morbidity level, and gender, use an unstructured mean for time, including all the interactions. Is there still a significant three-way interaction? Then explore a series of models where the higher order interactions include linear time, while the lower order terms

include the unstructured mean. Is there any need for an unstructured mean at all?

25. In the Pediatric Pain data, we fit the model with the interaction between coping style and treatment. Another way of defining the treatment is as a matched treatment and strategy (i.e., CS=A, TMT=A or CS=D, TMT=D) or mismatched strategy (CS=A, TMT=D or CS=D, TMT=A). Keeping the N treatment the same, fit the model with the A and D treatments redefined as matched intervention, and mismatched intervention. Does this change the need for an interaction?

26. Consider table 7.8 where we gave 3 different parameterizations for the Pediatric Pain covariates. There are a number of potential parameterizations.
    (a) Write out the X_i matrices for 3 different CS×TMT combinations for the very first parameterization.
    (b) Get your software to reproduce the estimates from the three parameterizations in table 7.8. Report the command line and any data management steps that you require to reproduce the results. If possible, do this without using any contrast-type statements.
    (c) Let α be the coefficients for the second parameterization and let γ be the coefficients for the third parameterization. Write formulae that show the relationship between the α and γ parameterizations.
    (d) Identify, create, or construct as many other *interpretable* parameterizations as you can. For each parameterization, up to 4, clearly identify the interpretation of the parameters and give the point estimates of the parameters by calculating them using the results of table 7.8 (in other words, do not fit them using software).
    (e) Get your software to reproduce the estimates from your choice of four additional parameterizations.
    (f) How many different parameterizations are there in total?

27. Suppose we have a model where neither time nor a function of time enters into the fixed effects. If we change the units of time, what happens to the fixed effects? Suppose that we have a covariate that changes with time, but is not a function of time. What happens to the coefficient if we change the units of time?

28. In the Small Mice data, fit the data to a quadratic model for the mean as a function of time. Now change the time units from days to weeks, and refit it. Compare the estimates, standard errors, t-statistics and p-values for the linear and quadratic terms. What is the relationship among them? Now change time to hours, and repeat.

29. Suppose we have a model with an intercept and a time slope with time measured in years. Suppose we now change time from years to months, and we change the baseline by subtracting 3 months from each date. Give the mathematical formulae for how the new parameters, intercept γ_1 and time coefficient γ_2, relate to our old parameters α_1 and α_2.

30. Consider a quadratic model for the mean as a function of time, $E[Y_{ij}|\alpha] = \alpha_0 + \alpha_1 t_{ij} + \alpha_2 t_{ij}^2$.

    (a) Now change time to $s_{ij} = c_2 t_{ij}$ giving a model $E[Y_{ij}|\gamma] = \gamma_0 + \gamma_1 s_{ij} + \gamma_2 s_{ij}^2$. What is the relationship of the γ parameters to the original α parameters?
    (b) Now change time to $s_{ij} = c_1 + c_2 t_{ij}$, and answer the previous question again.

31. Suppose that in our model with an intercept and a time slope, we change the time units to newtime $= c_1$oldtime $+ c_2$. How do the new γ parameters relate to the old parameters α?

32. Refit the Raven's analysis of age and time from section 7.4. This time, run the analysis twice, once for children under age 9 at baseline and a second time for those over age 9. Show that the first analysis is essentially the same as what we presented in table 7.11, while the analysis for the older kids gives an example where the difference in coefficients of age and time are not significant. What does this imply about the older children who enter school? Does schooling necessarily benefit them?

33. In the Pain data, test the treatment effects separately for attenders and distracters by running the analysis twice, once for each coping style. Write a short explanation of your conclusions. Try CS and UN covariance models. Is the analysis different with different covariance models?

34. Draw a figure like 7.5(a) for the Pain data with CS×TMT interaction. Does adding error bars to the plot improve it?

35. Fit a bent line model to the $\log_2$ BSI depression data. Try knots at 18 months and again at 36 months. What do you conclude? Plot the fitted mean over time.

36. Develop a cubic spline model for the BSI total data. What are your conclusions?

37. Carefully develop a cubic spline to the Wallaby weight data. How many knots are needed? How do the males and females differ in their growth over time?

38. Calculate fitted values for the BSI model with gender, season and the bent line for

    (a) Three male subjects who start in November, March, and July.
    (b) Three female subjects who start in January, May, and September.

    Plot the estimates on a single graph, labeling the lines appropriately.

39. Explore the seasonal effect in the BSI data.

    (a) Discuss the pros and cons of having a separate effect for each month over the current model.
    (b) Now consider using a sine and a cosine term, i.e., $\alpha_1 \sin(2\pi t_{ij}) + \alpha_2 \cos(2\pi t_{ij})$ for the seasonal effect.
    (c) How many sine and cosine term pairs can you add to get to a model as complicated as the monthly indicators? The way to add terms to the basic sine and cosine model is to add pairs of terms such as $\alpha_1 \sin(2\pi k t_{ij}) + \alpha_2 \cos(2\pi k t_{ij})$. We already have $k = 1$, the next terms have $k = 2$, then $k = 3$, and so on.
    (d) Fit the monthly models and separately the sine and cosine models. How many cosine and sine terms can you add until neither term in a pair is significant? As always, make a report of your findings.
    (e) Plot the fitted values, and decide whether you like the sines and cosines or the monthly parameters better.

40. Use a cubic spline model to model some other Wallaby measurement separately for males and females. Make a short report of your results. Is there a difference by gender? Is there a gender by time interaction?

41. For the Big Mice data, fit a quartic (fourth order) polynomial, a quintic (fifth order) polynomial, the two spline models mentioned in the text, and also try removing the knot at time 13.5 either instead of or in addition to the knot at time 8.5. Make a table of your results using ML fitting. Which model seems to be most satisfactory?

42. In the Wallaby data, analyze length of some body part over time. Restrict yourself to the first 100 days, and do the analysis separately for males and females. Report on your results.

43. Develop a model for the Dental data in table 2.10. Decide what the issue is that you are exploring with your models, and report on your conclusions.

44. In the multivariate model $Y_i \sim N_J(\mu, \Sigma)$ with J observations per subject, write out several sets of contrasts to test the hypothesis that all μ_j are equal to each other. Assume $J = 4$.

    (a) First write out the 3 contrasts that say that μ_1 is equal to each of the other μ_js in turn.

(b) Write out the 3 contrasts that have each μ_j equal to its successor.

(c) Write out a third set of contrasts.

45. Recreate table 5.1 and then produce the covariance matrix of these parameters to directly calculate the estimate and standard error of the linear combination $B'\mu$ for the following Bs.

(a) $B'_1 = (1, -2, 1, 0, 0, 0, 0)$ and $B'_2 = (0, 1, -2, 1, 0, 0, 0)$. Verify your results by having your software calculate the point estimate and standard error.

(b) $B'_3 = (1, 1, 1, 1, 1, 1, 1)/7$.

(c) $B'_4 = (-1, 0, 0, 0, 0, 0, 1)$.

(d) Describe B_1 through B_4 in words.

(e) Construct 95% confidence intervals for $B'_k\mu$ for $k = 1, 2, 3, 4$.

8
Modeling the Covariance Matrix

Our relations to each other are oblique and casual.
                                    – Ralph Waldo Emerson

So that men, thus at variance with the truth,
Dream, though their eyes be open; reckless some
Of error;

                                    – Dante

Overview

In this chapter, we discuss the covariance model specification. We study a large number of covariance models, and we discuss how to generate additional covariance models. Most of the discussion assumes balanced with missing data. Extension to random times are discussed. We discuss reasons for modeling the covariance.

1. Parameterized covariance models

   (a) Compound symmetry
   (b) Random intercept
   (c) Autoregressive
   (d) Independence
   (e) Random intercept and slope
   (f) Independent increments
   (g) Antedependence

    (h) Factor analytic

    (i) Toeplitz or banded

    (j) Unstructured

2. Non-constant variance

    (a) Naturally non-constant variance models

    (b) Extending constant variance models to have non-constant variance

    (c) Different variance parameters as functions of covariates

3. Extensions of various models to random times

4. Family relationships among models

5. Reasons for modeling the covariance

The choice of covariance model is important. For example, figure 8.1 illustrates four hypothetical data sets. In each data set, the population means are the same, namely $X_i\alpha = \mathbf{0}$ is a vector of zeros. This is represented by the solid line plotted in each figure. If we were to average all responses at a given time, the average value would be near zero up to sampling variability. Still, the profiles that individual subjects follow are quite different in each figure. These different profiles are entirely accounted for by the covariance model $\Sigma(\theta)$ that was used to generate the data in each plot.

It might seem desirable to use an unstructured covariance model to model the covariance matrix of longitudinal data as it would seem guaranteed to model the data correctly. Unfortunately, use of the unstructured covariance model is usually not feasible nor recommended. For random observation times there are too many variance and covariance parameters, and they cannot all be estimated. The variance of responses at time t_{ij} will not be estimable if there is zero or one observation at that time. When the data set is balanced with possibly missing data, when the sample size n is large and the number of repeated measures J is small or modest compared with n, then we can use the unstructured covariance model. Most data sets do not meet these constraints, and we need to use a *parameterized covariance model*.

A *parameterized covariance matrix* is one where all variances and covariances are functions of a small to moderate number of covariance parameters θ. The covariance model $\Sigma_i(\theta)$ defines a *family* of possible covariance matrices with members of the family indexed by θ. The parameters θ are estimable using appropriate software. The key issue is to choose a covariance model $\Sigma_i(\theta)$ that is *correct*, but *parsimonious*, meaning that (a) the true covariance matrix is in the family defined by the covariance model and that (b) vector parameter θ has as few elements as possible. In practice, we are happy if the true covariance matrix is close to some member of the family, although we do not formally define close.

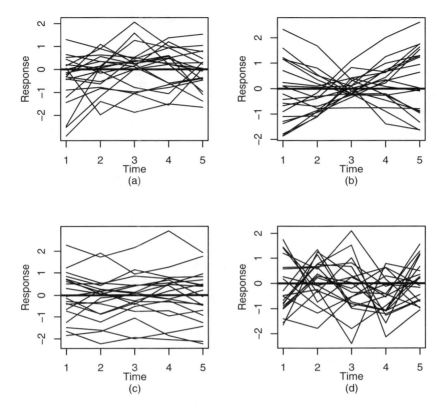

Figure 8.1. Constructed data with population mean zero. Twenty subjects contribute individual profiles for each of four different covariance matrices: (a) autoregressive, (b) random intercept and slope, (c) compound symmetry, (d) independence.

8.1 Parameterized Covariance Models

We discuss 10 different covariance models in this section. Each model has a motivating story; some have more than one. If a particular story applies to our data, the story specifies the covariance model we should use. Otherwise we use a combination of graphical display and model selection tools to choose a covariance matrix.

In the following subsections, we explain the story and implications of each model $\Sigma_i(\theta)$. For each model, we give the theoretical form of the covariance or correlation matrix, and we give numerical examples of the correlation and/or covariance matrix. We simulate illustrative data sets and plot the data in profile plots and in scatterplot matrices.

To illustrate the covariance and correlation matrices, we will assume data taken at equally spaced time points $t_i = t = c(1, 2, 3, 4, 5)'$ with no missing observations. All subjects Y_i then have identical covariance matrices $\Sigma_i(\theta) = \Sigma_i = \Sigma$. As we are concentrating on the covariance model, we set $\alpha = 0$ so that the population means are zero across time and $Y_i \sim N(0, \Sigma)$.

The parameter θ in $\Sigma(\theta)$ is generic. It refers to the distinct unknown parameters of a covariance matrix. For the autoregressive model of subsection 8.1.3, two parameters σ^2 and ρ determine all of the variances σ_{jj} and covariances σ_{jk} among observations, so $\theta = (\sigma^2, \rho)$. The names of the individual parameters in θ are model specific and the same parameter and name are used more than once in the different covariance models. In most models, parameters have multiple meanings; across models, some meanings may be shared, some may not. The parameter names for the compound symmetry model in subsection 8.1.1 have the same names as the parameters for the autoregressive covariance model. The variance parameter τ^2 shares a common definition $\tau^2 = \mathrm{Var}(Y_{ij})$ in both models. The correlation parameter ρ shares some but not all definitions as we will see. There is potential for confusion; context is required to keep parameter names matched with parameter definitions. The parameter names are more or less traditional and therefore suggestive to those already exposed to longitudinal data analysis. Besides: there aren't enough letters to give every parameter a unique name.

8.1.1 Compound Symmetry

The compound symmetry (CS) covariance model has two parameters, τ^2 and ρ. The first parameter is the variance of the observations

$$\mathrm{Var}(Y_{ij}) = \tau^2,$$

and under this model, observations Y_{ij} have constant variance across time. The parameter ρ is the correlation between any two observations from the same subject

$$\mathrm{Corr}(Y_{ij}, Y_{il}) = \rho,$$

for $j \neq l$. The times t_{ij} and t_{il} do not matter as long as $j \neq l$ or equivalently, $t_{ij} \neq t_{il}$. Thus this equi*covariance* covariance model also has equi*correlation* between observations. The algebraic form of the covariance matrix for data with 5 observations is

$$\mathrm{Var}(Y_i | \tau^2, \rho) = \tau^2 \begin{pmatrix} 1 & \rho & \rho & \rho & \rho \\ \rho & 1 & \rho & \rho & \rho \\ \rho & \rho & 1 & \rho & \rho \\ \rho & \rho & \rho & 1 & \rho \\ \rho & \rho & \rho & \rho & 1 \end{pmatrix}$$

and by ignoring the τ^2 out front we also have the correlation matrix. High values of ρ give very high correlation among all observations, low values of ρ give low correlation.

The key feature of the compound symmetry matrix is that for any time lag $t_{ij} - t_{il} \neq 0$ large or small, the correlation $\text{Corr}(Y_{ij}, Y_{il})$ is the same. This means that observations taken a few minutes apart and those taken a few years apart have the same correlation. This is unlikely for real data measured on human beings over long enough periods of time. In practice, measures that are very persistent over the data collection time frame may follow a compound symmetry covariance model. Weight might be an example; as adults, our weight often does not change over a period of years, other than changes due to typical daily energy consumption and expenditure.

Figures 8.2(a) and (b) shows 20 simulated subjects with a compound symmetry covariance matrix with $\rho = .95$, and similarly in 8.2(c) and (d) but with $\rho = .8$. Data in figures (a) and (b) have variance $\tau^2 = 1$, while data in (c) and (d) have variance $\tau^2 = 1$. Without inspection of the Y axis tick mark labels in the profile plots, we cannot tell that the variances are different in the upper and lower sets of figures. In figures 8.2(a) and (c), we see an approximately constant range in the responses as a function of time, and we conclude that the $\text{Var}(Y_{ij})$ are equal across time.

The greater correlations in (a) are noticeable because the profiles are more parallel in (a), while in (c), the paths cross quite a bit more. Similarly, stronger correlations at all time lags are visible in (b) than in (d). Also we notice that the off-diagonal plots in either (b) or (d) have the same approximate correlation. In particular, look at the plot in the $(1, 5)$ location, the lag 4 plot farthest from the main diagonal. The correlation appears to be the same as that in the lag one plots at the $(1, 2)$ and $(2, 3)$ down to the $(4, 5)$ locations. This is the clue that we have a equicorrelation or compound symmetry covariance matrix.

The population variances at times $t = (1, 2, 3, 4, 5)'$ for the data in figures 8.2(a) and (b) are

$$V = (1, 1, 1, 1, 1)',$$

and the population correlation matrix is

$$\text{Corr}(Y_i) = \begin{pmatrix} 1 & .95 & .95 & .95 & .95 \\ .95 & 1 & .95 & .95 & .95 \\ .95 & .95 & 1 & .95 & .95 \\ .95 & .95 & .95 & 1 & .95 \\ .95 & .95 & .95 & .95 & 1 \end{pmatrix}.$$

In 8.2(c) and (d), the variances are

$$V = (10, 10, 10, 10, 10)',$$

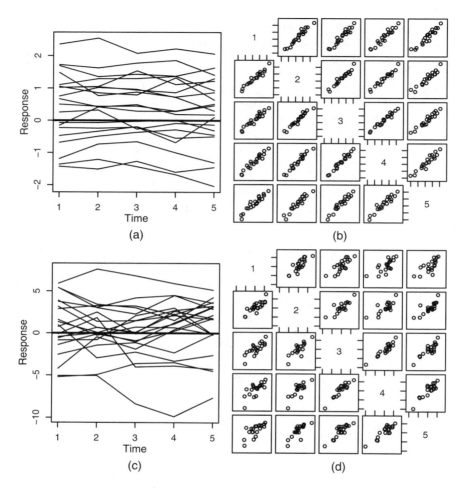

Figure 8.2. Constructed data, $\alpha = 0$, compound symmetry matrix. (a) Variance 1, correlation .95, (b) same data, scatterplot matrix. (c) Variance 10, correlation .8, (d) scatterplot matrix.

and the population correlation matrix is

$$\mathrm{Corr}(Y_i) = \begin{pmatrix} 1 & .8 & .8 & .8 & .8 \\ .8 & 1 & .8 & .8 & .8 \\ .8 & .8 & 1 & .8 & .8 \\ .8 & .8 & .8 & 1 & .8 \\ .8 & .8 & .8 & .8 & 1 \end{pmatrix}.$$

The compound symmetry matrix goes by several other names. It is also called *equicovariance*, (EC) because all the covariances are equal. It is also called the *equicorrelation* matrix. With equicorrelation, we potentially could have different variances at different times, a model we discuss in section 8.2. The compound symmetry model is also called a *random intercept* (RI) model. The RI model is identical to the CS model, but is arrived at from a different perspective. It allows for different generalizations and we discuss it in the next subsection.

8.1.2 Random Intercept Model

In figure 8.2(a), each subject's profile appears flat, at least approximately. Observations Y_{ij} vary around a different value γ_{i1} for each subject. These values γ_{i1} are the intercepts of the line each subject's responses vary around, since the true slopes are zero. The mean of $Y_{i1}, \ldots, Y_{i5}$ is an estimate $\hat{\gamma}_{i1}$ of the unknown subject-specific intercept γ_{i1}. The set of intercepts $\gamma_{i1}, i = 1, \ldots, n$ are a sample from the population of intercepts, paralleling our assumption that subjects are a sample from the population of subjects. We make a distributional assumption for the γ_{i1}; for convenience and in the absence of other information we often assume normality. The γ_{i1}'s are called *random intercepts*. We include the subscript 1 as well as i on the γ_{i1}'s because later on, when we discuss the random intercept and slope model, there will be two γ's, one, γ_{i1} for the intercept, as here, and γ_{i2} for the slope. We will then need the additional subscript to distinguish between the two random effects.

The random intercept model assumes observations bouncing around a flat line $Y_{ij} = \gamma_{i1} + \delta_{ij}$. The *random intercept and slope* model of subsection 8.1.5 is a generalization that allows for the lines to have a subject-specific non-zero slope. General random effects models are discussed further in chapter 9.

Let the mean of the γ_{i1}'s be α_1, which is zero in the constructed examples of this section. Let the variance of the γ_{i1}'s be D, and let the variance of the observations Y_{ij} around γ_{i1} be σ^2.

In the random intercept model, observations Y_{ij} are modeled as having two sources of variation. One is the variation of the means γ_{i1} around their population mean

$$\gamma_{i1} \sim N(\alpha_1, D),$$

and second is the variation of observations around the subject-specific mean

$$Y_{ij}|\gamma_{i1} \sim N(\gamma_{i1}, \sigma^2).$$

Alternatively, we may write

$$
\begin{aligned}
Y_{ij} &= \alpha_1 + \beta_{i1} + \delta_{ij} \\
\beta_i|D &\sim N(0, D) \\
\delta_{ij}|\sigma^2 &\sim N(0, \sigma^2)
\end{aligned}
\tag{8.1}
$$

where

$$\beta_{i1} = \gamma_{i1} - \alpha_1.$$

The random variables γ_{i1} and δ_{ij} are independent, both within and between subjects. What is the variance of Y_{ij} under this model?

$$\text{Var}(Y_{ij}) = \text{Var}(\gamma_{i1} + \delta_{ij}) = \text{Var}(\gamma_{i1}) + \text{Var}(\delta_{ij}) = D + \sigma^2. \qquad (8.2)$$

The covariance between Y_{ij} and Y_{il} simplifies because the various random terms in our model (8.1) are independent.

$$
\begin{aligned}
\text{Cov}(Y_{ij}, Y_{il}) &= \text{Cov}(\gamma_{i1} + \delta_{ij}, \gamma_{i1} + \delta_{il}) \\
&= \text{Cov}(\gamma_{i1}, \gamma_{i1}) + \text{Cov}(\gamma_{i1}, \delta_{il}) \\
&\quad + \text{Cov}(\delta_{ij}, \gamma_{i1}) + \text{Cov}(\delta_{ij}, \delta_{il}) \\
&= \text{Cov}(\gamma_{i1}, \gamma_{i1}) \\
&= \text{Var}(\gamma_{i1}) = D.
\end{aligned}
$$

Under model (8.1), the observations have constant variance $D+\sigma^2$ and constant covariance D. Thus this model is the same as the compound symmetry model. We can relate the τ, ρ parameters from the compound symmetry model to σ^2 and D via

$$\sigma^2 = \tau^2(1 - \rho),$$

and

$$D = \tau^2 \rho.$$

Or, solving for τ^2 and ρ in terms of σ^2 and D we have $\tau^2 = \sigma^2 + D$ and $\rho = D/(D + \sigma^2)$. The parameter ρ is called the *intraclass correlation coefficient*.

8.1.3 Autoregressive

The *autoregressive* or AR(1) covariance model also has two parameters, τ^2 and ρ. The parameter names are the same as for the compound symmetry matrix, and they have similar, but not identical meanings. Like the compound symmetry matrix, the variance of Y_{ij} is τ^2 for all i and j and the model assumes constant variance. In the AR model, the correlation between two observations Y_{ij} and Y_{il} depends on the absolute value of the time between them

$$\text{Corr}(Y_{ij}, Y_{il}) = \rho^{|t_{ij} - t_{il}|}.$$

The farther apart two observations are in time, the lower the correlation between them. In theory ρ can be negative when the data set is balanced, but in longitudinal data practice, negative correlations are rare, and as a routine matter we assume $0 < \rho < 1$.

In our simulated data with observations at $t_i = (1,2,3,4,5)'$, the covariance matrix is

$$\text{Var}(Y_i) = \tau^2 \begin{pmatrix} 1 & \rho & \rho^2 & \rho^3 & \rho^4 \\ \rho & 1 & \rho & \rho^2 & \rho^3 \\ \rho^2 & \rho & 1 & \rho & \rho^2 \\ \rho^3 & \rho^2 & \rho & 1 & \rho \\ \rho^4 & \rho^3 & \rho^2 & \rho & 1 \end{pmatrix}. \tag{8.3}$$

The correlation matrix is inside the parenthesis in (8.3). The lag 1 correlations are all equal to ρ. As we move farther from the main diagonal, the lag increases and the correlations decrease. Observations Y_{i1} and Y_{i5} have correlation ρ^4, the smallest correlation between observations in this example.

Figure 8.3 presents two simulated examples. Parts (a) and (b) have $\rho = .99$, while (c) and (d) have $\rho = .8$. In both sets of data the variances are 1 at all times. The population correlation matrix in 8.3(a) and (b) is

$$\text{Corr}(Y_i) = \begin{pmatrix} 1 & .99 & .9801 & .9703 & .9606 \\ .99 & 1 & .99 & .9801 & .9703 \\ .9801 & .99 & 1 & .99 & .9801 \\ .9703 & .9801 & .99 & 1 & .99 \\ .9606 & .9703 & .9801 & .99 & 1 \end{pmatrix},$$

and in 8.3(c) and (d), the population correlation matrix is

$$\text{Corr}(Y_i) = \begin{pmatrix} 1 & .8 & .64 & .512 & .4096 \\ .8 & 1 & .8 & .64 & .512 \\ .64 & .8 & 1 & .8 & .64 \\ .512 & .64 & .8 & 1 & .8 \\ .4096 & .512 & .64 & .8 & 1 \end{pmatrix}.$$

It is hard to distinguish between the AR(1) and the CS covariance models using a profile plot, particularly when the lag 1 correlation ρ is high. For example, the profiles in figure 8.3(a) are nearly parallel suggesting possibly a compound symmetry model – in parallel with the situation in figure 8.2(a). That there is less crossing of profiles in figure 8.3(a) than in 8.2(a), suggests that the correlations are higher in 8.3(a), but this does not help with the problem of determining the covariance model.

To decide between the AR(1) and CS covariance matrix, it is useful to look at a scatterplot matrix. In figure 8.3(b), we first look at the lag 1 plots $(1,2)$, $(2,3)$ and so on that plot the sequential observations $Y_{i(j+1)}$ against Y_{ij}. We see equal correlations in these figures. Next we look at the most distant of pairings, $(5,1)$, the plot with the largest lag plotting Y_{i5} against Y_{i1}. Here we see that while the correlation is still strong it has decreased from the lag 1 plots. Next we look along the rows and columns of figure 8.3(b). We see decreasing correlation as we get farther from the main diagonal. The decrease in correlation appears steady as we increase

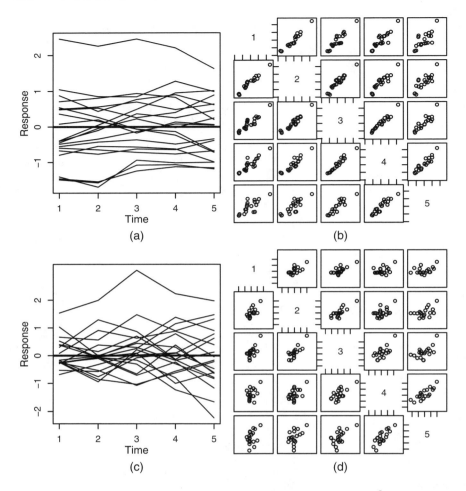

Figure 8.3. Constructed data, autoregressive covariance model, $\tau^2 = 1$. (a,b) $\rho = .99$, (c,d) $\rho = .8$.

lag. This indicates that an AR(1) type covariance matrix is more plausible than a compound symmetry matrix for this data set.

In figure 8.3(c) there is more crossing of the profiles than in 8.3(a), indicating a less strong correlation between consecutive observations. The variance appears constant across time in 8.3(c) as well. Due to sampling variability this is not as obvious as it is in 8.3(a), and learning to draw accurate conclusions about variance and correlation in the presence of sampling variability can be a relatively difficult part of interpreting statistical graphics. Figure 8.3(d) shows that the correlations are much weaker than in 8.3(b), and we see the correlations dropping to small values as we look across from the $(1, 2)$ to the $(1, 5)$ plot.

The AR(1) model can be written in terms of the increments from one observation to the next

$$Y_{ij} - x'_{ij}\alpha = \rho(Y_{i(j-1)} - x'_{i(j-1)}\alpha) + \delta_{ij} \qquad (8.4)$$

where $\delta_{ij} \sim N(0, \tau^2(1 - \rho^2))$. Because the variance of $Y_{i(j-1)}$ is τ^2, then $\rho Y_{i(j-1)}$ has variance $\rho^2 \tau^2$, and so δ_{ij} needs to have variance $(1 - \rho^2)\tau^2$ to make Y_{ij} also have variance τ^2. The first observation has no precursor. To complete the specification (8.4) we include the marginal distribution of Y_{i1}

$$Y_{i1} = x'_{ij}\alpha + \epsilon_{i1} \qquad (8.5)$$

where $\epsilon_{i1} \sim N(0, \tau^2)$. Equations (8.4) and (8.5) are completely equivalent to our original specification $Y_i \sim N(X_i\alpha, \Sigma_i(\theta))$ with $\Sigma_i(\theta)$ given by (8.3).

The difference between an AR(1) model and a compound symmetry model can be important scientifically. The autoregressive model says that the observations $Y_i(t_{ij})$ follow a path that wanders around with no discernible trend, other than a tendency to pull back toward the population mean $x'_{ij}\alpha$. If observations are taken close enough in time, the path will look continuous because of the high correlation $\rho^{|t_{ij}-t_{i(j+1)}|}$ between consecutive observations. If Y_{i1} is above the population mean, Y_{i2} will also tend to be above the mean. Given a long enough sequence $Y_{i1}, Y_{i2}, \ldots, Y_{in_i}$ with t_{in_i} large enough, Y_{ij} will pull back to near and then below the population mean. The population mean is zero in the simulations of this chapter, and is in general it is $x'_{ij}\alpha$. In contrast, the compound symmetry model (8.1) says that observations within a subject vary with variance $\sigma^2 = \tau^2\rho$ around a constant subject-specific mean $E[Y_{ij}|\gamma_{i1}] = \gamma_{i1}$ for all time. If γ_{i1} is fairly large and positive, then all observations Y_{ij} can be above the population mean α_1. Further, if Y_{ij} is above γ_{i1}, that tells us nothing about whether the next observation will be above or below γ_{i1} and the paths (t_{ij}, Y_{ij}) will not look continuous, as observations even close in time vary around γ_{i1} with independent errors.

The autoregressive model (8.4) says that ρ times the previous observation $Y_{i(j-1)}$ is the best predictor of Y_{ij}. The compound symmetry model says that a linear function of the average of all previous observations is the best predictor of Y_{ij}. In a scientific context, the compound symmetry model says that future observations are foreordained, up to measurement error, while the autoregressive model says that the observations will wander along a random path, like a drunk wandering in a park. The autoregressive model suggests that it might be possible to control the process perhaps by an occasional nudge to the drunk. The compound symmetry model suggests that the path is intrinsic to the subject, and therefore immutable, unless we are able to alter the subject.

The AR(1) is but one member of the AR(p) family of covariance models. Assuming equally spaced data, the AR(p) model relates the current observation Y_{ij} to the previous p observations.

$$Y_{ij} - x'_{ij}\alpha = \sum_{k=1}^{p} \rho_k (Y_{i\,j-k} - x'_{i(j-k)}\alpha) + \delta_{ij} \qquad (8.6)$$

Occasionally, we may have use of $p > 1$, in particular $p = 2$. Higher values of p require moderately long time series for each subject to fit the model and distinguish it from other covariance models.

8.1.4   Independence

The independence covariance matrix is the simplest of all covariance matrices and is a special case of all but one of the covariance models discussed in this section. The sole parameter is σ^2, and

$$\mathrm{Var}(Y_i) = \sigma^2 I,$$

with no correlation between observations. Simulated data is exhibited in 8.4(a) and (b). No correlation is visible in the scatterplot matrix and the lines in the profile plots have lots of crossings. This model has constant variance

$$V = (1, 1, 1, 1, 1)',$$

and the population correlation matrix in (a) and (b) is as stolid and boring as a covariance matrix gets

$$\mathrm{Corr}(Y_i) = \begin{pmatrix} 1 & 0 & 0 & 0 & 0 \\ 0 & 1 & 0 & 0 & 0 \\ 0 & 0 & 1 & 0 & 0 \\ 0 & 0 & 0 & 1 & 0 \\ 0 & 0 & 0 & 0 & 1 \end{pmatrix}.$$

The independence matrix is frequently useful as one component of a complex covariance matrix constructed as the sum of simpler covariance matrices. Subsection 8.5 discusses sums of covariance matrices. Independence by itself is usually not an appropriate covariance matrix for longitudinal data, but it is an important special case. Technically, independence does not require constant variance; we could have non-constant variance and independent observations but we use the IND model to mean constant variance as well.

8.1.5   Random Intercept and Slope

In subsections 8.1.1 and 8.1.2, we met the compound symmetry or random intercept model, the simplest *random effect* model with a single random effect and where observations within subject vary around a subject-specific mean γ_{i1}. The *random intercept and slope* (RIAS) model generalizes this model to observations Y_{ij} falling around subject-specific lines $\gamma_{i1} + \gamma_{i2}t_{ij}$ with unknown subject-specific intercepts γ_{i1} and subject-specific slopes γ_{i2}. Observations Y_{ij} for subject i will not fall exactly on the line, rather there is a residual variability σ^2 about the line. We may write

$$Y_{ij} = \gamma_{i1} + \gamma_{i2}t_{ij} + \delta_{ij}$$

with residual

$$\delta_{ij} \sim N(0, \sigma^2).$$

There is a different intercept and slope pair $\gamma_i = (\gamma_{i1}, \gamma_{i2})'$ for each subject. The γ_i, $i = 1, \ldots, n$ form a sample from the population of possible (intercept, slope) pairs. Like the RI model, and in the absence of other information, we assume a normal distribution for the random effects

$$\gamma_i \sim N_2(\alpha, D).$$

This is a bivariate distribution as there are two random effects. The mean $\alpha = (\alpha_1, \alpha_2)'$ of the γ_i's is the population mean of the intercept-slope pairs. The population mean time trend of the observations at time t_{ij} is the line

$$E[Y_{ij}|\alpha] = \alpha_1 + \alpha_2 t_{ij}.$$

In this section, our examples have population mean zero, and α is zero. For regular data, $\alpha \neq \mathbf{0}$ of course. For example, in the Weight Loss data, the population intercept is between 190 and 200 pounds, and the population slope is around -1 pounds per week.

The random intercept and slope model has four covariance parameters. The variance σ^2 is the error variance of the observations Y_{ij} around its subject-specific line $\gamma_{1i} + \gamma_{2i}t_{ij}$. The 2×2 matrix D has three unique parameters D_{11}, $D_{12} = D_{21}$, and D_{22}. The parameter D_{11} is the variance of the intercepts γ_{i1} in the population. Similarly, D_{22} is the population variance of the slopes β_{i2} and D_{12} is the covariance of the slopes and intercepts.

Unconditionally, that is, ignoring the person-specific intercept and slope parameters γ_i, and thinking in terms of the parameters σ^2 and D, the marginal variance of Y_{ij} is

$$\text{Var}(Y_{ij}|\sigma^2, D) = \sigma^2 + D_{11} + 2t_{ij}D_{12} + t_{ij}^2 D_{22}, \qquad (8.7)$$

and the covariance between Y_{ij} and Y_{il} is

$$\text{Cov}(Y_{ij}, Y_{il}) = D_{11} + (t_{ij} + t_{il})D_{12} + t_{ij}t_{il}D_{22}.$$

The marginal variance $\text{Var}(Y_{ij}|\beta_0, \sigma^2, D)$ is quadratic in time. The coefficient of t_{ij}^2 is D_{22}, the population variance of the slopes; the coefficient of $2t_{ij}$ is D_{12}, the population covariance between the intercepts and slopes; and the constant portion of the marginal variance $D_{11} + \sigma^2$ is the sum of the variance of the intercepts plus the residual variance.

If we continue to collect data over a long enough period of time, the marginal variance in this model must explode which is usually implausible. Over shorter periods, there is little problem. In particular, where the times are centered around the time of the minimum of the quadratic $t_{\text{minimum}} = -D_{12}/(D_{22})$, the change in variances across time may be slight. The subject-specific lines approximately cross at a single time point, which is t_{minimum}. If D_{12} is zero, the lines cross at $t = 0$; if D_{12} is positive, the subject-specific lines cross at time less than zero; and if D_{12} is negative, the lines cross at a positive time. If D_{12} is positive, the lines spread out starting from $t = 0$, and if $D_{12} < 0$, then the lines first converge starting from $t = 0$, and then diverge after time $t = -D_{12}/D_{22}$. The covariance between Y_{ij} and Y_{il} has a term $t_{ij}t_{il}D_{22}$ in it; as t_{ij} and t_{il} both increase, the correlation between observations gets quite large. As t_{ij} gets large, the correlation between observations Δ time units apart goes to one.

The RIAS model has an additional two parameters compared with the CS/RI and AR(1) models and the extra parameters give added flexibility to the RIAS model. A variety of different matrices are possible, including some which have negative correlations between observations. We present three examples to illustrate the variety of correlation matrices possible with this model.

Our first example mimics the weight data in certain respects. The parameters are $\sigma^2 = 1$, and

$$D = \begin{pmatrix} 100 & 0 \\ 0 & 2 \end{pmatrix}.$$

This has a very high intercept variance 100, relatively low variance 2 for the slopes, and no correlation between intercepts and slopes. This produces increasing marginal variances that are

$$V = (103, 109, 119, 133, 151)'.$$

The correlation matrix is

$$\text{Corr}(Y_i) = \begin{pmatrix} 1 & .982 & .957 & .923 & .882 \\ .982 & 1 & .983 & .963 & .935 \\ .957 & .983 & 1 & .986 & .970 \\ .923 & .963 & .986 & 1 & .988 \\ .882 & .935 & .970 & .988 & 1 \end{pmatrix}. \tag{8.8}$$

Simulated data is plotted in figures 8.4(c) and (d). Figure 8.4(c) does not allow us to easily see differences in slopes between subjects. Figure 8.5 remedies this by elongating and narrowing the figure as discussed in section 2.4.3. We can now see trends in some of the subjects, particularly those with largest negative or positive slopes. The correlation matrix in (8.8) is approximately *banded*, in that the correlations are similar for constant lags, although they are increasing slightly with increasing time. This correlation matrix is not quite like the perfect banding in the AR(1)

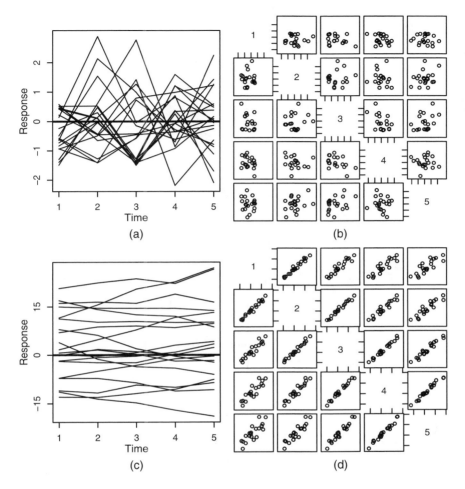

Figure 8.4. Constructed data, independence (a,b) variances 1. (c,d) Random intercept and slope, variance of intercepts is 100, variances of slopes is 2, no correlation between slopes and intercepts.

with its exactly constant correlation within lag. Nor does it quite have the geometric progression with increasing lag that the AR(1) model has. For example, $.98^2 \approx .96$ is approximately the lag two correlations in (8.8), however, $.98^3 \approx .94$ is higher than the lag three correlations of .92 and .93, and $.98^4 = .92$ is much higher than the .88 lag 4 observed in (8.8). Also unlike the AR(1) model, the variances are not constant.

Our second example has parameters $\sigma^2 = .75$ and

$$D = \begin{pmatrix} 1 & 1.2 \\ 1.2 & 2 \end{pmatrix}.$$

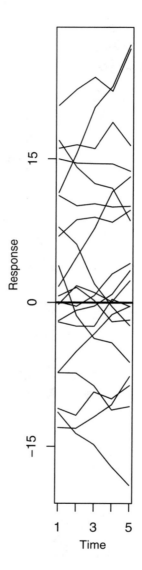

Figure 8.5. Constructed data, random intercept and slope, variance of intercepts is 100, variances of slopes is 2, no correlation between slopes and intercepts. Same data as in figure 8.4(c) but with a better choice of plot shape.

This produces variances that are starkly accelerating over time

$$V = (6.15, 14.55, 26.95, 43.35, 63.75)'.$$

The population correlation matrix is

$$\text{Corr}(Y_i) = \begin{pmatrix} 1 & .909 & .917 & .919 & .919 \\ .909 & 1 & .959 & .964 & .965 \\ .917 & .959 & 1 & .977 & .980 \\ .919 & .964 & .977 & 1 & .985 \\ .919 & .965 & .980 & .985 & 1 \end{pmatrix}.$$

Simulated data are plotted in figures 8.6(a) and (b). The main feature of this example is the positive .85 correlation between slopes and intercept causing the observations to immediately spread out over time. Figure 8.6(a) shows visible non-constant marginal variance over time. As compared with correlation matrix (8.8), we have a rather small population variance for the intercepts. The scatterplot matrix shows the increasing correlation between observations at later times, as compared to between similarly lagged observations at earlier times. Similar to (8.8), all correlations are quite high between all lags up to and including lag 4. Unlike (8.8), we do not have a banded matrix in (8.1.5); rather we have approximately a constant correlation along a row above the main diagonal or within a column below the main diagonal. The correlation between the first observation and the remaining observations is nearly constant.

The scatterplots in the scatterplot matrix 8.6(b) are constructed slightly different from the scatterplots in chapter 2. In chapter 2, each row or column of the scatterplot matrix is scaled so that the data fill up the entire plot. This means that if the data have non-constant variance over time, we do not see the change in variance in the plots. In the scatterplot matrices in this chapter, the plots are scaled so that each axis in each plot in the scatterplot matrix has the same scaling. When the marginal variance increases over time, the data at time t_1 will not fill up the entire range of the data plot while the data at later times will fill up increasing fractions of the axes.

The third example has parameters $\sigma^2 = .25$,

$$D = \begin{pmatrix} 9 & -2.5 \\ -2.5 & 1 \end{pmatrix},$$

This produces non-constant variances that decrease then increase

$$V = (5.25, 3.25, 3.25, 5.25, 9.25)'.$$

The correlation matrix is

$$\text{Corr}(Y_i) = \begin{pmatrix} 1 & .847 & .484 & .095 & -.143 \\ .847 & 1 & .769 & .484 & .274 \\ .484 & .769 & 1 & .847 & .730 \\ .095 & .484 & .847 & 1 & .933 \\ -.143 & .274 & .730 & .933 & 1 \end{pmatrix}.$$

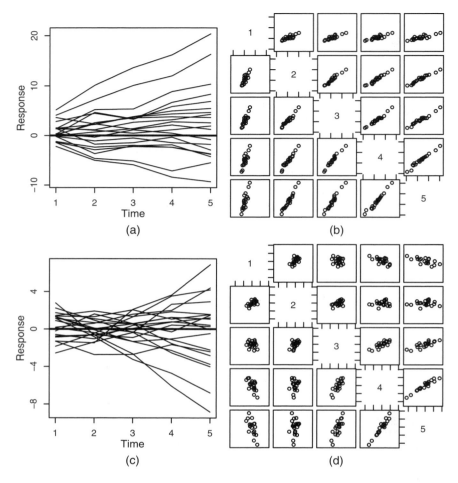

Figure 8.6. Constructed data, $\alpha = 0$, random intercept and slope. (a,b) Positively correlated intercept and slope. (c,d) Negatively correlated intercept and slope.

The key feature of this model is the strong negative correlation of the intercept and slope leading to a crossing pattern in the middle of the data set plotted in figure 8.6(c). Lag 1 observations have fairly high correlations, while lag 3 and 4 pairs have low or even negative correlation.

Occasionally, people will omit the intercept/slope covariance D_{12} from the random effects covariance matrix D, specifying independent intercepts and slopes. There are data sets where this may be a reasonable model. However, with longitudinal data and a RIAS model, it is important not to fix $D_{12} = 0$.

Suppose that we actually have a model where D_{12} happens to be 0. Then the marginal variance of the observations Y at time t_{ij} is

$$\text{Var}(Y) = \sigma^2 + D_{11} + t_{ij}^2 D_{22}. \tag{8.9}$$

The marginal variance is a minimum at time $t = 0$. Change the definition of time to $t_{ij}^* = t_{ij} + \Delta$, so that the new zero point $t^* = 0$ used to be at $t = -\Delta$. Plugging in $t_{ij} = t_{ij}^* - \Delta$ into (8.9) gives

$$\text{Var}(Y) = (\sigma^2 + D_{11} + \Delta^2 D_{22}) - 2\Delta D_{22} t_{ij}^* + t_{ij}^{*2} D_{22}. \tag{8.10}$$

Using the starred timescale t^*, we see in (8.10) that our data now has random intercepts with variance $D_{11} + \Delta^2 D_{22}$, found by plugging $t^* = 0$ into (8.10). Now the marginal variance is a minimum at $t^* = \Delta$; if we continue to force $D_{12} = 0$, then the model will not fit the data, as the model would say that the minimum variance was at $t^* = 0$.

From (8.7), the coefficient of $2t^*$ is the covariance of the intercepts and slopes, and the coefficient of $(t^*)^2$ is the variance of the slopes. The slopes in (8.10) still have variance D_{22}, but we now have a non-zero covariance between the intercepts and slopes of $-\Delta D_{22}$. Shifting the location of the timescale alters the population intercept variance and the population covariance between the intercept and slope. Shifting the location of time should not alter the scientific interpretation of our model, yet it can change the variance of the intercepts and the covariance of the intercepts and slopes. The only parameter of the D matrix that is invariant to the time shift is D_{22}. Essentially, the location shift argument says that if D_{12} is zero or near zero, that is a coincidence based on the choice of zero time for our model. We must include a population covariance parameter in the D matrix even if D_{12} is not significantly different from zero. The time shift argument is the same as the reason why we must include lower order polynomial terms in a p-order polynomial regression of y on x; by carefully changing from x to $x^* = x + \Delta$, we can make coefficients of the lower order terms $x^k, k < p$ be anything we want.

8.1.6   Independent Increments

Independent increments (II) is a *Markovian* model. *Markov* means that given the current observation, past observations are independent of the future observations. In particular, this means that some function of the current observation Y_{ij} is the best prediction of the next observation $Y_{i(j+1)}$. The AR(1) covariance matrix was a previous example of a Markov model. In the independent increments model, given Y_{ij}, the next observation is equal to Y_{ij} plus an additional increment $\delta_{i(j+1)}$. The $\delta_{i(j+1)}$'s have variances σ_{j+1}^2, for $j = 1, \ldots, 4$ and are independent of each other. At time 1, Y_{i1} has variance σ_1^2. The II covariance model has variances that increase with time. The correlation of consecutive observations depends on the values of the σ_j^2's.

The variance at time t_{ij} is

$$\text{Var}(Y_{ij}) = \sum_{k=1}^{j} \sigma_k^2,$$

the covariance between observations at times t_{ij} and t_{il} is

$$\text{Cov}(Y_{ij}, Y_{il}) = \sum_{k=1}^{\min(j,l)} \sigma_k^2.$$

In matrix form the covariance matrix is

$$\text{Cov}(Y_i) =$$

$$\begin{pmatrix} \sigma_1^2 & \sigma_1^2 & \sigma_1^2 & \sigma_1^2 & \sigma_1^2 \\ \sigma_1^2 & \sigma_1^2 + \sigma_2^2 & \sigma_1^2 + \sigma_2^2 & \sigma_1^2 + \sigma_2^2 & \sigma_1^2 + \sigma_2^2 \\ \sigma_1^2 & \sigma_1^2 + \sigma_2^2 & \sigma_1^2 + \sigma_2^2 + \sigma_3^2 & \sigma_1^2 + \sigma_2^2 + \sigma_3^2 & \sigma_1^2 + \sigma_2^2 + \sigma_3^2 \\ \sigma_1^2 & \sigma_1^2 + \sigma_2^2 & \sigma_1^2 + \sigma_2^2 + \sigma_3^2 & \sigma_1^2 + \cdots + \sigma_4^2 & \sigma_1^2 + \cdots + \sigma_4^2 \\ \sigma_1^2 & \sigma_1^2 + \sigma_2^2 & \sigma_1^2 + \sigma_2^2 + \sigma_3^2 & \sigma_1^2 + \cdots + \sigma_4^2 & \sigma_1^2 + \cdots + \sigma_5^2 \end{pmatrix}.$$

Rows and columns emanating right or down from any (j, j) diagonal cell are constant. In the examples, we report the correlation matrices; this model does not have constant correlation within rows, rather correlation decreases along rows or columns emanating from the main diagonal.

Our two examples are displayed in figure 8.7. In (a) and (b), we use parameters $\sigma_j^2 \equiv 1$. This means that the marginal variance $\text{Var}(Y_{ij}$ increases linearly with time j in figure 8.7(a), alternatively, the standard deviation increases as the square root of time. In particular, the variances are

$$V = (1, 2, 3, 4, 5)'.$$

and the correlation matrix is

$$\text{Corr}(Y_i) = \begin{pmatrix} 1 & .707 & .577 & .500 & .447 \\ .707 & 1 & .816 & .707 & .632 \\ .577 & .816 & 1 & .866 & .775 \\ .500 & .707 & .866 & 1 & .894 \\ .447 & .632 & .775 & .894 & 1 \end{pmatrix}.$$

We see a steady decrease in correlations with increasing time lag and a steady increase in lag 1 correlations as time progresses.

For our second example, we have parameters $\sigma_j^2 = j$. Now the variances increase quadratically, and the standard deviations increase approximately linearly with time.

$$V = (1, 3, 6, 10, 15)'.$$

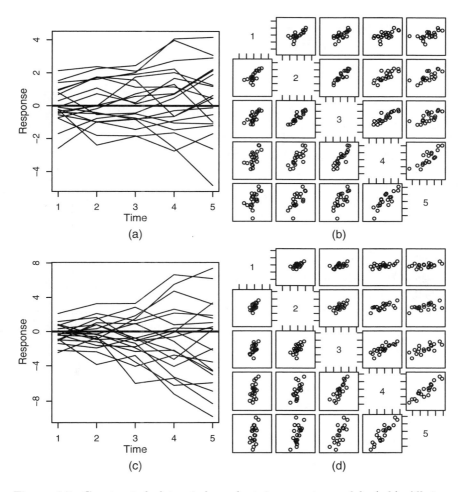

Figure 8.7. Constructed data, independent increments model. (a,b) All incremental variances σ_j^2 equal. (c,d) Linearly increasing incremental variances $\sigma_j^2 = j$.

and the correlations are

$$
\mathrm{Corr}(Y_i) = \begin{pmatrix}
1 & .577 & .408 & .316 & .258 \\
.577 & 1 & .707 & .548 & .447 \\
.408 & .707 & 1 & .775 & .632 \\
.316 & .548 & .775 & 1 & .816 \\
.258 & .447 & .632 & .816 & 1
\end{pmatrix}.
$$

8.1.7   Antedependence

The *antedependence* (ANTE) model is a generalization of the autoregressive model of section 8.1.3. It allows for non-constant lag 1 correlations over time. It has four correlation parameters $\rho = (\rho_1, \rho_2, \rho_3, \rho_4)$ for data with 5 repeated measures. For balanced data with J time points, there are $J - 1$ correlation parameters and a variance parameter σ^2. Each correlation parameter ρ_j is the lag 1 correlation between observations at times t_j and $t_{(j+1)}$. Higher lag correlations are the product of the intervening lag 1 correlations. The entire correlation matrix is

$$\text{Corr}(Y_i) =$$
$$\begin{pmatrix}
1 & \rho_1 & \rho_1\rho_2 & \rho_1\rho_2\rho_3 & \rho_1\rho_2\rho_3\rho_4 \\
\rho_1 & 1 & \rho_2 & \rho_2\rho_3 & \rho_2\rho_3\rho_4 \\
\rho_1\rho_2 & \rho_2 & 1 & \rho_3 & \rho_3\rho_4 \\
\rho_1\rho_2\rho_3 & \rho_2\rho_3 & \rho_3 & 1 & \rho_4 \\
\rho_1\rho_2\rho_3\rho_4 & \rho_2\rho_3\rho_4 & \rho_3\rho_4 & \rho_4 & 1
\end{pmatrix}$$

It should be clear that if the lag 1 correlations $\rho_1 = \rho_2 = \rho_3 = \rho_4$ are all equal, then we are right back at the AR(1) correlation model. An example is given in figures 8.8(a) and (b), with parameters $\rho = (.6, .75, .9, .99)$. We set the marginal variance $\text{Var}(Y_{ij}) \equiv \sigma^2 = 1$ in both of our examples. The correlation matrix is

$$\text{Corr}(Y_i) = \begin{pmatrix}
1 & .6 & .45 & .405 & .401 \\
.6 & 1 & .75 & .675 & .668 \\
.45 & .75 & 1 & .9 & .891 \\
.405 & .675 & .9 & 1 & .99 \\
.401 & .668 & .891 & .99 & 1
\end{pmatrix}.$$

The inverse of this matrix is

$$\begin{pmatrix}
1.562 & -.938 & 0 & 0 & 0 \\
-.938 & 2.848 & -1.714 & 0 & 0 \\
0 & -1.714 & 6.549 & -4.737 & 0 \\
0 & 0 & -4.737 & 54.514 & -49.749 \\
0 & 0 & 0 & -49.749 & 50.251
\end{pmatrix}.$$

which is *tri-diagonal*, meaning that only three diagonals are non-zero. Of course, the two non-zero, non-main diagonals are equal.

The reason for the name *antedependence* is that the parameterization of the inverse (hence ante) correlation matrix is quite simple, albeit perhaps less immediately interpretable than the correlations. In particular, the antedependence matrix we have discussed is called ANTE(1), with the one indicating that only one of the inverse's non-main diagonals is non-zero. ANTE(2) would mean that 2 off diagonals were non-zero, and so on. A quick check for antedependence for a given data set is to calculate the inverse of the empirical correlation or covariance matrix of the data set and check for zero or near zero values at the higher lags of the inverse matrix.

In our second example, we reverse the order of the correlations. The sequential lag 1 correlations are $(.99, .95, .75, .6)$. This correlation matrix shuffles the rows and columns of the previous example in an obvious fashion

$$
\mathrm{Corr}(Y_i) = \begin{pmatrix} 1 & .99 & .891 & .668 & .401 \\ .99 & 1 & .9 & .675 & .405 \\ .891 & .9 & 1 & .75 & .45 \\ .668 & .675 & .75 & 1 & .6 \\ .401 & .405 & .45 & .6 & 1 \end{pmatrix}.
$$

Simulated data are illustrated in figures 8.8(c) and (d).

Many data sets exhibit antedependence correlation matrices. *Growth curve* data starts from low values, grows rapidly up to a maximum, followed by a leveling off when growth stops. Growth curve data may be well modeled with an antedependence correlation matrix. During the era of rapid growth, correlations are lower, but as growth flattens out, changes are minimal, and a previous observation becomes an excellent prediction of the next observation. The antedependence correlation model can model this sort of data quite nicely. Suppose we have balanced, but unequally time-spaced data. If the AR(1) model would fit if we had had equally spaced data, then, if we happen to erroneously define our times with $t_j = j$, we will find that the ANTE model fits our data.

Like the AR(1) model, the ANTE(1) model can be written in terms of the increments from one observation to the next

$$
Y_{ij} - x'_{ij}\alpha = \rho_{j-1}(Y_{i(j-1)} - x'_{ij}\alpha) + \delta_{ij}
$$

for $j > 1$ where $\delta_{ij} \sim N(0, \sigma^2(1 - \rho_{j-1}^2))$. The variance of the increments δ_{ij} changes at each t_j to keep the marginal variance $\mathrm{Var}(Y_{ij})$ constant.

8.1.8 *Autoregressive Moving Average*

The autoregressive moving average or ARMA$(1, 1)$ model is a generalization of both the AR(1) model and the CS/RI model. The ARMA$(1, 1)$ model has three parameters, a variance parameter $\mathrm{Var}(Y_{ij}) = \sigma^2$ and two correlation parameters, γ and ρ. The first correlation parameter γ is the lag one correlation

$$
\mathrm{Corr}(Y_{ij}, Y_{i(j-1)}) = \gamma
$$

while ρ is the additional decrease in correlation for each additional lag. The lag k correlation is

$$
\mathrm{Corr}(Y_{ij}, Y_{i(j-k)}) = \gamma\rho^{k-1}.
$$

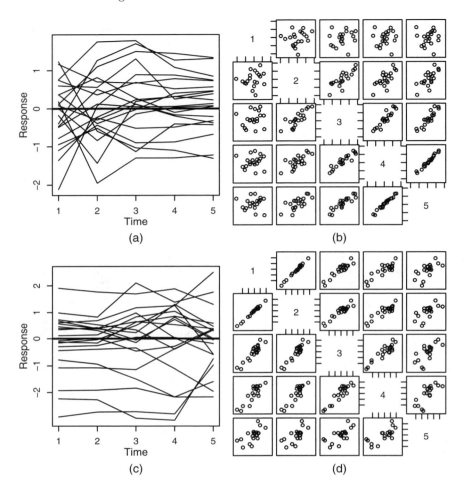

Figure 8.8. Antedependence models. (a,b) Increasing correlations, (c,d) decreasing correlations.

The full covariance matrix is

$$\text{Var}(Y_i) = \sigma^2 \begin{pmatrix} 1 & \gamma & \gamma\rho & \gamma\rho^2 & \gamma\rho^3 \\ \gamma & 1 & \gamma & \gamma\rho & \gamma\rho^2 \\ \gamma\rho & \gamma & 1 & \gamma & \gamma\rho \\ \gamma\rho^2 & \gamma\rho & \gamma & 1 & \gamma \\ \gamma\rho^3 & \gamma\rho^2 & \gamma\rho & \gamma & 1 \end{pmatrix}. \tag{8.11}$$

We give two numerical examples. The first has $\gamma = .65$ and $\rho = .95$, an example motivated by psychological scale data. The theoretical correlation matrix is

$$\text{Corr}(Y_i) = \begin{pmatrix} 1 & .65 & .6175 & .587 & .557 \\ .65 & 1 & .65 & .6175 & .587 \\ .6175 & .65 & 1 & .65 & .6175 \\ .587 & .6175 & .65 & 1 & .65 \\ .557 & .587 & .6175 & .65 & 1 \end{pmatrix}.$$

The initial drop-off in correlation from one observation to the next is high, but further drop-off at higher lags is small. In this example, one might mistake this for a compound symmetry model if one was not careful. The data is plotted in figures 8.9(a) and (b). A second example with $\gamma = .9$ and $\rho = .78$ is given in figures 8.9(c) and (d) with correlation matrix

$$\text{Corr}(Y_i) = \begin{pmatrix} 1 & .9 & .702 & .548 & .427 \\ .9 & 1 & .9 & .702 & .548 \\ .702 & .9 & 1 & .9 & .702 \\ .548 & .702 & .9 & 1 & .9 \\ .427 & .548 & .702 & .9 & 1 \end{pmatrix}.$$

There are at least three interesting special cases of the ARMA$(1,1)$ model.

1. $\rho = 1$ is the compound symmetry model,

2. $\gamma = \rho$ is the autoregressive model, and

3. $\rho = 0$ is the *moving average* or MA model.

The moving average has covariance matrix

$$\text{Var}(Y_i) = \sigma^2 \begin{pmatrix} 1 & \gamma & 0 & 0 & 0 \\ \gamma & 1 & \gamma & 0 & 0 \\ 0 & \gamma & 1 & \gamma & 0 \\ 0 & 0 & \gamma & 1 & \gamma \\ 0 & 0 & 0 & \gamma & 1 \end{pmatrix}. \tag{8.12}$$

Only the lag 1 correlations are non-zero. One way of representing the moving average model is

$$Y_{ij} = \beta \delta_{i\,j-1} + \delta_{ij} \tag{8.13}$$

where the δ_{ij} are a sequence of independent errors with

$$\text{Var}(\delta_{ij}) = \tau^2$$

Then $\text{Var}(Y_{ij}) = \tau^2(1 + \beta^2)$ and in (8.12), $\gamma = \beta/(1 + \beta^2)$ and $\sigma^2 = \tau^2(1 + \beta^2)$. Each δ_{ij} contributes to exactly two observations at times t_{ij} and $t_{i\,j+1}$.

The ARMA$(1,1)$ model is also created from the δ_{ij} by also including an AR(1) process on the right-hand side of (8.13)

$$Y_{ij} = \rho Y_{i\,j-1} + \beta \delta_{i\,j-1} + \delta_{ij}. \tag{8.14}$$

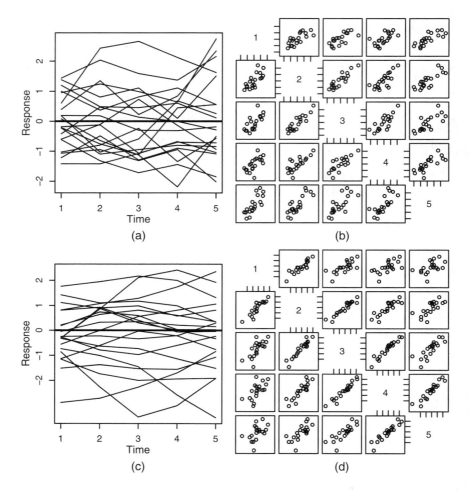

Figure 8.9. Autoregressive moving average example data. (a,b) $\gamma = .65$, $\rho = .95$, (c,d) $\gamma = .9$, $\rho = .78$.

If $\beta = 0$, we have an AR(1) process with correlation ρ, and $\mathrm{Var}(Y_{ij}) = \tau^2(1 + 2\rho\beta + \beta^2)/(1 - rho^2) \equiv \sigma^2$. In general, we can solve for γ and ρ of (8.11 in terms of ρ and β of (8.14). Fortunately, $\rho = \rho$ in both places, and for the record, we note

$$\gamma = \rho + \frac{\beta(1 - \rho^2)}{1 + 2\rho\beta + \beta^2}.$$

The ARMA(1, 1) models can be generalized. The ARMA(p, q) model is often fit to long time series. Like the AR(p) generalizing the AR(1), the ARMA(p, q) generalizes the ARMA(1, 1) by having the current observation depend on the previous p observations, the MA part has the previous q δ_{ij}s also contributing

$$Y_{ij} = \sum_{k=1}^{p} \rho_k Y_{ij-k} + \sum_{l=1}^{q} \beta_l \delta_{ij-l} + \delta_{ij}. \qquad (8.15)$$

8.1.9 Factor Analytic

Factor analytic, or FA covariance models are similar to the random intercept and random intercept and slope covariance models. FA is a generalization of RI and RIAS models, and those models are special cases of the FA model. In the RI model, we have one random effect, in the RIAS model, we have two. In the FA model, we may have one or two or p random effects and these are called the FA(1), the FA(2), and the FA(p) model. The difference between the RI or RIAS models and the FA(p) model is that in the FA(p) models, we must estimate the random effect patterns also. Each random effect pattern is a curve. No longer are observations Y_{ij} required to fall on a flat line (RI) or a straight line (RIAS). Now observations may fall around a curve (FA(1)), or around a linear combination of 2 curves (FA(2)) or a linear combination of p curves (FA(p)). The FA model patterns are parameterized by a vector of unknown parameters. In contrast, in the RI or RIAS models, the shapes of the curves are pre-specified (i.e., flat or lines of unknown slope and intercept) as part of the model.

The random intercept model (8.1) may be rewritten as

$$Y_{ij} = \alpha_1 + D^{1/2}\beta_{i1} + \delta_{ij}, \qquad (8.16)$$

along with the modified distributional assumption

$$\beta_{i1} \sim N(0,1)$$

and $\delta_{ij} \sim N(0,\sigma^2)$. In (8.16), we have made the random effect β_{i1} have known variance one, and instead we multiply it in the linear predictor (8.16) by unknown parameter $D^{1/2}$. The model is unchanged, as the variance of $D^{1/2}\beta_i$ is D. The FA(1) generalizes (8.16) by

$$
\begin{aligned}
Y_{ij} &= \gamma_{1j}\beta_{i1} + \delta_{ij} & (8.17)\\
\beta_{i1} &\sim N(0,1)\\
\delta_{ij} &\sim N(0,\sigma^2).
\end{aligned}
$$

The factor $D^{1/2}$, which was constant for different j, has been replaced by a vector of unknown parameters γ_{1j}, $j = 1 \ldots, J$. This specification assumes that the data are balanced, or balanced with missing data. If we had allowed β_{i1} to have variance D in (8.17), the variance of $\gamma_{1j}\beta_{i1}$ is $\gamma_{1j}^2 D$ and we could double the γ_{1j} and divide D by 4 and kept the same model. The D is an extra, undetermined parameter, so we fixed it equal to 1 to determine the model.

The covariance of Y_i in the FA(1) model is

$$\tau^2 \begin{pmatrix} \sigma^2 + \gamma_{11}^2 & \gamma_{12}\gamma_{11} & \gamma_{13}\gamma_{11} & \gamma_{14}\gamma_{11} & \gamma_{15}\gamma_{11} \\ \gamma_{11}\gamma_{12} & \sigma^2 + \gamma_{12}^2 & \gamma_{13}\gamma_{12} & \gamma_{14}\gamma_{12} & \gamma_{15}\gamma_{12} \\ \gamma_{11}\gamma_{13} & \gamma_{12}\gamma_{13} & \sigma^2 + \gamma_{13}^2 & \gamma_{14}\gamma_{13} & \gamma_{15}\gamma_{13} \\ \gamma_{11}\gamma_{14} & \gamma_{12}\gamma_{14} & \gamma_{13}\gamma_{14} & \sigma^2 + \gamma_{14}^2 & \gamma_{15}\gamma_{14} \\ \gamma_{11}\gamma_{15} & \gamma_{12}\gamma_{15} & \gamma_{13}\gamma_{15} & \gamma_{14}\gamma_{15} & \sigma^2 + \gamma_{15}^2 \end{pmatrix}.$$

If the γ_{1j} and γ_{1l} are large for given j and l, then Y_{ij} and Y_{il} will have high covariance. If a γ_{1j} is zero, then Y_{ij} will be uncorrelated with other responses. If γ_{1j} is negative while γ_{1l} is positive, then Y_{ij} and Y_{il} will be negatively correlated.

In our first example, we set $\gamma = (\gamma_{11}, \ldots, \gamma_{15}) = (0, 0, 3, 3, 3)'$. This produces marginal variances of

$$V = (1, 1, 10, 10, 10)'$$

where $10 = 1 + 3^2$, and a correlation model that is

$$\mathrm{Corr}(Y_i) = \begin{pmatrix} 1 & 0 & 0 & 0 & 0 \\ 0 & 1 & 0 & 0 & 0 \\ 0 & 0 & 1 & .9 & .9 \\ 0 & 0 & .9 & 1 & .9 \\ 0 & 0 & .9 & .9 & 1 \end{pmatrix}.$$

The correlation between the first two observations and any of the last three is zero, while the correlations are constant and high among the last three observations. Sample data is presented in figures 8.10(a) and (b).

In our second example, we set $\gamma = (0, 1, 3, 5, 3)'$. This produces profiles that look like a constant β_i times the vector $(0, 1, 3, 5, 3)'$, rising or falling steeply to time $t_{ij} = 4$, then changing direction and falling or rising slightly respectively, from time 4 to 5. Figure 8.10(c) illustrates data that looks something like an eggbeater. The marginal variances are

$$V = (1, 2, 10, 26, 10)'$$

and the correlation model is

$$\mathrm{Corr}(Y_i) = \begin{pmatrix} 1 & 0 & 0 & 0 & 0 \\ 0 & 1 & .67 & .69 & .67 \\ 0 & .67 & 1 & .93 & .90 \\ 0 & .69 & .93 & 1 & .93 \\ 0 & .67 & .90 & .93 & 1 \end{pmatrix}.$$

Because $\gamma_{13} = \gamma_{15}$, Y_{i3} and Y_{i5} have the same correlations with other observations.

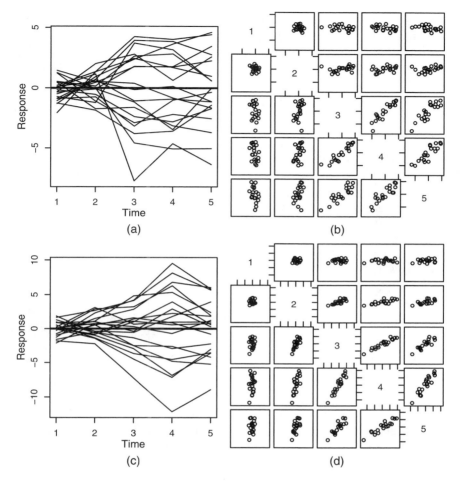

Figure 8.10. Factor analytic covariance example data. (a,b) $\gamma = (0, 0, 3, 3, 3)$ and $\sigma^2 = 1$. (c,d) $\gamma = (0, 1, 3, 5, 3)$ and $\sigma^2 = 1$.

The FA(2) model has two factors γ_{kj} and two random effects β_{i1} and β_{i2}.

$$
\begin{aligned}
Y_{ij} &= \gamma_{1j}\beta_{i1} + \gamma_{2j}\beta_{i2} + \delta_{ij} & (8.18) \\
\beta_{ki} &\sim N(0, 1) \\
\delta_{ij} &\sim N(0, \sigma^2).
\end{aligned}
$$

This model apparently has 11 parameters when $J = 5$, 10 for the γ_{kj} and one for σ^2. The random effects β_{ik} do not count when tabulating parameters in this model or in the random intercept or RIAS models. All of the β_{ik}, $k = 1, 2$, and the δ_{ij} are independent. The β_{i1} and the β_{i2} are independent because if they were correlated, several different sets of parameters give the same covariance model. Having redundant parameters is unnecessary and makes for problems when trying to estimate parameters. Second, there needs to be a constraint on the γ_{kj}. Suppose for example that $\gamma_{1j} = \gamma_{2j}$, then we have the FA(1) model again. An *orthogonality* constraint that is often used is

$$\sum_j \gamma_{1j}\gamma_{2j} = 0$$

guarantees that the γ_{kj} are well defined and leaves the model with 10 parameters. This constraint requires the vector of γ_{1j}s be *orthogonal* to the second vector of γ_{2j}s.

Factor analysis as a model for analyzing general multivariate data arose in psychometrics and factor analysis has its own language for the different parameters. The γ_{kj} are called the *factor loadings*, the random effects β_{ik} are called the *common factors*, σ^2 is the *unique variance* or *residual variance*, and the $\sum_k \gamma_{kj}^2$ are called the *communalities* or the *communal variances*.

We can write the factor analytic covariance model nicely in matrix form

$$\Sigma = \sigma^2 I + \sum_{k=1}^{K} \gamma_k \gamma_k'$$

where γ_k is the vector with elements γ_{kj}, $j = 1, \ldots, J$.

8.1.10   Toeplitz or Banded Covariance Model

The *Toeplitz*, or *banded* covariance model has constant covariance down the diagonals of the covariance matrix. It has constant variance, constant lag 1 covariance and correlation. The lag 2 correlation is different from the lag 1 correlation, but all lag 2 correlations are the same. We can write the general banded covariance model for longitudinal data with 5 time points in terms of the variance and the various lag k correlations ρ_k

$$\mathrm{Var}(Y_i | \tau^2, \rho_1, \rho_2, \rho_3, \rho_4) = \tau^2 \begin{pmatrix} 1 & \rho_1 & \rho_2 & \rho_3 & \rho_4 \\ \rho_1 & 1 & \rho_1 & \rho_2 & \rho_3 \\ \rho_2 & \rho_1 & 1 & \rho_1 & \rho_2 \\ \rho_3 & \rho_2 & \rho_1 & 1 & \rho_1 \\ \rho_4 & \rho_3 & \rho_2 & \rho_1 & 1 \end{pmatrix}.$$

Banded covariance model is mostly appropriate for equally spaced data. It would be odd if the difference in times $t_{i2} - t_{i1} \neq t_{i3} - t_{i2}$ were different, yet the correlations $\mathrm{Corr}(Y_{i2}, Y_{i1}) = \mathrm{Corr}(Y_{i3}, Y_{i2})$ were equal. The ARMA(1, 1) and AR(1) correlation models are special cases of this covariance model, and we do not illustrate a separate set of numerical examples for the banded covariance model.

8.1.11   Unstructured Covariance Model

The *unstructured covariance* (UN) matrix Σ is the most general covariance matrix possible. This model has $J(J+1)/2$ parameters when the data have J repeated measures. This is a large number of parameters, and they can be hard to estimate when the number of observations n is small relative to $J(J+1)/2$. Using the unstructured covariance matrix may lead to problems in estimation, as there can be too little data to estimate α and Σ accurately.

All of the examples in this section are special cases of the unstructured covariance matrix and we do not present additional specific examples here.

8.2   Non-constant Variance

Longitudinal data often exhibit *heterogeneity* or *non-constant variance*. The Big Mice data (figure 2.3) exhibit their lowest marginal variance at days 0, 1, and 2. As time increases, the individual mice grow and the variability between mice grows along with the mean weight. The variability of the Ozone data is low early in the day and rises to its highest variability in the early afternoon. The variance decreases slightly at the end of the day. The Pediatric Pain data in figure 2.7 showed skewness. Skewness is often associated with non-constant variance. Subjects with low average $\bar{Y}_i = m_i^{-1}\sum_{j=1}^{m_i} Y_{ij}$ tend to have lower within-person variability, whereas those with higher $\bar{Y}_i$ have higher within-person variability. Figure 2.9 showed that the within-subject standard deviations increase with the subject averages for the Pain data.

When our data exhibit skewness or non-constant variance, there are several modeling approaches we may take. We may (a) ourselves choose a transformation of the data, (b) let the data estimate the transformation, or we can (c) directly model the changing variance. We discussed the log transformation for the Pain data in chapter 2 and we briefly mentioned estimating transformations in section 6.6. When there is non-constant variance, but the data is symmetric around the mean, then transformation is not indicated and we will need to model the variance.

8.2.1   Non-constant Variance Over Time

When the non-constant variance is a function of time, we can model the changing variance in several ways. Some covariance models naturally incorporate non-constant variance. The unstructured covariance model, the random intercept and slope model, factor analytic and the independent increments covariance models all allow for changing marginal variances across time.

Our constant variance covariance models CS,, AR, IND, ANTE, and ARMA can be generalized to allow non-constant variance across time. In the unequal variance versions of these models, we keep the correlation matrix R equal to that from the equal variance model. The variances we allow to vary. For example, we allow the variance at time t_j to be σ_{jj}. The jkth element of Σ is the covariance term

$$\sigma_{jk} = \sigma_{jj}^{1/2}\sigma_{kk}^{1/2}r_{jk}$$

where r_{jk} is the jkth correlation element of matrix R. This defines non-constant variance across time with the same correlation matrix as before. The above formula can be compactly written in matrix notation. Let T be the diagonal matrix of standard deviations with diagonal elements $\sigma_{jj}^{1/2}$, and zeros off the diagonal. Then

$$\Sigma = TRT. \tag{8.19}$$

For the CS, AR, IND, ANTE, and ARMA models, the correlations ρ or ρ_j and the variance parameters σ^2 are distinct parameters and the decomposition (8.19) allows us to replace the original single variance parameter σ^2 with J variance parameters σ_{jj}^2. We abbreviate these models as CSH or ECH, ARH, INDH, ANTEH and ARMAH.

8.2.2 Constant Variance with Unstructured Correlations

We can reverse this generalization with the unstructured covariance matrix to produce a constant variance model with unstructured correlation model. What we did in the text after equation (8.19) for the constant variance models, we can also undo for non-constant variance models. The only covariance model we really wish to do this to is the unstructured covariance matrix. We define the constant variance unstructured correlation matrix as

$$\Sigma = \sigma^2 R$$

where R is as an unstructured correlation matrix. We use the abbreviation UNC for this covariance model. This model has $1 + J(J-1)/2$ parameters.

8.2.3 Non-constant Variance Component

RI, RIAS, and FA have an IND component with constant variance that can be generalized to have non-constant variance. For example, FA(1) naturally has non-constant variance $\sigma^2 + \gamma_{1j}^2$. However, we may generalize the σ^2 component to have non-constant variance. We replace $\sigma^2 I$ by a diagonal matrix $T = \operatorname{diag}(\sigma_{11}^2, \ldots, \sigma_{JJ}^2)$. In fact, in multivariate analysis, this is perhaps the most common form of the FA(p) model. We abbreviate this as the FAH(p) model. As with the FA(p) model, there may be more than 1 factor. The RI and RIAS model also have a $\sigma^2 I$ independence component and all could have a non-constant diagonal component $\operatorname{diag}(\sigma_{11}^2, \ldots, \sigma_{JJ}^2)$.

| Model | Cov | # | -2 Log | | | LRT vs. | |
#	model	parms	REML	AIC	BIC	IND	UN
7	ARH(1)	8	990.9	1006.9	1012.0	172.22	*26.2
12	ANTEH	13	983.1	1009.1	1017.4	180.00	*18.4
13	FAH(2)	19	977.1	1015.1	1027.2	185.98	*12.4
14	UN	28	964.7	1020.7	1038.6	198.41	–
2	AR(1)	2	1035.3	1039.3	1040.5	127.83	70.6
5	ARMA	3	1032.9	1038.9	1040.8	130.18	68.2
10	FA(2)	14	1004.0	1032.0	1041.0	159.05	39.3
8	FA(1)	8	1020.4	1036.4	1041.5	142.69	55.7
3	RS	2	1041.0	1045.0	1046.3	122.10	76.3
6	RIAS	4	1038.6	1046.6	1049.2	124.48	73.9
11	FAH(1)	14	1016.8	1044.8	1053.7	146.33	52.1
9	CSH	8	1050.7	1066.7	1071.8	112.43	86
4	RI	2	1105.3	1109.3	1110.6	57.8	140.6
1	IND	1	1163.1	1165.1	1165.7	–	198.4

Table 8.1. Results from fitting various covariance models to the Small Mice data set. Fixed effects include the intercept, linear, and quadratic time terms. Models are ordered by BIC from best to worst. Model # is based on the number of parameters, with ties ordered by BIC. RS stands for the random slope only model. Other covariance model names are standard. In a reversal of usual practice, the "*" identifies the three models that the UN model is *not* significantly better than. All models are significantly better than the IND model. AIC and BIC are in *smaller is better* form.

8.3   Examples

8.3.1   Small Mice Data

We illustrate fitting several covariance models with the Small Mice data. All models use the quadratic in day mean model that we identified in table 5.2. We found this quadratic form assuming a random slope covariance matrix. Now we consider whether this was the best choice for covariance matrix or not.

Table 8.1 shows results of model fitting for 14 covariance models. Rows are ordered by BIC from best to worst. Omissions are due to lack of software, inability of available software to fit some models to this particular data set, and lack of analyst imagination. The model number orders models by the number of covariance parameters. Columns include name, number of parameters, -2 times log REML likelihood, AIC (smaller is better), and BIC (smaller is better). The last two columns give the difference in log REML likelihoods between the given model and the independence covariance model and the given model and the unstructured covariance model. All covariance models are better than the independent model; even with 27 degrees of freedom, the .999 quantile of a chi-square is approximately

Time	ARH	ANTEH	FAH(2)	UN
2	32	30	30	30
5	55	48	49	47
8	67	58	58	59
11	103	101	101	103
14	124	132	133	134
17	125	134	134	134
20	111	119	121	120
# parms	8	13	19	28

Table 8.2. Estimated marginal standard deviations for the Small Mice data using four covariance models, ARH(1), ANTEH(1), FAH(2), and UN.

55, and every chi-square statistic comparing covariance models to the null model is significant.

Comparing models to the unstructured model gives more interesting results. The unstructured model is significantly better than most of the covariance models including the random slope model. The exceptions are three models with non-constant variance, ARH(1), ANTEH(1), and FAH(2). For those three models, the p-value comparing them to the unstructured model is around .15 to .20, indicating no need for the more complicated model.

BIC also selects these three models as the best models. Using BIC to compare just the top two, ARH(1) and ANTEH(1), we see that the posterior probability of ARH(1) is $1/[1+\exp(-5.4)] = .9955$; including any other model in the comparison is moot, the data strongly prefer the ARH(1) covariance model. We can also do the likelihood ratio test between ARH(1) and ANTEH(1). The difference is a chi-square statistic of 7.8 on 5 degrees of freedom, which we recognize as being so far from significant that we need not calculate a p-value. Similarly, AIC also prefers ARH(1). AIC orders models similarly to BIC, although there are a few inversions between models with many parameters such as the factor analytic models and models with few parameters, such as AR(1) and ARMA(1, 1). Since none of these models fit at all well, the reordering is not of great importance.

Our chosen model is the ARH(1) model. Now we would take the ARH(1) model, and, if there were more covariates, or other issues regarding the fixed effects, we would further explore different fixed effects models. The random slope model is not in competition for the best model. However, we see that there were a number of models much worse than the random slope. Surprisingly, it is not better than the AR(1) model, which we would not have considered due to the non-constant variance in the data; at least the random intercept model allowed for heterogenous and increasing marginal variances.

On occasion we may care about the exact differences among otherwise similar fitting models. We illustrate using the Small Mice data and the four

	2	5	8	11	14	17	20
ARH(1)							
2	1	.90	.80	.72	.65	.58	.52
5		1	.90	.80	.72	.65	.58
8			1	.90	.80	.72	.65
11				1	.90	.80	.72
14					1	.90	.80
17						1	.90
20							1
ANTEH							
2	1	.92	.69	.60	.56	.52	.48
5		1	.75	.65	.60	.57	.52
8			1	.87	.81	.76	.69
11				1	.93	.87	.80
14					1	.94	.86
17						1	.92
20							1
FAH(2)							
2	1	.93	.58	.31	.21	.22	.31
5		1	.74	.50	.40	.41	.49
8			1	.83	.80	.79	.81
11				1	.93	.91	.88
14					1	.94	.90
17						1	.88
20							1
UN							
2	1	.92	.50	.30	.18	.23	.35
5		1	.73	.51	.42	.42	.53
8			1	.87	.81	.75	.81
11				1	.93	.90	.87
14					1	.94	.89
17						1	.92
20							1

Table 8.3. Estimated correlation matrices for the Small Mice data using the four best covariance models.

best model fits. We concentrate first on the covariance model because this is the subject of the current chapter.

Table 8.2 presents the fitted marginal standard deviations for the 7 days for the four best models. The three more complex models are more flexible than the ARH model in their correlation model; they have quite similar standard deviations differing by only 1 or 2 units in the ones place; the three larger models are slightly different from the ARH variances, which appear higher at the earlier times and lower at the later times.

| | Covariance model | | | |
Parameter	ARH	ANTE	FAH(2)	UN
Intercept	78.8	81.3	84.7	89.6
(SE)	(6.9)	(6.1)	(5.7)	(4.6)
Day	67.2	66.4	67.1	64.2
(SE)	(3.1)	(2.4)	(2.3)	(2.3)
t	22	28	30	28
Day×day	-1.18	-1.18	-1.26	-1.11
(SE)	(.14)	(.12)	(.09)	(.09)
t	-8	-10	-15	-12

Table 8.4. Parameter estimates for the Small Mice data using the four best covariance models.

Table 8.3 presents the fitted correlations for the same four models. The ARH(1) correlation model is the simplest, and is not overly different from ANTEH(1). The FAH and UN differ slightly from ARH(1) and ANTE(1) in that the correlations between the first two days and the last three are lower than in the ARH(1) and ANTEH(1) models. This matches what we saw in the scatterplot matrix 2.1.

We are usually more interested in the mean model than the correlation model; we present parameter estimates in table 8.4 for the four models. The differences among the coefficients of day and day×day are small, roughly 1 unit of the corresponding standard error. The quadratic term is highly significant in all models. The difference among the intercepts is larger, almost 2 units of the standard errors. Are these differences important? Inspection the fitted means from each of the top four models at times from day 2 to day 20 shows that the four sets of fitted values differ most at day 20, with a range of approximately 26 milligrams. Plotting the four different mean functions shows that the differences are not large. See problem number 8.

The ARH(1) model is the best model by BIC, AIC, and traditional p-value methods, so our conclusion is to use the ARH(1) model here. Still, it is interesting to compare the four best models and see how they differ, and how they are similar. Tables 8.2, 8.3, and 8.4 illustrate the differences among the models, and we see that changes on the order of one standard deviation of parameters is quite common from different model specifications. It gives us pause in placing too much faith in the minute numeric details of any one analysis! In situations where several models are in quite close competition using BIC or AIC, we might well construct these and other tables to illustrate sensitivity of inferences to assumptions.

8.3.2 Pain Data

We try out a number of different covariance models for the Pain data. Table 8.5 presents a summary of the results. BIC picks the random intercept

Model	Cov	#	-2 Log			LRT vs.	
#	model	parm	REML	AIC	BIC	IND	UN
2	RI	2	408.2	412.2	416.5	142.9	23.1
4	ARMA$(1,1)$	3	406.1	412.1	418.6	145	21
5	RIAS	4	408.0	416.0	424.6	143.1	22.9
6	FA(1)	5	403.9	413.9	424.7	147.2	18.8
10	UN	10	385.1	405.1	426.7	166	–
9	FA(2)	8	396.7	412.7	429.9	154.4	11.6
8	FAH(1)	8	396.8	412.8	430.0	154.3	11.7
3	AR(1)	2	428.9	432.9	437.2	122.2	43.8
7	ANTEH(1)	7	426.3	440.3	455.4	124.8	41.2
1	IND	1	551.1	553.1	555.3	–	166

Table 8.5. Results from fitting several covariance models to the Pain data set. Fixed effects include coping style, treatment, and treatment by coping style interaction. Models are ordered by BIC from best to worst. Model # is based on the number of parameters, with ties ordered by BIC. All models are significantly better than the IND model, and all models are significantly worse than the UN model at the .01 level. AIC and BIC are in "smaller is better" form.

model slightly ahead of the ARMA$(1,1)$ model. Considering just the two models gives a posterior probability of RI being correct at around 90% with 10% probability accorded to ARMA$(1,1)$. In this example, the different criteria that we have select two different models. AIC, with its propensity toward larger models prefers the unstructured covariance model. Formal p-value testing also rejects RI in favor of the unstructured model, with a chi-square statistic of 23.1 and a p-value of .003.

The estimated correlation matrix from the unstructured covariance model is given in table 8.6. The range of the correlations .55 to .82 is a little large for us to comfortably conclude that the equal correlation model might be a reasonable model. The estimated variances at trials one through four are .48, .54, .64, and .52. The standard deviations on these estimates are approximately .1, suggesting that it would be of interest to fit an equal variance, unstructured correlation model.

We compare the fitted parameter values for the RI and UN models in table 8.7. We use a parameterization that sets the intercept to be the distracter baseline mean; the next parameter is the baseline difference attenders minus distracters. The last six parameters are the treatment effects A, D, and N first for the distracters then for the attenders. The table gives parameter estimates, standard errors, t-statistics, and two-sided p-values. Comparing the results, we see some interesting differences and similarities. The single most astonishing result is that the standard errors of the parameters are identical in both models, to the accuracy presented. In contrast, the parameter estimates themselves vary moderately. For the baseline effects, the differences are approximately .03 or .04. For the treatment effects,

Trial	Trial			
	1	2	3	4
1	1	.70	.82	.55
2		1	.71	.67
3			1	.73
4				1

Table 8.6. Estimated correlation matrix for the Pain data set from the unstructured correlation matrix.

differences are larger, as much as .1 units. In all cases, these differences are much less than 1 standard error.

Unfortunately, the differences do matter in terms of statistical significance, if we insist on declaring significance only for p-values exactly less than .05. The baseline difference is significant, $p = .04$ for the UN model, but barely not, $p = .0502$ for the RI model. For the DD treatment, the p-value is approximately .001 for the UN model, but only .006 for the RI model. The DN treatment effect is about the same in both models. Partly the problem is that p-values are continuous; .04 is not really importantly different from .06, except when we make a big deal about values being less than or greater than .05.

8.4   Non-constant Variance and Covariates

There is a more general covariance matrix than the unstructured covariance model! Consider a dichotomous baseline covariate G_i taking values zero or one. This may be used to define two classes of observations, each of which could have its own covariance matrix Σ_{G_i}. Now we have $J(J + 1)$ parameters, as each group requires its own Σ matrix. There is nothing special about two groups; there can be g groups and g different parameter sets each of size $J(J + 1)/2$. The number of parameters required escalates with the number of groups; a lot of data in each group is required to fit these models!

There is no requirement that we fit the unstructured covariance model to each group. We may use any parameterized covariance model that we desire in each group. The usual situation is to use the same covariance model with different parameters in each group. With g groups, and q covariance parameters, we will end up with gq total covariance parameters. Common choices might be the AR(1) and CS, with different parameters for each group. In principle we can even have a separate covariance model in each group. In practice we probably would not without large amounts of data and substantial scientific evidence for different covariance relationships among observations within groups.

UN model

Parameter		Est	SE	t	p
Baseline	D	3.51	.12		
	A-D	-.36	.17	-2.16	.035
Treatment	DA	-.14	.15	-0.92	.36
	DD	.49	.14	3.49	.0009
	DN	-.35	.15	-2.32	.024
	AA	.10	.15	.66	.51
	AD	.07	.15	.45	.66
	AN	-.02	.15	.12	.90

RI model

Parameter		Est	SE	t	p
Baseline	D	3.47	.12		
	A-D	-.33	.17	-2.00	.0502
Treatment	DA	-.24	.15	-1.66	.10
	DD	.40	.14	2.85	.0061
	DN	-.33	.15	-2.24	.029
	AA	.075	.15	.51	.61
	AD	.027	.15	.18	.85
	AN	-.075	.15	-.51	.61

Table 8.7. Parameter estimates for the Pain data set using the unstructured covariance model (upper) and the random intercept (lower). Parameterization uses distracter baseline as the intercept, A-D is the difference attenders minus distracters, and the last six effects are the treatment effects A, D, and N for the distracters and attenders. Columns are parameter names, parameter estimates, standard errors (SE), t-statistics, and two-sided p-values.

We may also set up regression models for covariance parameters. Most commonly, we want to do this for the variance σ^2. Consider the situation where we have increasing marginal variance over time. Rather than fitting a different variance at each time t_j, we can use a smooth function of time to model the variance

$$\log \sigma^2(t) = \gamma_1 + \gamma_2 t. \qquad (8.20)$$

We take a log of the variance before modeling because $\log \sigma^2$ takes values in minus infinity to infinity, and we need not put any restrictions on the parameters γ_1 and γ_2. If we did not take log, then we would need to restrict $\gamma_1 + \gamma_2 t$ to be greater than zero, which causes problems with maximum likelihood, both theory and practice. The log function is called the *link* function.

It can be that model (8.20) is not complex enough to properly model the variance. We can try a quadratic function of time $\log \sigma^2(t) = \gamma_1 + \gamma_2 t + \gamma_3 t^2$, and even higher order polynomials. With balanced data and J time points, a polynomial of order $J-1$ is the same model as we discussed in section 8.2 with a different σ_j^2 at each time t_j. Similarly, we may set up a categorical covariate $G_{ij} = j$ with J levels and use G_{ij} as a predictor of $\log \sigma^2$, again giving us back this same model. The covariate G_i may distinguish different times within a subject or it may distinguish different subjects or it may distinguish a combination.

In general, we can use any vector of covariates x_{ij} to model the variance σ^2

$$\log \sigma_{ij}^2 = x_{ij}'\gamma.$$

with parameters γ to be estimated. While it is possible, it is not required that we model σ^2 with the same covariates that we use to model the mean $x_{ij}'\alpha$.

In principle, other variance parameters may be modeled as functions of covariates. If ρ is a correlation parameter, then we may model it as a function of time-fixed covariates x_i using a logit link function

$$\text{logit}\rho = x_i'\phi$$

where ϕ is a vector of unknown parameters. This model restricts ρ to be between 0 and 1, which is the common case. To allow general correlation, we may set

$$\text{logit}[.5(\rho + 1)] = x_i'\phi$$

In all these various models, we test whether the coefficients are zero or not using t-tests equal to the estimate divided by its standard error or using a likelihood ratio test or by using AIC or BIC.

8.4.1 Pain Data Example

In analyzing the Pain data, we have often noticed differences between attenders and distracters, and we might well wonder whether these differences carry over to the covariance model also. We allow the covariance parameters to be different for attenders and distracters.

In section 8.3.2, we screened a number of covariance models assuming a single covariance model for all subjects. We determined that either the unstructured or the random intercept model would be best. Rather than reconsider all possible covariance models, we restrict attention to these two covariance models, with separate parameters for the attenders and the distracters. While it is entirely possible that a completely different model is appropriate once we fit the two groups separately, we hope, and we usually expect, that the information we gained from fitting the two groups together is generally pertinent when we expand our model to fitting the two groups separately.

Table 8.8 summarizes results. The first section of the table presents the results from fitting the entire data set as if it were being fit as a single data set with two parameter sets, one for attenders and one for distracters. RI-2 and UN-2 have RI and UN covariance models with separate covariance

M #	Cov	q	-2 Log REML	AIC	BIC	LRT vs. Single	LRT vs. UN
Two-parameter set							
11	RI-2	4	391.7	399.7	408.4	16.5, 2, .0003	30.5, 16, .02
12	UN-2	20	361.2	401.2	444.4	23.9, 10, .008	–
One-parameter set							
2	RI	2	408.2	412.2	416.5	–	23.1, 8, .003
10	UN	10	385.1	405.1	426.7	–	–
Attenders							
21	RI	2	151.3	155.3	158.2	–	16.5, 8, .04
22	UN	10	134.8	154.8	169.5	–	–
Distracters							
31	RI	2	240.5	244.5	247.4	–	14.1, 8, .08
32	UN	10	226.4	246.4	261.0	–	–

Table 8.8. Results from fitting various covariance models to the Pain data set. Fixed effects include coping style, treatment, and treatment by coping style interaction. M# is the model number; Cov is the covariance model; and q is the number of covariance parameters. REML is -2 log REML likelihood; AIC and BIC are in "smaller is better" form. The first likelihood ratio test (LRT) (χ^2 statistic, df, and p-value) compares the single covariance model to the separate coping style covariance models. The last column (χ^2 statistic, df, and p-value) compares the RI model to the UN covariance model. The first section is for the entire data set with difference covariance parameters for attenders and distracters. The second section fits a single covariance model for all subjects. The third and fourth sections present results for the attenders and distracters separately.

parameters for attenders and distracters. We continue to model the fixed effects with the coping style by treatment interaction. The second section reprises information from table 8.5 on fitting with a single covariance model and parameters for both groups. The third and fourth sections give results for attenders and distracters separately. We see that AIC and the -2 log REML likelihoods add between sections 3 and 4 to give the results of section 1; that is, the attender AIC plus distracter AIC equals the combined AIC, and similarly for -2 log REML. The calculation for BIC is not additive; the direct cause is the term $q \log n$, the number we plug in for n depends on whether we think of the data set as one unified whole $n = 64$ or two separate data sets with $n = 32$.

We see in table 8.8 that AIC and BIC both prefer the models with separate parameters for the attenders and distracters over the single parameter set; both RI-2 and UN-2 are significantly better than a single set of RI or UN covariance parameters for both attenders and distracters. This holds for the attenders and distracters analyzed separately as well. We can directly test model RI-2 as the null hypothesis with UN-2 as the alternative. The UN-2 model is significantly better with $p = .02$ at the .05 level, but the significance is much less than before. We see separately in the attenders and distracters that the UN model is barely significant (attenders, $p = .04$)

	Cov parm	Est	SE
Attenders			
	σ^2	.10	.02
	D	.30	.08
Distracters			
	σ^2	.23	.04
	D	.46	.13

Table 8.9. Parameter estimates for the RI model for attenders and distracters.

Group	Trial	Est	SE
Attenders			
	1	.38	.10
	2	.48	.13
	3	.40	.10
	4	.36	.10
Distracters			
	1	.59	.15
	2	.61	.16
	3	.88	.23
	4	.69	.18

Table 8.10. Variances for attenders and distracters from the unstructured covariance model.

or not (distracters $p = .08$). If before we preferred the RI model, we now have more reason to prefer it here, but again there is a choice to be made; even if one preferred UN before, one might well switch here. I prefer RI-2 to UN-2; partly due to BIC, and partly due to my preference for parsimony; the story attached to the random intercept model is easier to tell to other researchers. In spite of this preference, here, for this data set, I do fully expect that with enough subjects we would eventually prefer the UN-2 to the RI-2.

The distracter group shows less significance when comparing RI to UN than the attender group. This should not be too surprising, as they have higher variance than the distracters. Table 8.9 gives the variance parameters for the attenders and distracters from the RI model; attender variances are all smaller than for the distracters. The σ^2 variance is double in the distracters over the attenders, and D is 50% larger in the distracters than the attenders. Similarly, table 8.10 presents the fitted variances from the UN model for attenders and distracters. The average variance for the distracters is 70% larger than for the attenders.

We can convert the parameters from the RI (σ^2, D) parameterization to the (τ^2, ρ) CS parameters. This gives an attender correlation of .75 and a distracter correlation of .67. The marginal variance is $\tau^2 = D + \sigma^2$ is .4 for the attenders and a much higher .69 for the distracters.

Attenders

Trial	1	2	3	4
1	1	.84	.86	.61
2		1	.78	.71
3			1	.70
4				1

Distracters

	1	2	3	4
1	1	.61	.80	.52
2		1	.68	.66
3			1	.75
4				1

Table 8.11. Correlations from the unstructured covariance model for attenders and distracters.

Table 8.11 gives the fitted correlation matrices for attenders and distracters. As with the RI model, attenders have slightly higher correlations; as in the RI model, it is not overwhelming.

Table 8.12 presents the results for the fixed effects assuming different parameters in the two coping style groups, for the unstructured and random intercept models. The results are not greatly different, again unless one puts unusual faith in the magic .05 cutoff for significance for the baseline effect. The baseline significance level is .04 for the UN-2 model and .05 for the RI-2 model. Similarly, the DD treatment effect is significant at $p = .005$ for UN-2 and at $p = .02$ for the RI-2 model. There are slight differences from the one covariance parameter fits. The distracter treatment effects have higher standard errors than the attender treatment effects, and the the null counseling intervention given to distracters effect is not quite significant now.

8.5   Sums of Covariance Matrices

The variety of covariance models presented so far are not enough to model some data sets. One way to add flexibility to the covariance models is to take two familiar models and add them together. For example, we might have

$$Y_i = X_i \alpha + \epsilon_i + \delta_i$$

where

$$\epsilon_i \sim N(0, \Sigma_1(\theta_1)),$$

and

$$\delta_i \sim N(0, \Sigma_2(\theta_2)),$$

UN model

Parameter		Est	SE	t	p
Baseline	D	3.53	.12		
	A-D	-.35	.16	-2.16	.035
Treatment	DA	-.14	.16	-0.86	.39
	DD	.48	.16	3.05	.005
	DN	-.33	.17	-1.97	.057
	AA	.10	.13	.83	.41
	AD	.07	.13	.59	.56
	AN	.00	.13	.00	.99

RI model

Parameter		Est	SE	t	p
Baseline	D	3.47	.13		
	A-D	-.33	.17	-1.99	.0507
Treatment	DA	-.24	.17	-1.41	.16
	DD	.40	.16	2.45	.018
	DN	-.34	.18	-1.93	.059
	AA	.073	.11	.64	.53
	AD	.030	.11	.284	.79
	AN	-.074	.11	-.64	.53

Table 8.12. Parameter estimates for the Pain data set using the unstructured covariance matrix (upper) and the random intercept (lower) with different parameter sets for attenders and distracters. Parameterization uses distracter baseline as the intercept, A-D is the difference attenders minus distracters, and the last six effects are the treatment effects A, D, and N for the distracters and attenders. Columns are parameter names, parameter estimates, standard errors (SE), t-statistics, and two-sided p-values.

are two independent variance terms. The model for the covariance of Y_i is

$$\mathrm{Var}(Y_i) = \mathrm{Var}(\epsilon_i + \delta_i) = \Sigma_1(\theta_1) + \Sigma_2(\theta_2).$$

The RI model can be thought of as the sum of two covariance model components. One is the intercept term β_i, and the other is the δ_i measurement error component. Similarly, the RIAS and the FA models also have a measurement error component δ_i and they also have an additional random component; for RIAS it is a line with random slope and intercept. With a little algebra we can make the RIAS have three independent components; in our usual formulation there are also three components but the intercept and slope terms are not independent, and in this section we wish to restrict ourselves to independent components.

These complex covariance models are usually used by researchers with strong prior opinions about the covariance matrix for their data. Actually figuring out from plots of the data alone that one should add a random intercept and slope and (say) an antedependence matrix is not easy. Typically we would have some thought that makes us wish to expand our covariance model. We would then entertain the sum of the two models and would compare the combined model to the two individual models separately to see if the additional complexity was necessary and then we would compare it to the unstructured model to see if we needed still additional flexibility in our model.

8.5.1 Autoregressive Plus Random Intercept Plus Independent Error

A popular covariance model combines the random intercept and AR(1) model. Let the parameters be $\theta_1 = (\tau_1^2, \rho_1)$ for the RI model and $\theta_2 = (\tau_2^2, \rho_2)$ for the AR(1) covariance model. The variance at time j is then

$$\text{Var}(Y_{ij}) = \tau_1^2 + \tau_2^2,$$

the covariance is

$$\text{Cov}(Y_{ij}, Y_{ik}) = \tau_1^2 \rho_1 + \tau_2^2 \rho_2^{|k-j|},$$

and the corresponding correlation is

$$
\begin{aligned}
\text{Corr}(Y_{ij}, Y_{ik}) &= \frac{\tau_1^2 \rho_1 + \tau_2^2 \rho_2^{|k-j|}}{\tau_1^2 + \tau_2^2} \\
&= \rho_1 \frac{\tau_1^2}{\tau_1^2 + \tau_2^2} + \rho_2^{|k-j|} \frac{\tau_2^2}{\tau_1^2 + \tau_2^2}.
\end{aligned} \tag{8.21}
$$

At a lag of 1, the correlation is a linear combination $w_1\rho_1 + w_2\rho_2$ where $0 < w_1 = \tau_1^2/(\tau_1^2 + \tau^2) < 1$ and $w_1 + w_2 = 1$. As the lag increases, the correlation decreases smoothly to $w_1\rho_1$.

Homework problem 19 discusses further aspects of this model.

8.5.2 Local Effect

A useful and common example of combining two covariance models into one occurs when we add additional independent noise $\tau_2^2 I$ to an already defined covariance model.

$$\text{Var}(Y_i) = \tau_2^2 I + \tau_1^2 R(\rho), \tag{8.22}$$

where $R(\rho)$ is a correlation matrix that is a function of parameters ρ, such as that given by the AR(1) model (8.3) or the ANTE model. This independent error $\sigma^2 I$ is called the *local* effect. In spatial data analysis, the $\tau_2^2 I$ component is called a *nugget* effect. It is the extra variability of an observation beyond that expected if the covariance between $Y_i(t)$ (an observation at time t) and $Y_i(t + \Delta)$ (an observation at time $t + \Delta$) were calculated and extrapolated as the time between them Δ goes to zero. In (8.3),

$$\text{Cov}[Y_i(t_0), Y_i(t_0 + \Delta)] = \tau_1^2 \rho^\Delta.$$

As Δ goes to zero, this converges to $\mathrm{Var}[Y_i(t_0)]$. If $\mathrm{Var}(Y_i)$ is as in (8.22) with the local effect, then the variance $\mathrm{Var}[Y_i(t_0)] = \tau_2^2 + \tau_1^2$, but the covariance still converges to τ_1^2, and τ_2^2 is the extra local variance. The AR(1) covariance model says that if we were to take two observations at the exact same time t_0, then those two observations would be identical. This may be moot when we never take two observations at identical or nearly identical times, but if we can and we do, then it can be a major reason for an AR model not fitting the data.

When we add the local effect $\tau_2^2 I$ to the AR(1) model it decreases the lag one correlation to

$$\gamma = \rho \frac{\tau_1^2}{\tau_1^2 + \tau_2^2}$$

and lag k correlations are

$$\rho^k \frac{\tau_1^2}{\tau_1^2 + \tau_2^2}$$

The AR(1) plus local effect is identical to the ARMA(1, 1) covariance model. The ARMA variance parameter σ^2 is equal to the sum of the two variances $\tau_1^2 + \tau_2^2$ in the AR plus local effect model, and the ARMA γ parameter is defined above. The parameter ρ is the same in both models.

8.6   Parameterized Covariance Matrices and Random Times

All of the parameterized covariance models of section 8.1 extend in principle to random times.

8.6.1   Models Easily Extensible to Random Times

The first four models of section 8.1, CS/RI/EC, IND, AR, and RIAS are readily available in software; the others are not. A key characteristic of these five models is that the extension to random times requires no new parameters and no new model specification.

In the compound symmetry model, all variances and covariances are functions of the two parameters, τ^2 and ρ. The time or times of the observations are not involved; whether observations are balanced or not is irrelevant. For the AR(1) model, the correlation between two observations at times t_j and t_l is

$$\rho^{|t_j - t_l|}.$$

Both the CS and AR(1) models with constant variance can handle even wildly unbalanced data. The RIAS model covariance model can also be written down in closed form no matter the times; this has long been programmed into software and many programs can fit RIAS to data observed at random times.

8.6.2 Extensions to Random Times Requiring Additional Specifications

This section uses additional mathematics and may be omitted.

In the previous subsection, we required no new parameters to define the variance and covariances of data observed at random times. In this section, we discuss extensions that require additional or different parameters. Usually we must replace a discrete set of parameters, for example the variances $\text{Var}[Y_{ij}] = \sigma_j^2$ with a continuous function, namely $\text{Var}[Y(t)] = \sigma^2(t)$. Although often there are no restrictions on these functions, typically we would want the functions to be continuous functions of time. To make these models practical, we would also have to create parametric functions that depend on only a few parameters and then estimate those parameters. This last extension we do not discuss. Because the software for these models is not generally available, these models may be moot. In discussing extensions of these models to random time data, our discussion will be more mathematical than in most other parts of the book.

For independent increments, the discrete parameters σ_j^2 are replaced by a positive function of time $\psi^2(t_{ij}) > 0$. The variance of an observation at time t_{ij} is now an integral

$$\text{Var}(Y_{ij}) = \sigma^2(t) = \int_0^{t_{ij}} \psi^2(t)dt.$$

The covariance between two observations with $j < l$ is

$$\text{Cov}(Y_{ij}, Y_{il}) = \sigma^2(t_{ij}).$$

For the ANTE and ANTEH correlation models, we must replace the ρ_j parameters with a function $\phi(t_1, t_3)$ that tells us the correlation between two observations at times t_1 and t_3. To keep to the Markov nature of these models, the function must satisfy

$$\phi(t_1, t_3) = \phi(t_1, t_2)\phi(t_2, t_3)$$

for any time t_2 satisfying $t_1 < t_2 < t_3$. Let us insist on $0 < \phi(t_1, t_3) < 1$. One way to do this is to let $\psi(t) < 0$ be a function that is always negative. Then define

$$\phi(t_1, t_3) = \exp\left(\int_{t_1}^{t_3} \psi(t)dt\right)$$

For the CSH, ARH(1), and ANTEH(1) models, the variances σ_j^2 are replaced by a function $\sigma^2(t) > 0$ that defines the variance as a function of time.

Fitting these models to data with random times is currently problematic because they are not commonly available in software yet.

8.6.3 Spatial Correlation Models

Another class of correlation models are available for for unbalanced time data. These are called *spatial correlation* models. Essentially, we specify the correlation function as a continuous function of the absolute difference in times between two observations $|t_{ij} - t_{il}|$. One example is the AR(1) model

$$\text{Corr}(Y_{ij}, Y_{il}) = \rho^{|t_{ij} - t_{il}|}.$$

The power AR (PAR) model is a one-parameter generalization of the AR(1) model

$$\text{Corr}(Y_{ij}, Y_{il}) = \rho^{|t_{ij} - t_{il}|^{\gamma}}.$$

This model has three important special cases. When $\gamma = 1$, we have the AR(1) again. When $\gamma = 0$, we have the compound symmetry model. Now assume balanced equal-spaced data and that the smallest non-zero value of $|t_{ij} - t_{il}|$ is one. As γ gets very large, the PAR correlation model converges to the moving average model! This model is not the same as the ARMA(1, 1) model, yet it is close, in that both have the same three special cases and the same number of parameters. When γ is less than one, $|t_{ij} - t_{il}|^{\gamma}$ is closer to 1 than $|t_{ij} - t_{il}|$ by itself, and the model is in between the AR(1) and the CS model. When $\gamma > 1$, the times are made larger, and the correlations at lags greater than zero converge to zero more rapidly than the AR(1) model.

A third model is the normal kernel (NK) correlation

$$\text{Corr}(Y_{ij}, Y_{il}) = \exp(-c_1 |t_{ij} - t_{il}|^2)$$

which has one parameter $c_1 > 0$. A generalization, (GNK) replaces the power 2 with a second constant $c_2 > 0$.

8.7 Family Relationships Among Covariance Models

There are a number of situations among our covariance models where simpler models are special cases of the more complex models. The models fall into two main families: the random effects family and the autoregressive/moving average family. We include the IND, UNC, UN models as the smallest and largest covariance models in both families. Table 8.13 identifies the nesting relationships for the random effects family, which includes the RI, RIAS, and FA models. The RI is nested in all the other models.

M	Name	# parms	1	2	3	4	5	6	7	8	9	10	11
			Model number										
1	IND	1	—	*	*	*	*	*	*	*	*	*	*
2	EC/CS/RI	2		—	*	*	*	*	*	*	*	*	*
3	RIAS	4			—	*		*		*			*
4	RIASAQ	7				—							*
5	CSH	$J+1$					—			*			*
6	FA(1)	$J+1$						—	*	*	*		*
7	FA(2)	$2J$							—		*		*
8	FAH(1)	$2J$								—	*		*
9	FAH(2)	$3J-2$									—		*
10	UNC	$.5J(J-1)+1$										—	*
11	UN	$.5J(J+1)$											—

Table 8.13. Family relationships among random effects models. First three columns are model number, name, and the number of parameters in the model. Asterisks * in the remaining columns indicate that the row model is nested in the column model. RIASAQ is the random intercept, slope, and quadratic model.

M	Name	# parms	1	2	3	4	5	6	7	8	9	10	11	12
			Model number											
1	IND	1	—	*	*	*	*	*	*	*	*	*	*	*
2	AR(1)	2		—	*			*	*	*	*	*	*	*
3	AR(p)	$p+1$			—				*				*	*
4	MA(1)	2				—	*	*	*				*	*
5	MA(q)	$q+1$					—		*				*	*
6	ARMA(1,1)	3						—	*				*	*
7	ARMA(p,q)	$p+q+1$							—				*	*
8	ARH(1)	$J+1$								—		*		*
9	ANTE(1)	J									—	*	*	*
10	ANTEH(1)	$2J-1$										—		*
11	UNC	$.5J(J-1)+1$											—	*
12	UN	$.5J(J+1)$												—

Table 8.14. Family relationships among random effects models. First three columns are model number, name, and the number of parameters in the model. Asterisks * in the remaining columns indicate that the row model is nested in the column model.

Table 8.14 shows the nesting relationships among the AR(p), MA(p), ARMA(p, q), and the ANTE models. The UNC model fits more comfortably here than in the random effects where the random effects models cause ubiquitous heterogeneity of variance. The banded covariance model is a special case of UNC and a generalization of AR(1) and ARMA(1, 1).

8.8   Reasons to Model the Covariance Matrix

In the chapter introduction, we mentioned that it is often not possible to use an unstructured covariance matrix in modeling longitudinal data. There are other reasons for wanting to model the covariance matrix as a function of θ.

8.8.1   The Science

In many problems, uncovering a model for the variances and correlations is an important part of the science. Consider the modeling of CD4 blood cell counts in subjects suffering from the human immunodeficiency virus (HIV). HIV attacks the immune system; as the immune system is compromised, CD4 counts decrease. When CD4 counts become low enough, people are subject to numerous diseases that if untreated, lead to a quick death. A model that says that log CD4 counts fall on or about a subject-specific straight line with a negative slope implies that the eventual end is foreordained. In contrast, a model that says that CD4 counts follow an AR(1) model says that they follow a random walk pattern, sometimes up, sometimes down, and this suggests that short term or even long term recoveries might be possible, even when the general population trend is down.

8.8.2   Estimating Fixed Effects

In many analyses, we are primarily interested in estimating the fixed effects α. The covariance parameters θ are termed *nuisance parameters*, meaning that while the covariance parameters are necessary to properly specify the model, they are not the parameters we are interested in estimating or drawing conclusions about. Doing a good job in choosing the covariance model improves the *efficiency* of the fixed effect estimates and will allow for more accurate confidence intervals and hypothesis tests. A good job in modeling the covariance matrix is as discussed early in this chapter: selecting a covariance matrix model that is close to, or includes the actual covariance matrix as a special case, or possibly is slightly too large, but not way too large. Getting the covariance matrix wrong, meaning that the true covariance matrix is not in or near the family of covariance matrices we are using, may cause our confidence intervals to be incorrectly wide or narrow and our tests to not have proper level and to potentially reduce power.

8.8.3 Prediction of Individual Future Observations

Given one or a few observations from a subject, we may wish to predict their future observations. It is the covariance matrix that lets us make predictions. For predictions to be accurate, the covariance matrix should be correctly modeled.

8.8.4 Prediction of Missing Observations

We usually have *missing data* in longitudinal data sets – subsets of observations not observed on one or another subject. It is data that we intended to collect but did not. Reasons can be many: lost data forms or corrupted data files; an inability to find or follow subjects in our study; study participants may skip or delay data collection meetings; death of the patient or termination of the study. There is sometimes a need to *fill in* the missing values in a data set. For example, some researchers are in the business of supplying data sets to other research groups for analysis, and a balanced complete data set is much easier for other researchers to analyze than an unbalanced one. To fill in or *impute* the missing values well, it is again necessary to have a good model of the covariance matrix between the observed and the unobserved observations.

8.8.5 Variability

Finally, for statisticians, looking at and thinking about variation and co-variation is good plain fun. Our interest in, and comfort with variability is why some of us become statisticians.

8.9 Historical and Future Notes

Not all models we have discussed may be available in software. But even if they are not, it is important to know what we ought to do, and software will eventually catch up to theory. ANTE originally came in non-constant variance form, you may only find software that allows for ANTEH and not the ANTE form. The various model extensions that we sketched for random times are currently not available in software.

In the future, we expect that additional models will be implemented that allow the covariance model to vary as a function of covariates. Models that allow correlations between observations to vary as a function of covariates have just begun to be explored in detail and are not yet available in software. You can also expect to see models that allow the covariance model to be unstructured but close to a specified parameterized covariance matrix.

8.10 Problems

1. For the random intercept and slope (RIAS) model, find the time t where the marginal variance of the Y_{ij} is 10 times the variance of the minimum variance.

2. In the independent increments model, show that the correlation between Y_{ij} and $Y_{i(j+k)}$ in the independent increments model is decreasing as k increases. What condition on the variances of the increments causes the correlation to go to zero?

3. In the AR(1) model and 8.4, show that the variance is constant with time, assuming $Y_{i1} \sim N(x'_{i1}\alpha, \sigma^2)$.

4. Use the Dental data for the next three problems. For the fixed effects, use the model with intercept, gender, time, and time×gender.

   (a) Fit all the covariance models that you can to this data. This should include at least 10 different covariance models. Be sure to include the independence and the unstructured covariance models.

      i. Create a table summarizing the results of the fitting. Include covariance name, number of parameters, log likelihood, AIC, and BIC. Do a likelihood ratio test of each model against the independence model and against the unstructured covariance model.

      ii. Which covariance model(s) fit(s) best? Discuss.

   (b) Draw an inference plot for the Dental data. Using the covariance model you identified as best in problem 4a, estimate the average response and its standard error for the boys and the girls at each time point and plot them over time.

      Explain the difference between this plot and the similar one you drew in chapter 2 problem number 26e.

   (c) Write a summary paragraph of your Dental data analysis results for the doctor who collected this data. Discuss your covariate modeling from problem 43 in chapter 2 and the covariance modeling from this time. Give your conclusions and also give an interpretation for the doctor of your choices for the fixed effects and the covariance model.

5. Discuss more general models for the Dental data.

(a) For the fixed effects, are there any additional models that could be considered for this data? Discuss. You do not need to fit them yet, if they exist.

(b) What is the most general fixed effect model that can be fit to this data?

(c) Fit the most general fixed effect model using your choice of covariance model.

(d) Is this a better model than the one from problem 4a? Report a formal test and also your opinion.

(e) For the covariance model, are there any models that are more general than the ones you have already fit to this data? Discuss. You do not need to fit them.

(f) What is the most general covariance model that can be fit to this data?

6. Consider using a stronger transformation for the Pediatric Pain data. In particular, consider the negative inverse square root $-\text{time}^{-.5}$ transformation.

(a) Why do we take the negative in this transformation?

(b) Draw plots of the data. Is the skewness improved over the log transformation? Is there non-constant variance as a function of coping style?

(c) What covariance model now best fits the data? Is it the same as before?

(d) Use your chosen covariance model and present results for the fixed effects. Do you get the same qualitative results as before?

(e) Back-transform the results to provide predictions on the original seconds scale for the different groups.

7. Use the Small Mice data for this problem.

(a) Using the ARH(1) covariance model, explore to see if a more complex mean model is indicated. In particular, check on whether the unstructured mean model is better than the quadratic model.

(b) Using the best model that you found in the previous step, refit all the correlation models to see which one fits best.

(c) Create a table of your results. Calculate the fitted correlation model and see how it compares with the results presented above.

(d) Does your fitted mean function differ substantially (meaning significantly and also meaning practically) from the quadratic mean?

8. More Small Mice data.

(a) Draw plots of the fitted values from the quadratic day model for the four best Small Mice covariance models. Draw them on one plot; the purpose is to investigate how different the four sets of fitted values are.

(b) Where are the biggest differences?

(c) What happens if you extend your plot out to day=0 or to days greater than 20? Display the plot and describe how the curves differ.

(d) Because the differences are hard to inspect, one way to expand the y axis is to subtract some standard value from the fitted values. Take the average fitted response across the four models at each day and subtract that from the four sets of fitted means, and plot the differences.

(e) Now describe your graph. How do the different quadratics differ? Are the differences of substantive import? Are they of statistical import?

9. In table 8.3, which covariance model gives estimates closest to the raw data correlations?

10. Use the Big Mice data for this problem.

(a) Pick an initial covariance model that you can fit to this data that does a reasonable job of describing the raw data. Explain your reasoning.

(b) Develop a mean function that describes the time trend of the mice data. Explain the path through model space that you take and your reasoning. Produce a table describing the various models.

(c) Now do a thorough job exploring various covariance models for this data. Produce a table describing the fits of the various models. Briefly explain your reasoning behind various inclusions or exclusions.

(d) Draw an inference plot with the fitted values and error bars showing plus or minus 2 standard errors.

(e) Draw a prediction plot showing the fitted values for plus or minus 1 sample standard deviation at each time.

(f) Are you happy with your final model?

11. Practice identifying covariance models. Write a program that picks a covariance matrix model at random and picks parameters at random for a fixed J. Generate 20 cases and have the software display a profile plot and a scatterplot matrix, and have it output the empirical covariance and correlation matrices. Your goal is to guess the actual covariance model and, for a real challenge, to estimate the parameters as well.

12. For the Dental data, determine if boys and girls follow the same or different covariance models and what the parameter values are.

13. Drug and grocery stores often have machines that will measure your blood pressure and heart rate. Use the RI model $Y_{ij} = \alpha_1 + \beta_i + \delta_{ij}$ for this problem.

   (a) Measure your systolic blood pressure 4 times. From these numbers, which of α_1, $\alpha_1 + \beta_i$, σ^2, D or σ^2/D can you estimate? What are your estimates for those that you can estimate?

   (b) Make a profile plot of your observations. On the vertical axis put the blood pressure, on the horizontal axis, put the order of the observations 1, 2, 3, and 4. Connect the dots with line segments.

   (c) Have every one in your class or other group measure their systolic blood pressure 4 times. Now which of α_1, $\alpha_1 + \beta_i$, σ^2, D, and σ^2/D can you estimate? Make simple estimates of these quantities from inspecting plots of the data.

   (d) Make a profile plot of everyone's observations. Can you explain how changes in μ, σ, and $D^{1/2}$ would affect the nature of this plot? In particular, explain

      i. If μ were higher (lower) than it is in your data, how would the plot change?

      ii. If σ were larger (smaller), how would the plot change?

      iii. If D were larger (smaller), how would the plot change?

14. Repeat problem 13 for

   (a) Diastolic blood pressure.

   (b) Heart rate.

15. Take your measurements on four consecutive days in problems 13 or 14 rather than at a single visit. Which parameters μ, σ^2, D should stay approximately the same as for problem 13 and which should change? How should they change? Why?

16. Take your measurements on four consecutive weeks in problems 13 or 14 rather than at a single visit. Which parameters μ, σ^2, D should stay approximately the same as for problem 13 or problem 15 and which should change? How should they change? Why? Is there a continuous change in any parameter from 4 consecutive measures, 4 daily measures to 4 weekly measures?

17. Would repeating problem 13 for height or weight be at all interesting? How about daily measures? Weekly? Monthly?

18. Use data from problems 13, 14 or 15. Identify the best fitting covariance model to this data. Is there an increasing or decreasing trend in the data? Would you expect there to be one?

19. The AR(1) plus RI covariance model can be thought of as having three error terms

$$Y_{ij} = x'_{ij}\alpha + \epsilon_{ij} + \beta_i + \delta_{ij}$$

where $\epsilon_i = (\epsilon_{ij})$ has the AR(1) covariance model with variance and correlation parameters (τ^2, ρ), β_i is a random intercept with variance D and δ_{ij} has variance σ^2, and ϵ_i, β_i, and δ_{ij} are all independent.

(a) Derive the variances, covariances, and correlations of Y_i for this model.

(b) Show that your formulae are the same as in subsection 8.5.1.

(c) There are several interesting special cases of this model.

   i. If the AR(1) correlation parameter is zero, show that we have the RI model back again. What are the parameters?

   ii. Suppose that $\sigma^2 = 0$; what is the model now? Is it like any other model discussed in this chapter?

   iii. Suppose that $D = 0$. What model is this now? What is the difference between this model and the previous model?

20. Show that the generalized normal kernel is the same as the power AR model.

21. Can the spatial correlation models be used if the data is balanced? What would the advantages and disadvantages be?

22. An exploration of what happens when software has trouble fitting a particular model to a data set. Use the Small Mice data set, and for the fixed effects use an intercept and a linear time trend. Do NOT use any other fixed effects. Fit the data using the top 5 covariance models from table 8.1 and also the random slope (RS) model. For each model, calculate the fitted values at the times where we have data, and plot the fitted values for each model and also plot the means of the observations at each time.

(a) Which of the fitted lines actually goes near the data points?

(b) Check the output of the software, does it give some warning that the output may not be valid?

Note:

It is dangerous to state that a particular software package does not fit a particular model and data set correctly. First, this may even be true for a particular software version, but software publishers have a habit of upgrading their software to correctly handle more and more situations. Second, software often allows users control over details of the function maximization, and intelligent choice of these options can aid the software in finding the best fit to the data.

23. Suppose that we have a model with a fixed effect with an intercept and slope and with time measured in years. Suppose we change time from years to months. Explain how the covariance parameters change if the covariance model is

(a) A random intercept model?
(b) A random intercept and slope model?
(c) An AR(1) model?

In particular, explain the relationship between the old parameters and the new parameters. How will the standard errors of the covariance parameters change?

24. Suppose that we have a model with an intercept and a time slope with time measured in years. Suppose we change time from years to months, and we change the baseline by subtracting 3 months from each date. Explain now how the covariance parameters change if the covariance model is

(a) A random intercept model?
(b) A random intercept and slope model?
(c) An AR(1) model?

In particular, explain the relationship between the old parameters and the new parameters. You will need to develop a notation for the old parameters and for the new parameters, then write a formula for the new parameters in terms of the old parameters. Suppose that you fit these three models to data twice, first with time measured in years, and second with time measured in months and with a change in the zero time point. How will the standard errors of the covariance parameters change?

25. Use the Pediatric Pain data for this problem, and fit the full CS×TMT interaction fixed effects model. Try a range of different power transformations for the response, and see how the different transformations affect inferences.

(a) Use a CS covariance model in this part. Consider a range of power transformations, say from −1 up to 2 in increments of .5. What is the estimated correlation between observations within a subject? Plot the correlations against the transformation parameter.
(b) Use an AR(1) correlation model and repeat the previous question.
(c) Use the UN correlation model and repeat the first item's tasks for each correlation parameter.
(d) What conclusion do you draw about correlation estimates and transformation?

26. For the Weight Loss data as illustrated in table 7.4, is the finding that estimated weekly means are lower than the sample means *robust* to the choice of covariance model?

(a) That is, do we get the same result for other choices of covariance model? Create another table like table 7.4 for another choice of covariance model.

(b) Suppose that we found another covariance model where the weekly estimates were the same as the sample means, but the covariance model did not fit the data nearly as well as other covariance models that did show estimates below the sample means. Would you conclude that there was uncertainty in the conclusion about weight loss over time? Justify your answer.

(c) Show that the independence covariance model meets the criteria of estimated weekly means matching the sample means and not fitting the data nearly as well as any other covariance model.

(d) Plot the data in a way that shows how subjects missing data at the last few time points are likely to have lower weights than those subjects who did not have missing data. Consider plotting those subjects with missing data in a darker line type and those with complete data with a lighter line type. Then try either plotting the weight loss data as in figures 2.16 or 2.17 or even 2.10 or some combination of the figures.

(e) State your conclusions about how the subjects with missing data differ from those with fully observed data.

27. For each of the numerical examples in section 8.1, plot the theoretical correlogram. For a subset of examples there will be only one (possibly bent) line on your plot. What feature do these examples have in common?

28. When fitting many responses from a single data set, we often analyze a single response very carefully, then generalize those conclusions to the other responses from the data set. Explore this approach for three of the Cognitive data responses. Use the same fixed effects for all three responses: age_at_baseline, gender, time, and time×intervention interaction. The original investigators used the RIAS covariance model for these data.

(a) Consider the Raven's response. Is the RIAS model the best possible covariance model for the Raven's data? Find the best covariance model that you can.

(b) Now explore height from the Anthropometry data. What is the best covariance model for the height data? Is this the same as the best covariance model for the Raven's?

(c) Do the same for the weight response from Anthropometry.

(d) Do you get the same covariance model in each case? If you do, then consider second and possibly third best choices that you get as well as your first choices so that you have three covariance models for the next parts.

(e) Consider the three different covariance models that you have chosen, and fit them to each of the three responses. How far off are they from being the best choices for each response? Is the difference on the log likelihood scale similar each time?

(f) For each of the three responses, do your conclusions regarding the fixed effects change? By how much?

9
Random Effects Models

I abhor averages. I like the individual case.

— Louis D. Brandeis

There never was in the world two opinions alike, no more than two hairs or two grains; the most universal quality is diversity.

— Michel Eyquem

Overview

In this chapter, we discuss random effects models for longitudinal data.

1. Random effects models are hierarchical models with two levels.

2. Data sets may have three or more levels! We include additional random effects into the data set for each level.

3. The marginal model is the random effect model written in the multivariate linear regression model form.

4. Random effects estimates are shrinkage estimates.

The random effects model is convenient to work with. Further, the story underlying the model has force for many data sets: that each subject's observations fall near a subject-specific line or other subject-specific curve. In the basic random effects model, we assume that subject-specific effects account for all of the correlation between observations on a single subject.

Any remaining variability is modeled as normally distributed error, often assumed to be independent and identically distributed as well. The decomposition assists in modeling by allowing us to develop three nearly separate models, one for the fixed effects, one for the subject-specific effects, and a third simple model for the remaining variability within a subject. The subject-specific effects are called random effects. In chapter 8, we studied two random effects models, the random intercept (RI) model and the random intercept and slope (RIAS) model. In table 8.13, we even considered a random intercept, slope, and quadratic (RIASAQ) model. Each subject has their own quadratic trend, with observations varying around that quadratic trend, and each subject-specific quadratic trend varies around the population trend.

Sets of subjects may be related to each other, for example children in a family or students in a classroom. When subjects belong to identifiable families or classrooms or other identifiable groups, subjects in the same group *cluster* together. Children are *nested* inside families and students are *nested* in classrooms. This requires introduction of random effects in our model to account for the clustering. Observations may also cluster. For example, subjects may sometimes weigh themselves on a home scale. The home scale requires a random effect of its own, as well as the subject random effects. We discuss how to include random effects to account for clustering of subjects and observations.

9.1 Random Effects Models as Hierarchical Models

Random effects models are examples of *hierarchical* models. Hierarchical models have many applications in statistical modeling and are particularly valuable for modeling data with a hierarchical structure. In hierarchical data structures, there is typically more data at the lower (observation) level and less data at higher levels. For longitudinal data, we have several observations (lowest level) per subject (higher level), and we have more total observations N than subjects n. In the next section, we will add another level to the hierarchy; subjects may be nested in families or schools.

Hierarchical models have one statistical model for the variability from each level in the hierarchy, and there is a model for the fixed effects as well. In section 8.1.2, on the random intercept model, we set out a model for observations Y_{ij}, $j = 1, \ldots, n_i$ from a single subject given a subject-specific parameter γ_{i1} and population-level parameter σ^2

$$Y_{ij} = \gamma_{i1} + \delta_{ij}, \qquad (9.1)$$

where

$$\delta_{ij} \sim N(0, \sigma^2). \qquad (9.2)$$

The δ_{ij}'s are independent and identically distributed residual errors with variance σ^2. The standard deviation σ is a measure of how far each subject's observations may vary around that subject's constant mean γ_{i1}. A second statistical model describes the distribution of the subject-specific parameters γ_{i1} as functions of population parameters α_1 and D.

$$\gamma_{i1} = \alpha_1 + \beta_{i1} \qquad (9.3)$$

where

$$\beta_{i1} \sim N(0, D). \qquad (9.4)$$

This second part of the model is the top level or population-level model. The γ_{i1}'s are subject-specific parameters. As before, we include the subscript 1 in γ_{i1} because there may be more than one γ per subject in some models, and this will keep notation consistent.

If we were able to observe the γ_{i1}'s as data, then α_1 would be a regression parameter and β_{i1} would be an unobserved subject level residual error. In contrast, in equation 9.1, the δ_{ij} are observation level residual errors. In this hierarchical model we have two sets of errors, one set from each level of the model. We call the δ_{ij}'s residual errors, and we will call the β_{i1}'s the random effects. If there are additional levels such as classroom or family in the model, each level will have a set of errors or random effects and these are identified by the level. The δ_{ij} errors are usually called residual errors or residuals. All higher levels are generically called *effects* or *random effects* and identified by the name of the level, for example, student random effects, or, more simply, student effects, class effects, and school effects.

At the top level of our two level hierarchical model (9.1), (9.2), (9.3) and (9.4), we have population parameters α_1 and D. Additionally, σ^2 is a population parameter because it is the variance of all observations around subject intercepts γ_{i1}. The parameter α_1 is the average of all subject intercepts γ_{i1}, and D is the variance of all subject intercepts. The population intercept α_1 is the average of subject intercepts. It is not the average of all observations. Suppose that some subjects contribute twice as many observations as other subjects. An average of observations would weight subjects contributing many observations more than subjects with fewer observations. If the subjects with more observations are all high on average, we will not estimate the population mean correctly.

There are a number of ways to write this model, all interchangeable. For example

$$Y_{ij} = \alpha_1 + \beta_{i1} + \delta_{ij}$$

omits the γ_{i1} and writes Y_{ij} in terms of all the random effects and residual errors in one equation. Combined with the distributions of β_{i1} and δ_{ij} in (9.2) and (9.4) given above, we have a complete model specification. This model collapses the hierarchies into a single equation. The γ_{i1}'s are still there, but hidden.

As we saw in section 8.1.5, the random intercept and slope covariance model has $q = 2$ random effects. In the absence of covariates, there is still a population intercept α_1 and population slope α_2 with $\alpha = (\alpha_1, \alpha_2)'$. In the population-level model, each subject has their own intercept γ_{i1} and slope γ_{i2}, with

$$
\begin{aligned}
\gamma_{i1} &= \alpha_1 + \beta_{i1} \\
\gamma_{i2} &= \alpha_2 + \beta_{i2}
\end{aligned}
$$

meaning that each individual intercept γ_{i1} differs from the population average intercept α_1 by the random effect β_{i1}. Similarly, the individual slope γ_{i2} differs from the population average slope α_2 by β_{i2}. With two random effects, β_i is an $r = 2$-vector and it has a bivariate normal distribution $N_2(\mathbf{0}, D)$. The matrix D is now a 2×2 matrix

$$
D = \begin{pmatrix} D_{11} & D_{12} \\ D_{21} & D_{22} \end{pmatrix}
$$

and D_{11} is the variance of the intercepts β_{i1}, D_{22} is the variance of the slopes β_{i2}, and $D_{12} = D_{21}$ is the covariance of the intercepts and slopes. We use the letter r to denote the number of random effects; for the RI model, $r = 1$.

The RIAS observation-level model describes the distribution of observations Y_{ij} as functions of σ^2 and the subject effects $\gamma_i = (\gamma_{i1}, \gamma_{i2})'$, where the subject-specific intercept is γ_{i1} and subject-specific slope γ_{i2}.

$$
\begin{aligned}
Y_{ij} &= \gamma_{i1} + \gamma_{i2} t_{ij} + \delta_{ij} \\
\delta_{ij} &\sim N(0, \sigma^2).
\end{aligned} \tag{9.5}
$$

The $\delta_{ij}, j = 1, \ldots, n_i$ and γ_i are all independent, both within subject and across subjects for all i and j.

We may write the random effects model in matrix form, specifying the model for all observations from a single subject. Define the covariate matrix $X_i = \mathbf{1}$ for the RI model and $X_i = (1, t_i)$ for the RIAS model, assuming no additional covariates here. Also define a second covariate matrix $Z_i = \mathbf{1}$ or $Z_i = (1, t_i)$ for the RI or RIAS models respectively. The jth row of Z_i is z'_{ij}. Let α be the vector of regression coefficients. Let $\delta_i = (\delta_{i1}, \ldots, \delta_{in_i})'$ be a vector of all observation level errors. The general form of the longitudinal random effects model is

$$
\begin{aligned}
Y_i &= X_i \alpha + Z_i \beta_i + \delta_i \\
\beta_i &\sim N_r(0, D) \\
\delta_i &\sim N_{n_i}(0, \sigma^2 I).
\end{aligned} \tag{9.6}
$$

We have set up this notation so that it can be used with many different data sets and many different models. Subscripts are put on the normal $N(\cdot, \cdot)$ distribution to remind us of the dimension of the random vectors β_i and δ_i. In the RI model, r is one, in RIAS, $r = 2$ and in RIASAQ, $r = 3$.

In general, we have $X_i, n_i \times p$ and α is $K \times 1$, and columns of the X_i are covariates as in chapter 7. The Z_i is also a matrix of covariates! It is similar to the X_i matrices, but usually they are not identical. The Z_i here are equal to X_i because we have no other covariates. With other covariates in the model, the X_i matrix would get larger; the Z_i will stay as already specified. Always Z_i will have the same number of rows as X_i, and every column of Z_i must be in X_i. Any column in the X_i matrix can potentially go in Z_i, subject to interpretability and appropriateness. In practice, we limit the number of columns of Z_i and always $r \leq K$ and usually r is much less than K.

When $K = r$, we have $X_i = Z_i$ and almost the only models then considered are the random intercept model where $X_i = Z_i = \mathbf{1}$ or the RIAS model where $X_i = Z_i = (\mathbf{1}, t_i)$. These models are typically used to introduce random effects models; they are not too interesting in practice; there are no groups to compare, no covariates other than time whose effect on responses we can describe.

Practical restrictions apply to the number of random effects. Adding an additional random effect to our model increases r to $r + 1$ and adds many parameters to our model. At the observation level model, this adds n parameters, one per subject. At the population-level, D goes from a $r \times r$ matrix to a $(r+1) \times (r+1)$ matrix, an increase of $r + 1$ parameters. As a practical matter, we do not want to have r come close to N/n, the average of the n_i, as the model becomes so flexible that multiple sets of parameter values may provide the same good fit to the data, and we will not be able to estimate the parameters α, D, σ^2 accurately.

In the random effects model (9.6), the α parameters are called the *fixed effects* as we have called them all along. We have generalized the name for use with all of our covariance models, but the name originated with the random effects model. The γ_i describe the distribution of observations from a single subject and vary from subject to subject. In contrast, α describes the population of subjects and not a single subject. Parameter α is fixed across subjects, hence the term fixed effects. The γ_i or β_i are described with a model that allows them to vary randomly subject to subject, hence the name random effects.

There are two mean functions for an observation Y_{ij}, depending on whether one is thinking about the population-level model or the subject-specific model. The population mean function is

$$E[Y_{ij}] = x'_{ij}\alpha.$$

The subject-specific mean function adds the random effect term to the population mean function

$$E[Y_{ij}|\beta_i] = x'_{ij}\alpha + z'_{ij}\beta_i.$$

We make the expectation conditional on β_i when we wish to indicate the subject-specific mean instead of the population mean.

The population consists of individuals with their own random effects. We can construct a prediction interval for a random effect from the population. A 95% interval for the intercepts γ_{i1} is

$$(\alpha_1 - 2D_{11}^{1/2}, \alpha_1 + 2D_{11}^{1/2}),$$

and similarly, for the slopes, the same formula, but replacing the subscript 1's with 2's

$$(\alpha_2 - 2D_{22}^{1/2}, \alpha_2 + 2D_{22}^{1/2}).$$

Because we do not know the parameters, we plug in parameter estimates into the formulae, although that will make the intervals only approximate 95% intervals in practice. More complex formulas can be developed to adjust the intervals for the fact that we have only estimated α and D.

9.1.1 Some Examples

Many different random effects models are possible. Random polynomials of any order, particular low orders are popular models to use. We could have a random bent line model: observations for a given subject follow a single line up to a particular time t_0. After t_0, the slope changes by a random amount, and the observations continue on a different, but still linear path over time.

Figure 9.1 graphically illustrates four random effects models. In each figure, data from four subjects is plotted. Each observation $Y_{ij} = x_{ij}'\alpha + z_{ij}'\beta_i + \delta_{ij}$ is plotted against t_{ij} using a unique plotting symbol for subject i. For each subject, the true subject-specific mean value $x_{ij}'\alpha + z_{ij}'\beta_i$ is drawn in over time using different line types. Finally, the population mean response $x_{ij}'\alpha$ is also drawn in with a bold solid line. Each subject is distinct from the others; few observations are close to the population mean trend, this is mainly because so few subjects are plotted. In all of the examples, the observations fall fairly close to the subject means; σ^2, the variance of the δ_{ij} errors, was set to be low.

All of the models illustrated have the Z_i and X_i covariate matrices equal to each other, $X_i = Z_i$. The simple random intercept model that we illustrate in figure 9.1(a) has but a single column for the X_i and Z_i matrices, both matrices are a column of ones. Different subjects i and k have different times t_i and t_k so $X_i \neq X_k$ and $Z_i \neq Z_k$. The RIAS model has two columns, $X_i = (1, t_i) = Z_i$. The RIASAQ model has $X_i = (1, t_i, t_i^2) = Z_i$ with three fixed and three random effects. For the RIASAQ model, the variance D of the random effects is a three by three matrix. The elements of D down the diagonal, D_{jj}, for $j = 1, 2$, and 3 are the variances of the random intercept, random slope, and random quadratic terms. The three off-diagonal elements D_{12}, D_{13}, and D_{23} are the covariances among the three random effects.

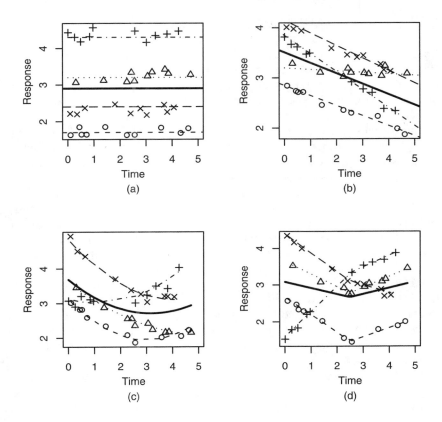

Figure 9.1. Examples of data from four different random effects models. In each plot, data from four subjects are plotted. Each subject uses a different symbol. The average response for each subject as a function of time is drawn in using distinct line types, and the population average response over time is drawn in a heavy solid line. (a) Random intercept model; (b) random intercept and slope; (c) random intercept and slope and quadratic; and (d) random intercept and slope and random change in slope at $t = 2.5$.

The random bent line model has three fixed and three random effects, α is a three-vector, and D is a three by three matrix. The first two columns of X_i and Z_i are the same as for the RIAS or RIASAQ model. The third column is $(t_i - \mathbf{1}t_0)_+$ as discussed in section 7.6, where $t_0 = 2.5$ is the location of the bend in the lines and $(t_{ij} - t_0)_+ = 0$ if $t_{ij} \le t_0$ and $(t_{ij} - t_0)_+ = t_{ij} - t_0$ for t_{ij} after $t*$.

9.2   Three-Level Hierarchical Data Sets

Longitudinal observations are *nested* or *clustered* inside subjects; observations on a given subject are more similar to each other than observations on different subjects. Longitudinally collected data sets are hierarchically organized with two levels: the observation level and the subject level.

The hierarchy need not stop at two levels. It is common in longitudinal studies to have three or more levels of nesting or clustering. Subjects from a single cluster are naturally more similar to each other as compared with subjects from different clusters. Researchers often find it expeditious to enroll groups or clusters of similar subjects.

For example, in the BSI study, researchers study HIV+ parents. Parents are identified and enrolled in the study. A second portion of the study collects data on their adolescent children. All adolescents of the parents are then enrolled into a second portion of the study. Each parent in the study may contribute one or more adolescent children to the study. Adolescents from a single parent are nested inside parent; each parent is a cluster.

We may apply an educational intervention to a classroom. Typically the intervention is actually applied to the teacher, who then teaches students in classes. Data is collected on the students. Students in a particular class share a teacher, classmates and environment and will be more similar to each other than to students from another class. Each class is a cluster of students.

This similarity among students in a class or children in a family needs to be accounted for in our models. We do this by including family or class as a covariate in some fashion in the analysis. Typically we model the cluster effect as a random effect. Each cluster k has its own random effect η_k (pronounced eta sub k), which is added to the fixed effects (plus random effects if any) for each subject in the cluster.

Suppose we have a random intercept model for subjects, a random intercept for clusters, and no additional covariates. The population mean is α_1. Each cluster k has its own intercept $\alpha_1 + \eta_k$, which differs by an amount η_k from the population mean level α_1. Subject i who happens to belong to cluster $k(i)$ has its own average level $\alpha_1 + \eta_{k(i)} + \beta_i$ which differs from the kth cluster average level by an amount β_i. Observations Y_{ij} from subjects in cluster $k(i)$ follow the model

$$Y_{ij} = \alpha_1 + \eta_{k(i)} + \beta_i + \delta_{ij} \qquad (9.7)$$

and vary from the subject mean by an amount δ_{ij}. We make the usual normality assumptions on β_i and δ_{ij}. Additionally, we assume normality for $\eta_{k(i)}$. All random effects and residuals are independent of each other, both within and across subjects. The full model includes model (9.7) and

$$\eta_{k(i)} \sim N(0, D_\eta)$$
$$\beta_i \sim N(0, D_\beta)$$
$$\delta_{ij} \sim N(0, \sigma^2). \qquad (9.8)$$

The model (9.7) – (9.8) has three levels, observation j, subject i (children, students), and cluster $k(i)$ (family, classroom). Each level has an associated residual error δ_{ij} or random effect β_i or $\eta_{k(i)}$. The data set is hierarchical with three levels, and our model reflects this structure.

The cluster $k(i)$ is not identified in the subject and observation subscripts i and j. We use only two subscripts to identify an observation. Cluster $k(i)$ is a *covariate* in the data set identifying the cluster that the particular subject belongs to.

Each level in the hierarchy has a sample size associated with it. There are still $N = \sum_i n_i$ observations, with n_i observations per subject. We have n, the number of subjects in the data set, and we also have n_c, the number of families or classes or clusters in the data set with $N > n > n_c$. Each sample size corresponds to the number of residual errors or random effects associated with that level. We refer to N, n, and n_c and so on as the number of *units* whether observations, subjects, or classes at that level.

More generally, suppose that we have covariates X_i, fixed effects α, and subject residual ϵ_i with covariance matrix $\Sigma(\theta)$, which may be any of the models discussed in chapter 8 including possibly the random intercept model. The general three-level hierarchical model with a random intercept model for the cluster effect is

$$Y_i = X_i\alpha + \mathbf{1}\eta_{k(i)} + \epsilon_i \qquad (9.9)$$

with

$$\epsilon_i \sim N(0, \Sigma(\theta))$$
$$\eta_{k(i)} \sim N(0, D_\eta). \qquad (9.10)$$

The hierarchical model (9.9)–(9.10) has many variations. The model here for the clusters is a random intercept model, but more complicated random effects models are possible; with enough data we could consider fitting a random intercept and slope $\eta_{k1} + \eta_{k2}t_{ij}$ model for clusters. This sort of model would usually only be fit if the subject level covariance model $\Sigma(\theta)$ was also a RIAS covariance model or possibly a more complex random effects model such as a RIASAQ. If classes are nested inside schools or families inside apartment buildings, then we might add random effects for school or apartment building, creating a four level hierarchical model.

We have discussed time-fixed and time-varying covariates, these are respectively subject-level and observation-level covariates. With the introduction of a third level to our model, we now have a new type of covariate: cluster-level covariates are constant for all subjects in a given cluster. One can consider interactions between cluster level covariates and subject level covariates or between cluster level and observation level covariates. Models and their interpretation can get quite complex.

Effects of subject level covariates are well estimated by clustered data sets, but cluster level covariate effects may not be well estimated. Care must be taken when estimating coefficients of covariates at the cluster level. The key sample size for deciding how much information is in the data set for estimating subject level or cluster level covariates is the sample size at the subject or cluster level, n or n_c. In an educational study, one of two teaching methods is used at the class level. We assign $n_c/2$ teachers to each of two teaching methods. To assess the impact of the teaching methods, we longitudinally assess students with a class. Whether there are 20 or 50 students in the classroom and whether there are 3 or 30 followup measures, we have $n_c/2$ pieces of data on each teaching method. The classroom average test score increase is the most relevant single datum from each class, not the set of individual student increases. If n_c is small, there may be little power for estimating classroom covariate effects such as teaching method or teacher gender.

In the same study, suppose we wish to assess the main effect of student gender on learning. The 50 students in a class with a near equal mix of males and females will provide quite a lot of information about the effects of gender. The class or cluster level random effects will not affect the estimated difference between boys and girls in each class. The student gender estimate is affected only by the student random effect and by the residual errors; cluster level random effects affect the boys and girls equally and cancel when we subtract the girls average score from the boys average score within a single cluster.

Hierarchical models with multiple levels of random effects are very powerful tools for analyzing hierarchical data sets. Trouble can arise because there is increasing uncertainty at higher levels where there are fewer units. Classroom random effects can be hard to estimate, and software can fail to produce any results, or worse, may produce results that are outright incorrect. As a rule, no matter how complicated the model is at the observation levels of our hierarchical model, we are restricted to simpler and simpler models at the higher levels (classrooms, schools, school districts, states, countries) with fewer and fewer units at each increasing level. If there are only a few units at a higher level, then we may treat the units as fixed effects. If there are too few clusters, software will have trouble estimating the cluster effect; we then might fit the data using cluster as a fixed effect only.

What makes the difference between a fixed effect and a random effect? In the United States, many studies have only one or two racial/ethnic groups or at most perhaps four. We routinely treat race/ethnicity as a fixed effect in analysis of US data. In some countries, there can be hundreds of ethnic groups. If we get members from multiple tens of groups into a single study, we might then treat race as random rather than fixed.

There are many potential sources of clustering or nesting in statistical data sets. Interviewers who go out to survey subjects may have their own impact on responses, and so may merit their own random effect. Lab technicians may routinely measure samples high or low and so require their own effects in the analysis. Measuring devices such as weighing scales, voltmeters, and so on may have their own random effects that introduce variation into the analysis. In agricultural studies, we may assess the crop yield from plots of land over many years. Each plot of land is slightly different, and requires its own random effect. In the BSI data, one parent may contribute up to 5 adolescents to the analysis. A proper analysis includes a parent random effect in the analysis. In the Weight Loss data, some weight measurements are taken in clinic, assumed to be the gold standard, and some weight measurements are taken at home. Because the home scales are all different, presumably cheaper and more prone to error, these measurements should have a home scale random effect for each subject. For the Weight Loss data, the extra random effect is nested inside the subject, rather than having several subjects nested inside scale.

9.2.1   Parent Random Effects in the BSI

In the BSI data, both adolescents and their parents are assessed over time. Fully 111 parents contributed more than one adolescent to our data set; one parent contributed 5 adolescents. Overall there are 423 adolescents from 280 parents. We might reasonably presume that adolescents that share a parent are more similar than adolescents with different parents. This suggests that we need to include a parent random effect in our analysis.

We fit two models to the adolescent log base 2 total BSI scores over time. A small constant $1/53$, equal to the smallest non-zero possible response, was added to BSI score because the minimum value is zero before taking logs. Both models used an $ARMA(1,1)$ for the adolescent responses over time. The difference in the two models was the inclusion or not of a parent random effect with variance D. Covariates for both models were the same and included an intercept; time in months; slope changes at 18 and 36 months; indicators for season effects, spring months 3–6, summer months 7–10 and winter months 11–2; adolescent gender; an indicator of recent parental alcohol use; an indicator of recent marijuana use; and two indicators for the three-category variable that parent (i) never reported hard drug use or reported hard drug use and either (ii) was a recent user or (iii) was not a recent user.

Estimates and standard errors for the covariance parameters for both models are given in table 9.1. The -2 log REML increases by 10.5 units when we add parent random effect to the model. Compared to a chi-squared random variable with 1 degree of freedom, this is very significant and the

Model	Cov parm	Est	SE	t
No	D_η	–	–	–
Parent	ρ	.94	.011	88
Effect	γ	.59	.024	24
	σ^2	4.04	.20	21
Random	D_η	.72	.23	3
Parent	ρ	.90	.02	40
Effect	γ	.50	.04	14
	σ^2	3.33	.22	15

Table 9.1. Covariance parameter estimates from fitting a random parent effect. Top part is without the parent random effect, bottom part is with the parent random effect, and D_η is the parent random effect variance. Both models have an ARMA(1, 1) covariance for the across time adolescent covariance model with correlation parameters γ and ρ. P-values are all highly significant $p < .0001$ except for the parent random effect, which is significant at $p = .001$.

data support use of the parent random effect. The sum of $D_\eta + \sigma^2$ from the parent random effect model is equal to σ^2 alone from the model without the parent random effect. Un-modeled variation in the data does not disappear. When we omit a variance term like the parent variance from the model, that variation shows up in some other variance term, often σ^2.

Comparing the two results for the fixed effects shows that including the parent random effect has little impact on the fixed effects inferences. There are slight changes in estimates and in significance, but no major changes in results. The estimated difference between boys and girls decreased, with concomitant decrease in the significance of the gender effect. Most of the covariates in our analysis change from observation to observation, that is, they vary over time within subject. Even the parent drug variables are time-varying within a subject and are not parent level fixed variables. Including or omitting the parent random effect could potentially affect parent level fixed effect variables, such as parent gender, parent race, or parent health status at baseline, but none of these variables are included, and so there is little practical influence of including the parent random effect in this analysis.

9.2.2 Weight Loss Data

In the discussion of figure 10.5(b), we saw an odd pattern over time that included a decrease at weeks 4 and 5 and an increase at week 6 in the residuals. In table 7.4, we see that the estimated population weight actually increases between week 5 to week 6 before continuing its downtrend at weeks 7 and 8. The Weight Loss data has one important piece of additional information beyond the day or week of the measurement. Recorded for each observation is whether the subject came into the clinic to be weighed or not. Most measurements at weeks 1, 2, 3, and 6 were in-clinic visits. At

Estimates	Parameter	Model			
		1	2	3	4
	Intercept	193.72	193.46	193.45	193.45
	Day	-.27	-.17	-.17	-.17
	Day*day	.21	.11	.11	.11
	Visit	–	-2.38	-2.38	-2.37
Standard errors					
	Intercept	4.20	4.19	4.20	4.20
	Day	.03	.03	.03	.03
	Day*day	.06	.06	.05	.05
	Visit	–	.30	.42	.43
Fit statistics					
	σ^2	4.52	3.44	2.52	2.50
	# parms	4	4	5	7
	-2log REML	1474.1	1420.7	1399.4	1397.3
	AIC	1482.1	1428.7	1409.4	1411.3
	BIC	1488.7	1435.3	1417.6	1422.7

Table 9.2. Parameter estimates, standard errors, and fit statistics for four models involving visit for the version 2 of the Weight Loss data. The number of covariance parameters in the model is given in the row marked # parms. Model 1 has no visit effect, model 2 has a fixed visit effect, model 3 has a random visit effect, and model 4 allows the random visit effect to be correlated with the random intercept and slope.

the other times, subjects were phoned at home, asked to go step onto their bathroom scales, and report the result. The indicator variable for visit type can be added to our model.

We used version 2 of the data set. Time is measured using the exact day of each visit rather than week and we fit a quadratic model in day to the data along with a random intercept and slope covariance model. This is model 1. Table 9.2 gives selected results (parameter estimates, standard errors and goodness of fit statistics) for this model in the column headed 1. Model 2 adds in a fixed visit effect. We see that the home visit effect is −2 pounds. Apparently an easy way to lose two pounds is to weigh yourself at home rather than in the clinic! The standard error is quite small compared to the estimate; visit effect is very significant. After adding in the visit effect, the estimate for the quadratic terms decreases substantially, although it is still borderline significant. We see that σ^2 decreases substantially from model 1 to model 2.

Scales are all different; we assume the clinic scale is the gold standard. We might expect each subject's scale to differ from the clinic scale by varying amounts. Some may be right on, some may weigh high, and more probably weigh low. When we have many measuring devices, we expect that measuring device is likely a random effect. We incorporate both a fixed and random scale effect in model 3. In model 3 we include the random scale effect as a separate random effect, uncorrelated with the other random effects. Model 3 improves substantially on model 2, with a decrease in -2 log REML of over 20 for adding one parameter and a continuing decrease of σ^2 from 3.4 to 2.5.

The variance of the random effects was estimated to be 4, so the standard deviation is 2. An approximate 95% interval for the amount that home scales add to the weight ranges from $-2.4 - 2 \times 2 = -6.4$ pounds up to $-2.4 + 2 \times 2 = 1.6$ pounds. So most scales weigh low, but a few may weigh higher than the truth.

We can get estimates for the individual random scale effects and add them to the estimated main effect -2.38. The range of the actual estimated scale effects ranges from a low of -6.6 pounds up to $+.3$ pounds. Only one subject had a scale that was estimated to be positive.

Model 3 assumes independence between subject's random scale effect and their random intercept and slope. Because there is one scale per subject, we can consider whether the scale effect is correlated with the subject's random intercept or random slope. For example, suppose heavier subjects weighed themselves lower at home, or that subjects who lost weight more rapidly weighed higher at home.

In model 4, we estimate the covariances between the scale effect and the random intercept and slope. This adds two parameters to model 3. In model 4, the covariance between the visit effect and the intercept is -13 with a standard error of 12; the covariance between the visit effect and the slope is $-.027$ with standard error .054. Neither covariance is remotely significant. The -2 log REML improves only by 2 units, which is a very modest improvement for adding two parameters. We decide to stay with model 3. Our conclusion is that subjects' scales at home weigh about 2.4 pounds lighter than the clinic scale, and the variance around this estimate is 4.09 or a standard deviation of 2 pounds.

9.3 Marginal Model

This section uses some linear algebra.

The model (9.6) is the hierarchical version of the random effects model. We can also write the model in the form we used in previous chapters

$$Y_i = X_i \alpha + \epsilon_i \tag{9.11}$$

$$\epsilon_i \sim N_{n_i}(0, \Sigma(\theta)). \tag{9.12}$$

This is called the *marginal model*; it does not involve the β_i's, at least not directly. Instead, the covariation among observations that the β_i are designed to accommodate has been placed inside the covariance matrix $\Sigma(\theta)$. What is $\Sigma(\theta)$ and what are the unknown parameters θ? We can calculate this by equating the right-hand sides of equation (9.11) and (9.6).

We get

$$X_i\alpha + \epsilon_i = X_i\alpha + Z_i\beta_i + \delta_i.$$

Canceling $X_i\alpha$ gives

$$\epsilon_i = Z_i\beta_i + \delta_i.$$

Taking the variance of both sides, $\text{Var}(\epsilon_i) = \Sigma(\theta)$, and on the right-hand side

$$
\begin{aligned}
\Sigma(\theta) &= \text{Var}(Z_i\beta_i + \delta_i) \\
&= \text{Var}(Z_i\beta_i) + \text{Var}(\delta_i) \\
&= Z_i\text{Var}(\beta_i)Z_i' + \text{Var}(\delta_i) \\
\Sigma(\theta) &= Z_i D Z_i' + \sigma^2 I,
\end{aligned}
$$

where I is as usual the identity matrix of appropriate size, in this case n_i by n_i. This is the same calculation that we did for the random intercept model in equation 8.2 generalized to matrix form rather than one observation at a time. Matrix Z_i can have any number of columns r and D is $r \times r$ rather than 1 by 1. The unknown parameters θ are the unique elements of D and σ^2. We use two standard facts about variances of sums and matrix products in producing the above formula. The first was that the variance of a sum of independent terms is the sum of the variances. The β_i and δ_i are independent, and the variance of the sum is the sum of the variances. Second, the variance of $Z_i\beta_i$, a known matrix Z_i times a random vector β_i is equal to $Z_i\text{Var}(\beta_i)Z_i'$. This is the matrix generalization of the formula $\text{Var}(ax) = a^2\text{Var}(x)$ where x is a random variable and a is known.

We can also think of the errors ϵ_i as having covariance matrix which is the sum of two pieces

$$\epsilon_i = Z_i\beta_i + \delta_i.$$

This is an example of a covariance model from section 8.5 on sums of covariance matrices with one piece $Z_i D Z_i'$ and one $\sigma^2 I$.

9.4   Estimation of Random Effects and Shrinkage

Suppose we have estimates $\hat{\alpha}$, $\hat{D}$, and $\hat{\sigma}^2$, the estimates may be either ML or REML. How might we estimate β_i? There are two pieces in our model that contribute information for estimating β_i. One is the statement that $\beta_i \sim N(0, D)$. This is called the *prior* of β_i. The second piece is the *likelihood* of the data Y_i which says that $Y_i|\beta_i \sim N(X_i\alpha + Z_i\beta_i, \sigma^2 I)$. There are many other parts to our model, involving other observations Y_l and other random effects β_l, but no part that involves β_i. We combine information about β_i from the prior and the likelihood by multiplying the two densities involved. We write out the likelihood of β_i times the prior after substituting the estimates $\hat{\alpha}$, $\hat{D}$ and $\hat{\sigma}^2$ for α, D, and σ^2, then take logs. Additive terms not involving β_i may be dropped. This leaves us with the sum of two terms that are quadratics in β_i

$$-.5\hat{\sigma}^{-2}(Y_i - X_i\hat{\alpha} - Z_i\beta_i)'(Y_i - X_i\hat{\alpha} - Z_i\beta_i) - .5\beta_i'\hat{D}^{-1}\beta_i. \qquad (9.13)$$

The first term comes from the likelihood, and the second term comes from the prior.

If one has not seen formulae like this before, the information about β_i in the formula is not readily apparent; but those that have worked with similar equations, the best estimate of β_i is but a few algebraic steps away. The usual approach to producing an estimate of β_i is to find the value of β_i that maximizes the above formula. To do this most easily, one differentiates 9.13 with respect to β_i and sets the resulting formula equal to zero. Solving for β_i then gives us a conditional maximum likelihood estimate for β_i, conditional on the values of $\hat{\alpha}$, $\hat{D}$ and $\hat{\sigma}^2$. The result is

$$\hat{\beta}_i = (Z_i'Z_i + \hat{D}^{-1}\hat{\sigma}^2)^{-1}Z_i'(Y_i - X_i\hat{\alpha}). \qquad (9.14)$$

This is the estimated random effects vector. It is an *empirical Bayes* estimate of β_i. It is *Bayes* because there is a prior and a likelihood. It is *empirical* because we have substituted in estimates $\hat{\alpha}$, $\hat{D}$, and $\hat{\sigma}^2$ for the true unknown values, rather than performing a full Bayesian analysis.

We explore this formula for $\hat{\beta}_i$ with a number of special cases. First, suppose that $\hat{D}$ was arbitrarily large, then the inverse $\hat{D}^{-1}$ is arbitrarily small: essentially zero, and we replace it with 0. The formula becomes

$$\hat{\beta}_{i,\infty} = (Z_i'Z_i)^{-1}Z_i'(Y_i - X_i\hat{\alpha}).$$

This formula should look familiar, although the notation is unusual. It is in the form of $(X'X)^{-1}X'Y$, the least squares estimate of the regression parameter for linear regression. The familiar covariate matrix X has instead been replaced by Z_i, and the response Y has been replaced by $(Y_i - X_i\hat{\alpha})$. In (9.14), for our response, we adjust Y_i by subtracting off the estimated fixed effect $X_i\hat{\alpha}$. If $\hat{D}^{-1}$ is zero, then β_i is the least squares estimate from regression $(Y_i - X_i\hat{\alpha})$ on Z_i.

Suppose instead that $\hat{D}$ is zero, so that $\hat{D}^{-1}$ is infinitely large, and $Z_i'Z_i + \hat{D}^{-1}\hat{\sigma}^2$ is also arbitrarily large, and $(Z_i'Z_i + \hat{D}^{-1}\hat{\sigma}^2)^{-1}$ is then zero, and our estimate $\hat{\beta}_{i,0} = 0$! If D is zero, our assumption $\beta_i \sim N(0, D)$ implies that all of the β_i's are zero because their mean is zero and their variance is zero, and this is built directly into formula (9.14).

In the formula for $\hat{\beta}_i$, $\hat{D}$ is neither zero nor arbitrarily large, but is rather somewhere in the middle. Approximately, you can think of $\hat{\beta}_i$ as falling in between $\hat{\beta}_{i,\infty}$ and $\hat{\beta}_{i,0}$, depending on the size of $\hat{D}^{-1}\hat{\sigma}^2$. In truth, only for $r = 1$, with one random effect does $\hat{\beta}_i$ fall on the line connecting $\hat{\beta}_{i,\infty} = (Z_i'Z_i)^{-1}Z_i'(Y_i - X_i\hat{\alpha})$ and $\hat{\beta}_{i,0} = 0$. In more than 1 dimension, then $\hat{\beta}_i$ falls on some arc connecting the two extreme estimates. The arc is a straight line only in an implausible special case when $r > 1$. For $r = 1$, we typically have a random intercept model with Z_i equal to a vector of n_i ones. Then $Z_i'Z_i = n_i$ and

$$\hat{\beta}_i = \frac{n_i}{n_i + \hat{\sigma}^2 \hat{D}^{-1}} \bar{R}_i$$

where

$$\bar{R}_i = \frac{1}{n_i} \sum_{j=1}^{n_i} Y_{ij} - x_{ij}'\hat{\alpha}$$

is the average value of $Y_{ij} - x_{ij}'\hat{\alpha}$. If $\hat{D}^{-1} = 0$, then $\hat{\beta}_i = \bar{R}_i$, but as $\hat{D}^{-1} > 0$,

$$|\hat{\beta}_i| = \frac{n_i}{n_i + \hat{\sigma}^2 \hat{D}^{-1}} |\bar{R}_i| < |\bar{R}_i|.$$

This formula says that $\hat{\beta}_i$ is closer to zero than $\bar{R}_i = \hat{\beta}_{i,\infty}$. Because of this, we call $\hat{\beta}_i$ a *shrinkage* estimate, it is shrunk toward zero from the estimate $\hat{\beta}_{i,\infty}$.

The variance of $\hat{\beta}_i$ is

$$\text{Var}(\hat{\beta}_i) = (Z_i'Z_i\sigma^{-2} + D^{-1})^{-1}.$$

The inverse of this formula has two terms, $Z_i'Z_i/\sigma^2$ and D^{-1}. The first term is the inverse of the variance from the likelihood, the regression of R_i on Z_i, and the second term is the inverse of the prior variance from $\beta_i \sim N(0, D)$. The inverse variances add to give the inverse of the variance of the resulting estimate.

There is a fixed effects version of model 9.6 that omits the assumption $\beta_i \sim N(0, D)$.

$$
\begin{aligned}
Y_{ij} &= x_{ij}'\alpha + z_{ij}'\beta_i + \delta_{ij} \\
\delta_{ij} &\sim N(0, \sigma^2).
\end{aligned}
\tag{9.15}
$$

The β_i's are now fixed effects parameters, just like α. There is no correlation among observations, even within a person, and all observations are independent according to this model. This model can be fit using least squares like in linear regression. There are now K parameters in α and nq parameters $\beta_i, i = 1, \ldots, n$. The total number of independent fixed effects parameters is $K + (n-1)r$. The reason for the -1 times r is that the X_i matrix has r columns that are the same as columns of Z_i and so there are r free parameters in α that can be set to anything before finding the least squares solution. Easiest perhaps is to set them equal to 0, effectively removing the redundant columns from the X_i matrices before fitting. The problem with the fixed effects model 9.15 is that the number of parameters $K + (n-1)r$ increases linearly with the sample size n. This leads to poor properties of the estimators. There is a theorem in statistics that to have a good estimator of a large number of regression parameters like the collection $\alpha, \beta_1, \ldots, \beta_n$, we must use a prior distribution on the parameters. We mostly, albeit not entirely, satisfy the requirements of this theorem when we set a prior $\beta_i \sim N(0, D)$ for the random effects.

In the random effects model (9.6), the parameters α, σ^2, D at the top level are estimated using the model specification (9.11) and (9.12) without specific reference to the random effects. There are only $K + 1 + r(r+1)/2$ parameters to estimate, which does not grow with the sample size. We say that the random effects are *integrated out* of the model. If the random effects are of interest, we estimate them with equation (9.14).

One way of thinking about the fixed effects model is to let D in 9.6 be arbitrarily large. As D gets larger and larger, there is less and less information in the assumption $\beta_i \sim N(0, D)$; it has less and less impact or influence on our inferences. This lack of influence is not a good thing. The $\beta_i \sim N(0, D)$ assumption is information about the β_i's; it says that the β_i's are near zero on average, and that the mean square $\mathrm{E}[\beta_{ij}^2] = D_{jj}$. If this assumption is accurate, then it helps with the estimation of the β_i's and in turn this helps with the estimation of α. Assuming that the β_i's have a variance at all, then the only issue is whether the distribution of the β_i's follows a normal distribution. As a general statement, some unimodal bell-curved shape is likely a good guess, and the normal makes for easy enough computation. Second, it is very difficult to determine the true sampling distribution of the β_i's as they are difficult to estimate. We do not have much information to estimate any one β_i; we have but n_i observations in Y_i to help estimate β_i.

Because the normal assumption is useful, substantially better than no assumption at all, and because it is quite difficult to identify a different distribution from the data, we usually will stick to the normal assumption. A modestly popular alternative is a t distribution because the t distribution allows for outliers far from the mean of the β_i's. This has the net effect of decreasing the influence of cases that fall far from the mean of the β_i's on the inferences.

9.5   A Simple Example of Shrinkage

We explain shrinkage another way, in a simple situation equivalent to the random intercept model with a single observation per subject and known parameters α, D and σ^2.

Consider a test maker creating a test. To study the properties of the test, they give the test J times to n subjects. The test results from a given subject are assumed to be constant over time, up to σ^2 sampling variation. Thus each subject has a mean test score, γ_i, which varies about the population mean test score α_1 by $\gamma_i = \alpha_1 + \beta_{i1}$. At the jth test taking, subject

i gets a score $Y_{ij} = \alpha_1 + \beta_i + \delta_{ij}$. The random effects β_i are distributed $N(0, D)$ and the residuals δ_{ij} are distributed $N(0, \sigma^2)$.

The test maker gives the test many times to many subjects, and with enough testing, and some modifications to the test, the scoring of the test is designed so that α_1, D, and σ^2 are known. The test is published, and psychologists and educators use the test, administered just once, to estimate γ_i for different subjects.

Now allow only $n_i = 1$ observation per subject, only 1 fixed effect α_1 and known values for $\alpha_1 = 100$, $\sigma^2 = 10^2$ and $D = 10^2$. Consider an IQ test. We suppose IQ is normally distributed with a mean 100 in and a standard deviation of 10 IQ points. So for example, 68% of subjects have IQs between 90 and 110, and 95% of subjects have IQs between 80 and 120. Measuring a subject's IQ precisely requires a much too long of a test. To be practical we give a short test to subjects, and so we measure each individual subject's IQ with error. Because of extensive studies by the test manufacturer, we know that the standard deviation σ^2 of the test is 10 IQ points and is normally distributed around the true IQ score $\gamma_i = \alpha_1 + \beta_i$. Our assumptions are first

$$\text{true IQ} \sim N(100, 10^2)$$

and second, that

$$\text{test score} \sim N(\text{true IQ}, 10^2).$$

The first distribution is *prior* information about the population of IQs. The second assumption is information about the test properties for a specific subject. IQ is equal to $\gamma_i = 100 + \beta_i$, and we wish to estimate $100 + \beta_i$ for a specific subject.

A subject selected at random takes the IQ test and scores 120. Suppose you are only allowed (for the moment) to report their IQ as either 110 or 130. Which score is a better estimate of the subject's IQ?

From the test manufacturer's results, we know that the test has a standard deviation of 10, and so that both 110 and 130 are one standard deviation away from the 120 that we observed. It would seem to be a toss up as to whether 110 or 130 is preferable to report, and we would prefer to report 120. Essentially this is the fixed effects model's conclusion without using the population information true IQ $\sim N(100, 10^2)$ assumption.

But this apparent indecision between 110 and 130 does not take into account the population information. We know that subjects in the population are distributed with IQ$\sim N(100, 10^2)$. Given the choice of 110 or 130, a person with an IQ 110 is as likely to score 120, 1 test sd above their true IQ as a person with an IQ of 130 is likely to score 120, 1 test sd below their true IQ. However, we know that 110 is 1 population sd above the population mean, whereas 130 is 3 population sds above the mean. There are more people with IQs of 110, and relatively few people with IQs of 130. So given the choice between 110 and 130, why would we guess 130, when

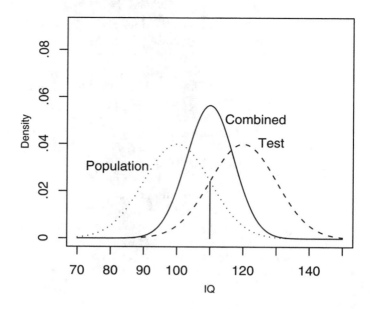

Figure 9.2. IQ example 1. The dotted density on the left is the density of IQ scores in the population, normal with mean 100 and sd 10. The dashed density on the right is the likelihood information coming from the test score of 120 and test sd of 10. The solid line is the inference from combining both sources of information. It is a normal density with mean 110 and sd $\sqrt{50}$. The vertical line shows the height of the population density at the proper estimate of 110.

there are very few people that far out on the scale? We would of course guess 110 as there are many more people with IQs of 110.

The score 110 is the correct estimate in the choice between 110 and 130. In fact, given the information provided, it is the correct estimate given our choice of any potential conclusion. What we have is formula (9.14) with $Z_i = 1$, $Z_i' Z_i = 1$, $\hat{\alpha} = 100$, $\hat{\sigma} = 10^2$, $\hat{D} = 10^2$, and $Y_i = 120$. Then $\beta_i = (1/2) \times (120 - 100) = 10$, and we estimate the true IQ as $100 + \beta_i = 110$.

This is called *shrinkage*. The fixed effects model estimates 120, whereas the random effects model estimates 110. We shrink the raw test score toward the population mean of 100. Further, the standard error from the random effects model is $\sqrt{50}$, whereas the standard error from the fixed effects model is 10. The formula for the standard error of the IQ estimate is given in problem 6. On average, the random effects model estimate 110 is closer to the subject's true IQ than the fixed effects model, and the standard error can be smaller because we have included the population information in our estimate. Figure 9.2 illustrates the situation. The left-hand dotted density is the density of IQ scores in the population, $N(100, 10^2)$. The right-

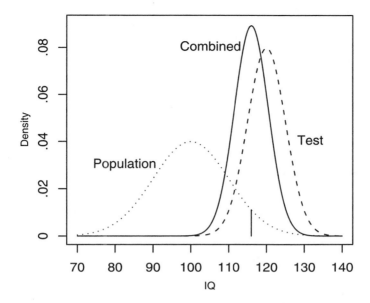

Figure 9.3. IQ example 2. The dotted density on the left is the density of IQ scores in the population, normal with mean 100 and sd 10. The dashed density on the right is the likelihood information coming from a test score of 130 and test sd of 5. The solid line is the inference from combining both sources of information. It is a normal density with mean 124 and sd $\sqrt{20}$. The vertical line shows the height of the population density at the proper estimate of 124.

hand $N(120, 10^2)$ dashed density is the subject-specific information that we get from the observation $Y_i = 120$; it is called a likelihood. The center solid-lined density is the conclusion we draw from combining both population and subject-specific information. The density is more peaked than either source density, indicating that the standard deviation of our conclusion is smaller than both of the input densities. The center solid-lined density is the density that we get from taking formula 9.13, exponentiating it (i.e., undoing the log) assuming it is proportional to a density in β_i, and *standardizing* it so that it integrates to 1 like a true density, and that is what we plotted in figure 9.2.

Figure 9.3 presents a second example. Again the population of IQ scores is normal with mean 100 and standard deviation 10. This time, the subject takes a more precise test with a standard deviation of 5. This could be accomplished, perhaps, by making a longer IQ test that has 4 times as many questions. Alternatively and equivalently, we could have given the subject 4 independent IQ tests with standard deviation 10 and averaged the results. This subject scores 130 and the density on the right has mean

130 with standard deviation 5. The compromise distribution drawn in solid line is closer to the subject density than to the population mean because there is more information from the test in this example than in the previous example; the test standard deviation is 5 around the true IQ score rather than 10 in the previous example. We estimate the subject's IQ at 124 with a standard deviation of $20^{1/2}$.

9.6 Problems

1. Fit the two BSI models described in section 9.2 and confirm the minimal changes to the fixed effects from including the parent random effect. The clustering in this data set is not very heavy, and the consequences of omitting the clustering are small. In many data sets, you ignore the clustering at your own peril.

2. Several parent level variables are available in the BSI data set. Include a few of them in the two BSI analyses described and see how including or omitting the parent random effect affects the inference.

3. In the Weight Loss data, see if you can estimate an additional fixed and random slope for the times when subjects are at home. Is the fixed slope estimate different from the slope when subjects come into the clinic? Do we need two random slopes?

4. In the Weight Loss data, subjects do not start losing weight immediately. They are introduced to the weight-loss regime at week 2. Model the time trend as being constant from week 1 to week 2, then following a linear or quadratic trend. Does taking this into account in the modeling of the time trend affect the conclusions about the home scale effects?

5. In the Pediatric Pain data, at each trial where there is a pain tolerance measure, subjects were asked to rate the pain of the trial on a 1 to 10 scale. Higher ratings indicate greater pain. Decide on a random effects model and then analyze the pain rating data. Is there a difference between coping styles in baseline pain rating? Do any treatments cause an increase in pain rating for any coping styles?

6. In the IQ example, the variance of the IQ estimate is $1/(D^{-1} + \sigma^{-2})$. Suppose college students have an average IQ of 110 with standard deviation 10, and a college student takes a test with standard deviation $\sigma = 10$ and scores 130. What IQ do you estimate and with what standard error? Suppose the student scores 90, what is the estimate and standard error? Does the standard error depend on the score?

7. Consider random intercept and slope models with a randomized intervention/control delivered at $t = 0$. The intervention does not change the observations at $t = 0$ but can affect the slope over time.

   (a) Describe a random effects model where the treatment effect changes the random effects variance.

   (b) The intervention makes the random slope variance larger, although it will leave the intercept variance the same. Describe a model that includes this constraint.

10
Residuals and Case Diagnostics

Good fortune lieth within bad, and bad fortune within good.
— Lao-tzu

Luck is the residue of design.
— Branch Rickey

Overview

We define outlier and influence statistics for the multivariate linear regression model. For random effects models, we define several additional residuals and discuss residual plots at length.

A *residual* is a measure of how far an observation Y_{ij} is from its predicted value $\hat{Y}_{ij} = X_{ij}\hat{\alpha}$. An *outlier statistic* is used to formally identify observations or sets of observations that are unusually far from their predicted value, usually with an associated test statistic. Residuals can be plotted against each other or against time or covariates to suggest lack of fit of the current model and to identify changes to the model that will improve the fit. An *influence statistic* identifies observations or subjects that have a large impact on the conclusions of an analysis. It does not necessarily identify outliers, although outliers are often influential.

Residuals, outlier statistics, residual analysis, and influence assessment are well developed in linear regression. Statisticians are only beginning to develop corresponding methods for longitudinal data.

10.1   Residuals and Outlier Statistics

In chapter 2, we discussed two types of empirical residuals: empirical within-subject residuals in subsection 2.4.4 and empirical population residuals in subsection 2.4.5. For the empirical within-subject and population residuals, there was no formal statistical model specified, but taking the residuals allowed us to see something that was not as easily visible in the profile plot of the original data. A *residual* is a difference between an observation Y_{ij} and a fitted value $\hat{Y}_{ij}$. Typically, the fitted value is calculated from a model, although informal methods as in subsections 2.4.4 and 2.4.5 are common.

The basic residual

$$e_i = Y_i - X_i\hat{\alpha}$$

is an estimate of ϵ_i in the multivariate regression model (5.6)

$$Y_i = X_i\alpha + \epsilon_i$$

with

$$\epsilon_i \sim N(0, \Sigma(\theta)). \tag{10.1}$$

These $e_i = (e_{i1}, \ldots, e_{in_i})'$ residuals are called *population residuals*. They can be plotted in profile plots or scatterplot matrices to help determine if the mean and variance model are properly specified. These are the model based generalizations of the empirical population residuals defined in section 2.4.5.

If our current model is a good model, the e_{ij} should have mean zero at each time point. If there appears to be a trend over time or a non-zero mean at some time point, that indicates that the population mean has not been well specified as a function of time. The ideas of subsection 2.4.1 can be applied to the e_{ij} instead of the Y_{ij} to see if a covariate predicts e_{ij}. This would indicate that the covariate could be usefully added to the current model.

The distribution of e_i, given that we know the covariance parameters θ is normal with mean zero and variance

$$\text{Var}(e_i) = \Sigma_i(\theta) - X_i\text{Var}(\alpha)X_i'.$$

The variance of e_{ij} is the jth diagonal element of this matrix. The variance $\text{Var}(e_i)$ is slightly smaller than $\Sigma_i(\theta)$ because we subtracted off an estimate $X_i\hat{\alpha}$ of the population mean rather than the true population mean $X_i\alpha$. We estimate $\text{Var}(e_i)$ by substituting $\hat{\theta}$ for θ.

We identify observation Y_{ij} as an outlier if the outlier statistic

$$\Psi_{ij}^2 = \frac{e_{ij}^2}{\text{Var}(e_{ij})}$$

is greater than $\chi^2_{(1-a)}(1)$ or equivalently if

$$\Psi_{ij} = e_{ij}/\text{Var}(e_{ij})^{1/2}$$

is less than the normal quantile $-z_{1-a/2}$ or greater than $z_{1-a/2}$.

We can similarly construct multivariate outlier statistics that identify sets of e_{ij} as being outlying. The main use for multivariate outlier statistics would be to identify outlying subjects, and we call these statistics Ψ_i^2. If Ψ_i^2 is larger than $\chi^2_{1-a}(n_i)$, we identify subject Y_i as outlying at the $a100\%$ level.

Technically, Ψ_i^2 is a quadratic function in e_i

$$\Psi_i^2(\theta) = e_i'[\text{Var}(e_i)]^{-1}e_i.$$

We estimate $\Psi_i^2(\theta)$ by plugging in $\hat{\theta}$ for θ, which we then write Ψ_i^2. Function $\Psi_i^2(\theta)$ is distributed as a chi-square with n_i degrees of freedom and we approximate Ψ_i^2 as chi-square with n_i df as well.

We can also use a likelihood ratio test to test the hypothesis that a set of observations are outlying. The Pain data analysis of section 10.4 illustrates.

10.2 Influence Statistics

An influence statistic identifies when a particular observation, set of observations, or subject has a large impact on our inferences. Because we are usually mostly interested in α, we use Cook's distance, a statistic that assesses influence on $\hat{\alpha}$. Let $[I]$ denote a set of observations whose influence we wish to assess. Cook's distance is

$$C_{[I]} = K^{-1}(\hat{\alpha} - \hat{\alpha}_{[I]})'[\text{Var}(\hat{\alpha})(\hat{\theta})]^{-1}(\hat{\alpha} - \hat{\alpha}_{[I]})$$

where K is the number of coefficients in α, and $\hat{\alpha}_{[I]}$ is the estimate of α where we omit the set of observations $[I]$ from the analysis. Let C_{ij} be the influence statistic when $[I]$ contains only the jth observation on the ith subject, and let C_i be the statistic when $[I]$ contains all the observations from subject i.

Observations or subjects with the largest values of C_{ij} or separately the largest values of C_i are identified as influential. We do not compare C_{ij} to C_i. We would, at a minimum, wish to confirm that influential subjects are indeed members of the population under study and that there were no transcription errors in the recording of their data.

Cook's distance is a general purpose influence statistic. Often we are most concerned with a specific inference from our analysis, such as whether treatment is significantly better than control. In this case, we would check whether omitting the subjects or observations identified by C_i or C_{ij} from the data set changes the specific conclusion of our analysis. If deleting them does change our conclusions, then we say that the conclusions are not robust to deleting those observations and that our conclusions depend strongly on the validity of the influential data points. Our conclusions are fragile at best. We must note in our conclusions that the conclusions are sensitive to the presence of the influential observations. If deleting the influential observations does not change the fundamental conclusions, then we report the analysis based on the full data set.

The usual approach to assessing influence proceeds subject by subject. We might first inspect the n subject influence statistics C_i. For aberrant subjects, we would also calculate the n_i observation-level influence statistics C_{ij} and inspect the e_{ij} residuals for that subject.

There are two approaches to calculating $\hat{\alpha}_{[I]}$. One re-estimates the estimate $\hat{\theta}_{[I]}$ of θ after set deletion as well as re-estimating $\hat{\alpha}$. This takes a little bit of computational time although these days that is not much of a problem. A common shortcut does not bother with recalculating $\hat{\theta}$ and deletes the observations while keeping $\hat{\theta}$ fixed. Calculation of $\hat{\alpha}_{[I]}$ can then be done in closed form.

10.3 Additional Residuals for Multivariate Regression

The residuals e_i from longitudinal data often have substantial variation in them, so that plotting them in profile plots is not of great additional value beyond the insight gained from plotting Y_i. Several additional residuals for longitudinal data have been proposed although little has been said about how to use them in further plotting. In longitudinal data, the e_i are correlated; these additional proposals basically create a set of uncorrelated residuals from the e_i.

The first set of residuals l_{ij} are called *next step prediction residuals*. The first element l_{i1} is e_{i1} divided by its standard deviation. Each successive l_{ij} is constructed so that it (i) is uncorrelated with the previous l_{ik} for $k = 1$ to $j - 1$ and (ii) has variance one. One plots the l_{i1} against baseline covariates, and considers plotting all l_{ij} against time-varying covariates x_{ij}. One also plots the l_{ij} in a scatterplot matrix and in profile plots. Any correlations or patterns in these plots indicate that the current model is incorrect.

Principal component (pc) residuals linearly transform the e_{ij} into a different set of uncorrelated residuals m_{ij}. The first m_{i1} is the linear combination $b'_1 e_i = \sum_j b_{1j} e_{ij}$ of the e_{ij} with the largest variance λ_1. The largest variance is over all possible vectors b_1 such that $1 = b'_1 b_1 = \sum_j b_{1j}^2$. The second pc residual m_{i2} is the linear combination $b'_2 e_i = \sum_j b_{2j} e_{ij}$ with the second largest variance λ_2 but with $b'_2 e_i$ uncorrelated with $b'_1 e_i$, that is $b'_2 \Sigma b_1 = 0$, and also with length $b'_2 b_2 = 1$. The jth pc residual has $b'_j e_i$ with the jth largest variance λ_j such that $b'_j e_i$ is uncorrelated with the first $j - 1$ pc residuals and so on up to $j = J$. The vectors b_j are called the *principal components* or the *eigenvectors* of the Σ matrix, and the variances λ_j are the *eigenvalues* of Σ. As we have constructed them, the pc residuals have different variances λ_j. A constant variance version of the m_{ij} is $m_{ij}/\lambda_j^{1/2}$, called *standardized pc residuals*.

Both the principal component and the next step prediction residuals can be plotted in profile plots and scatterplot matrices. All should be mean zero at any time point, so non-zero means indicate a lack of fit of our model. A scatterplot matrix of the $m_{ij}/\lambda_j^{1/2}$ or the l_{ij} should show uncorrelated scatterplots without any patterns. Any deviation from random normal noise indicates a lack of fit of the model.

Our discussion of pc residuals and next step prediction residuals assumes balanced data. The literature does not explain how to deal with unbalanced data at this time. At best we must assume that most subjects have complete data. For next step prediction residuals, we can produce the residuals separately for each subject based on the data at hand. Consider a subject with observations at times 1 and 3 plus perhaps later observations but without an observation at time 2. We can construct the residual at time 3 by predicting it from the observation at time 1. For pc residuals, we can create separate standardized residuals for each subject from the estimated covariance matrix $\Sigma_i(\hat{\theta})$ for that subject.

A potential use of the pc residuals is to calculate

$$f_{ij} = e_{ij} - b_{1j}m_{1i}.$$

The f_{ij} can be plotted in profile plots. Like e_{ij}, m_{ij} and l_{ij}, these are mean zero. Because the first pc residual has the largest variance of all the pc residuals, the f_{ij} have substantially smaller variance compared to the e_{ij}. It may well be possible to see lack of fit, such as non-zero means at particular times of the f_{ij} residuals even when those non-zero means were not visible in the e_{ij} residuals. Non-zero means indicate that the population mean $X_i\alpha$ as a function of time is not correctly specified. We can plot the m_{i1} against baseline covariates. We may also split subjects into two groups according to a group indicator and then plot the f_{ij} residuals in two profile plots to see if the covariate is predictive of the residuals. This would indicate that the covariate should be added into the X_i matrix in our regressions.

In the next section, we will meet up with random effects residuals. These are another way, particular to random effects models, of decomposing the e_i residuals into several components.

10.4   Residuals and Residual Plots in Random Effects Models

In this section, we identify formal residuals associated with our random effects models that correspond to the empirical within-subject residuals that we saw in subsection 2.4.4. The Wallaby tails analysis presented in figure 7.11 of section 7.6 presented an example of these residuals.

In section 9.1, we identified two *error terms* or *residual terms* δ_{ij} and β_i in the hierarchical model 9.1 – 9.4. In the marginal model, we had one residual ϵ_i that was the sum of two pieces, $Z_i\beta_i + \delta_i$. We have three interrelated residuals δ_i, ϵ_i, and β_i in the random effects model. Each residual has its uses. We shall call the corresponding estimated residuals d_i, e_i, and $\hat{\beta}_i$

$$d_{ij} = Y_{ij} - x'_{ij}\hat{\alpha} - z_{ij}\hat{\beta}_i$$
$$e_{ij} = Y_{ij} - x'_{ij}\hat{\alpha}$$

and $\hat{\beta}_i$ was defined in formula (9.14). The d_{ij} and e_{ij} residuals have the same hierarchical structure as the original data Y_{ij} and can be plotted just as the Y_{ij}'s can be plotted, for example in scatterplot matrices and profile plots. We can plot summary statistics of the d_{ij} and e_{ij} and plot them as we plotted mice means and standard deviations within day in figures 2.6(a) and (b) or we can plot empirical correlograms as in figure 2.22. The $\hat{\beta}_{ik}$'s, $k = 1, \ldots, r$ can be plotted against each other and against baseline covariates.

The e-residuals $e_i = Y_i - X_i\hat{\alpha}$ are population-level residuals. We may plot them in profile plots and for balanced data, in scatterplot matrices. The interpretations of these plots are similar to the interpretations that we gave to these plots when we plotted the raw data Y_i. The ϵ_i errors have mean zero at any given time or at any given value of a covariate. We can check that the e_i residuals also have these properties. The variance-covariance matrix of ϵ_i is $\Sigma(\theta)$. We can check that the e_i have a variance-covariance matrix of approximately $\Sigma(\hat{\theta})$, although as mentioned in section 10.1, the correct covariance is $\Sigma(\hat{\theta}) - X_i\text{Var}(\alpha)X'_i$. We look to see in the profile plots that the variances of the e_i over time follow those estimated from the model, and we check that the correlations in the scatterplot matrix also follow those estimated from the model.

The δ_{ij}'s are within-subject residuals and are assumed to have mean zero, constant variance σ^2, and to be independent of each other. The corresponding d_{ij} may be easier to work with than the e_i, as the theoretical properties of the δ_i are simpler than those of the ϵ_i. We plot the d_{ij} in profile plots and scatterplot matrices. We have to be a bit careful in assessing correlation, as the d_{ij} are likely to have residual correlation in them, even though the δ_{ij} do not. When we estimate the d_{ij}, the estimation process introduces correlation into the residuals even when none exists in the δ_{ij}.

Figure 7.11(b) of section 7.6 analyzing the Wallaby tail length data illustrates profile plots of the d_{ij} residuals. We see a very clear pattern particularly in the first 300 days. There are several time intervals where all of the d_{ij} residuals are entirely positive or entirely negative. This indicates at a minimum that our model does not fit. We were fitting a spline model to the tail data, and the pattern in the residuals indicates that more knots are needed to better approximate the tail growth trends in the first few hundred days. The range on the vertical axis from -3 to $+2$ indicates the range of the differences between the fitted values and the data. The original data ran from 0 up to 50, and now the errors are plus or minus 2 or 3. More specifically, around $t \approx 200$, or just a little earlier, the average error appears to be between -1.5 and -2, and the original data appears to be around 20, suggesting that the error in the model is approximately $(1.5/20)100\% = 7.5\%$ to $(2/20)100\% = 10\%$ at day $t = 200$.

The $\hat{\beta}_i$ are the estimated random effects. They are also called empirical Bayes residuals. The $\hat{\beta}_{ik}$ may be plotted against each other, as in a scatterplot for $r = 2$ or in a scatterplot matrix if $r > 2$. They can be plotted against baseline covariates. If the $\hat{\beta}_{ik}$ appear correlated with a baseline covariate G_i, when we would add a column $Z_{ik}G_i$ as a column in the X_i matrix, where Z_{ik} is the kth column of the Z_i matrix.

It is also of interest to plot the subject-specific fitted values, $x'_{ij}\hat{\alpha} + z'_{ij}\hat{\beta}$ in profile plots. These plots can be compared to the raw data to see if they display the same structure as the raw data do. In the Wallaby tail length data, the model fit in figure 7.11(c) does not display exactly the same curves at the earliest times as the original data in plot 7.11(a). The fitted values from the more complicated model in 7.11(d) did a better job of imitating the Wallaby tail length data.

We look at several examples.

10.4.1   Small Mice Data

When I first analyzed the Small Mice data, I saw that the mice weights were very similar at $t = 2$; they all started at about the same weight, indicating a fixed intercept in the model. They grew at a fast rate, but the rate differed for different mice, indicating both a fixed effect for time plus a random time effect. There appeared to be a kink in the data around day 14 but I could not tell what this indicated for the model. So I fit a first model with a fixed intercept and slope and a random slope as well and calculated the subject-specific fitted values $\hat{Y}_{ij} = \hat{\alpha}_1 + \hat{\alpha}_2 t_{ij} + \hat{\beta}_{i1} t_{ij}$ and residuals $d_{ij} = Y_{ij} - \hat{Y}_{ij}$.

Figure 10.1 plots (a) the data for the Small Mice, (b) the fitted values from this random slope, fixed intercept model. In figure 10.1(b), the fitted values continue linearly at all times. The variance of the fitted values at a given time is small for early days, and large at later times. The fitted values do not seem to capture the patterns we see in figure 10.1(a), where after day 14 the lines do not spread further apart. The residuals are plotted in 10.1(c). The residuals d_{ij} at $t = 2$ and at $t = 20$ are all negative. Nearly all residuals at times $8 \leq t \leq 14$ are positive, and the few negative residuals are close to zero. We see a very clear quadratic pattern over time in this plot. The quadratic curves seem to be the same for each mouse and we

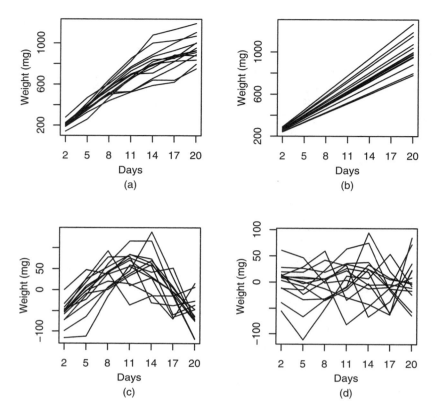

Figure 10.1. Small mice data. (a) Profile plot of raw data again. (b) Fitted values from random slope model. (c) E-residuals from random slope model. (d) E-residuals from random slope model with fixed quadratic term.

conclude that we need to add a fixed quadratic time trend t_i^2 to our model. The residuals at each time t_i do not have mean zero; including a fixed t_i^2 term in our model should theoretically remove this quadratic structure from the residuals and move it into the fitted values.

Figure 10.1(d) plots the residuals from the model with the quadratic fixed effects and random slope model

$$Y_{ij} = \alpha_1 + \alpha_2 t_{ij} + \alpha_3 t_{ij}^2 + \beta_{i1} t_{ij} + \delta_{ij}.$$

Now there is no time where the residuals are predominantly negative or predominantly positive. The mean structure seems acceptable now.

There is still one problem that the residual plot 10.1(d) shows. At the left-hand side of the plot, the lines are strongly parallel, indicating a strong positive correlation between the residuals at $t = 2$ and $t = 5$. Between $t = 5$ and $t = 8$, the correlation is not as strong, but is still positive. The remaining correlations appear small, and are hard to judge, but between the last

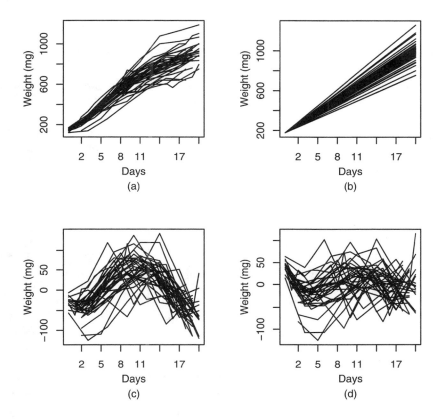

Figure 10.2. Big mice data. (a) Profile plot of raw data again. (b) Fitted values from random slope model. (c) E-residuals from random slope model. (d) E-residuals from random slope model with fixed quadratic term.

two time points, the lines cross and we see moderate negative correlation between the residuals. Overall, this indicates that our correlation structure is not well modeled by the random slope model. We explored the correct correlation structure for the mice data in section 8.3.1, and indeed we saw in table 8.1 that a number of covariance models fit better than the random slope model.

We consider the same models for the Big Mice data. The range of the times extends earlier to $t = 0$ and $t = 1$. The raw data, presented in figure 10.2(a), gives the same general impression as we had with the Small Mice data, and similarly the fitted values in figure 10.2(b) from the random slope model do not seem to adequately describe the data. In both figures 10.1(b) and 10.2(b), the fitted values seem too straight and do not quite seem to imitate the raw data in 10.1(a) and 10.2(a) very well. The residuals d_{ij} from the fixed intercept and slope plus random slope model is plotted in

10.2(c). We again see the general quadratic trend in the residuals showing the need for a t_{ij}^2 term. However, with this data set, there is additional mean structure for $t \leq 2$, where we see a twist that is not part of a quadratic time trend. Still, the quadratic is the strongest feature of the residual plot, and we add the quadratic term to the model and recalculate the d_{ij} residuals plotted in figure 10.2(d). Unlike 10.1(d), there is still a modest time trend, particularly between days 0 and 8 where there is a small quadratic trend, and possibly a cubic or even quartic over the entire time. We mentioned briefly at the very end of section 7.2 that the Big Mice data are fit by a quintic (fifth-order) polynomial in time. The residual plot indicates that at least a cubic polynomial in time is needed, and we would continue building up to still more complicated models for the mean of the weights over time.

10.4.2 Pediatric Pain Data

We fit a random intercept model with a single set of covariance parameters to the log base two pain tolerance data. The fixed effects are coping style (CS), treatment (TMT), and CS×TMT. We plot a profile plot of the d_{ij} residuals in figure 10.3 separately for attenders and distracters. We see that the spread of the attender residual profiles runs from -1 to 1 except for a single outlier at around $+3$ that belongs to subject 21. The spread of the distracter residuals spreads evenly from below -1 up to $+3$. The variance of the distracter residuals seems much higher than for the attender residuals, and in section 8.4 we did fit a model to the Pain data that allowed for different covariance parameters for the attenders and distracters. Here we continue the analysis with just one set of covariance parameters.

Maximum likelihood (ML) fitting can be used to give a test that a subject is an outlying subject. We fit the full data set using maximum likelihood, and then we delete subject 21 and refit the data. The minus 2 log likelihood for the model with subject 21 is 570.7. Without subject 21, the -2 log likelihood is 550.4. The difference is 20.3, and, approximately, the difference is distributed as a chi-square distribution on 4 degrees of freedom, as by deleting subject 21 we have removed 4 observations. This difference has a p-value of .00044; the .999 percentile of a chi-square(4) is 18.5. Subject 21 is clearly an outlier. However, recall that we picked subject 21 from among 64 subjects as the worst outlier. We probably should adjust our assessment of significance for the fact that this chi-square test is quite possibly the worst of 64 chi-square test statistics. We may use the Bonferroni correction to adjust our p-value to account for the implicit multiple comparison that we did when looking at figure 10.3. The Bonferroni correction to the p-value multiplies the observed p-value of .00044 by 64 to give a Bonferroni corrected p-value of .028, which is still significant.

Figure 10.4 shows a stacked histogram of the REML $\hat{\beta}_i$'s for attenders and distracters. Each bar of the histogram is constructed from blocks of height one corresponding to one observation. Clear blocks correspond to

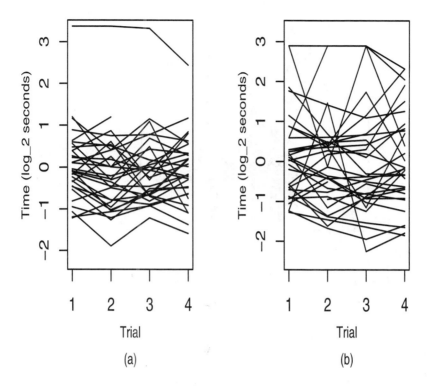

Figure 10.3. Pediatric Pain data. Profile plots of the (a) attender and (b) distracter d_{ij} residuals.

distracter observations and solid blocks correspond to attender $\hat{\beta}_i$'s. For example, between .75 and 1, there is 1 distracter and 3 attender $\hat{\beta}_i$'s. We see that attenders generally have smaller variance than distracters. There is a single outlying high attender, again subject 21, giving further impetus to our removing that subject from the analysis.

When we compare the analyses with and without subject 21, there is no major change in the general findings regarding the effects of treatment. There is however a change in the inference regarding the difference between the attenders and distracters at baseline. The estimated baseline difference distracter minus attender goes from .48 to .59, while the standard error stays about the same and the p-value goes from .04 to .007. This is expectable, as subject 21 is the highest attender, and deleting subject 21 lowers the baseline attender average.

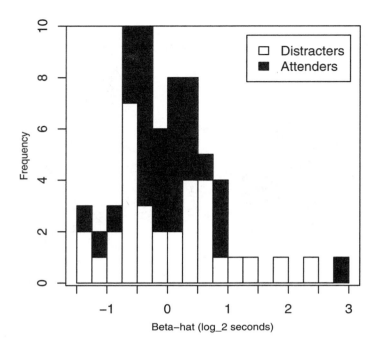

Figure 10.4. Pediatric Pain data. Stacked histogram of estimated random intercepts by coping style (a) attenders (dark blocks), (b) distracters (clear blocks).

10.4.3 Weight Loss Data

When we first plotted the Weight Loss data in figure 2.15, we saw only that each subject had their own weight and that each subject's observations appeared to vary around their own subject mean weight. Call this fixed plus random intercept model model 1.

$$Y_{ij} = \alpha_1 + \beta_i + \delta_{ij}.$$

We fit this model, calculate d_{ij} residuals and re-plot the residuals in figure 10.5(a). This is a more principled version of figure 2.4.4 that we plotted in chapter 2. As we saw there, the residuals are mostly positive at early times and mostly negative at later times, indicating that subjects are losing weight over time and that we need to include a fixed time slope in our model.

Let model 2 be model 1 with the addition of a fixed time slope

$$Y_{ij} = \alpha_1 + \alpha_2 t_{ij} + \beta_i + \delta_{ij}.$$

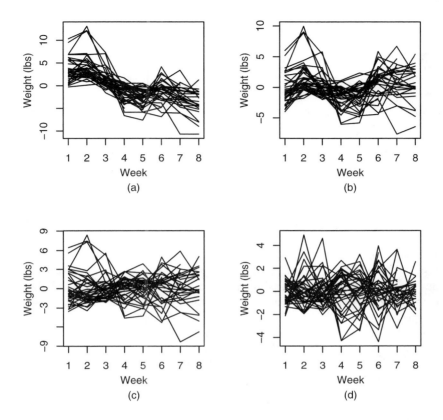

Figure 10.5. Weight Loss residuals from models 1 through 4. (a) Model 1, random intercept; (b) model 2, random intercept, fixed slope; (c) model 3, unstructured mean, random intercept; (d) model 4, unstructured mean, RIAS. The vertical axes in each plot are different; read the labels to see the range.

Profiles of residuals for this model are plotted in figure 10.5(b). The range of the residuals, from −5 to +5 with a few exceptions, is smaller than the range from model 1, and this is reflected in the smaller range on the vertical axis in figure 10.5(b) as compared to figure 10.5(a). In this figure, we see that there is still a trend over time. The trend is not simple to describe, but the mean residual value seems to undulate in a regular fashion, low at week 1, high at weeks 2 and 3, low again at weeks 4 and 5, high again at week 6, and unclear for the last two weeks. If one explores various polynomials, one will find that a quartic polynomial does a fair job of describing the data. Not liking a quartic, we might move to an unstructured mean model. Model 3 has an unstructured mean model, and the covariance model is a random intercept. Residuals from this model are plotted in figure 10.5(c).

There is no longer a time trend for the mean in figure 10.5(c), but there is still a pattern for the variances over time. The variance starts high at

Model	-2 LL	$\hat{\sigma}^2$	$\hat{D}_{11}$	$\hat{D}_{12}$	$\hat{D}_{22}$
1	1690.9	15.8	637	–	–
2	1532.6	7.8	646	–	–
3	1478.5	6.5	644	–	–
4	1383.9	3.1	678	-5.7	.61

Table 10.1. Summary information from fitting models 1 through 4 to the Weight Loss data: Model number, -2 log likelihood, $\hat{\sigma}^2$, $\hat{D}_{11}$, $\hat{D}_{12}$, $\hat{D}_{22}$.

week 1, is smallest right before or at week 4, continues small at week 5, and then is large again for the final weeks. This pattern is also visible in figure 10.5(b), but we paid attention first to the non-constant mean function. For both 10.5(b) and (c), there is an important detail that can be best brought out by looking at the individual profiles from model 2 or 3, and that is that the non-constant variance is not caused by increasing and decreasing σ^2 variance but because profiles that start high end up low, whereas those that start low tend to end high. And plenty of residual profiles start in the middle and end in the middle too. Those subjects high at the beginning and lower at the end suggests that some subjects have time trends that are steeper than the population trend. Similarly, subjects starting low and ending high suggests that other subjects have slopes that are less steep compared to the population trend. This means that we need to add a random slope to our model as well as the random intercept producing model 4. The residuals from this model are shown in figure 10.5(d). The range of the residuals from this model are the smallest of the four models, indicating that each model is better than the previous in fitting our data.

Table 10.1 gives some summary results from the four fitted models. The second column gives the -2 log likelihood. Each improvement in the model results in substantial increases in the -2 log likelihood. The third column gives $\hat{\sigma}^2$, which is decreasing each time we add a term to the model that explains variation in Y_{ij}. The last three columns give the one or three elements of the variance $\hat{D}$ of the random effects. The estimated variance of the random intercepts $\hat{D}$ increases slightly from model 1 to model 4. As we add terms into the model, the terms explain the δ_{ij} errors and so σ^2 decreases. But the additional terms in the model do not explain the variation in the intercepts, and so D_{11} does not also decrease.

As mentioned in section 6.1, the likelihood ratio test between the RI and the RIAS models is distributed as a 50-50 mixture of chi-square random variables with 1 and 2 df. The difference in likelihoods is 94.6 which is very significant even for a chi-square with 2 df. We conclude that model 4 fits much better than model 3.

10.5   Problems

1. Calculate outlier and influence statistics for subjects in the Pain data. Use the unstructured covariance model. For the fixed effects use coping style, treatment, and coping style by treatment interaction.

2. Calculate outlier and influence statistics for all subjects in the Weight Loss data for models 1, 2, 3, and 4. Are there more or fewer outliers as the model gets more complex? Are there more or fewer influential observations as the model gets more complex? For your preferred model, plot the outlier and influence statistics against n_i. Is there any pattern?

3. Fit the log pain tolerance data using maximum likelihood and a random intercept model with and without subject 21 and show that the -2 log likelihood is as stated in section 10.4. By deleting an attender subject that was very high at all times, how would you expect the baseline parameter estimates to change? The treatment estimates? And is this indeed what happens?

4. Fit the log pain tolerance data with different covariance parameters for attenders and distracters.

    (a) Is subject 21 still outlying? Why do you say so? Plot residual profile plots and histograms of the estimated random intercepts.
    (b) What happens to the variance parameter estimates for attenders when you delete subject 21?
    (c) What happens to the variance parameter estimates for the distracters when you delete subject 21?

5. Develop the best model that you can for the Weight Loss data that includes either a fixed or random visit effect. What weight loss do you estimate over the period of time in the study?

11
Discrete Longitudinal Data

The opposite of a correct statement is a false statement. But the opposite of a profound truth may well be another profound truth.

– Niels Bohr

A second such blow would have killed him, but the horse kicked at random ...

– Thomas Bulfinch

Overview

In this chapter, we discuss random effects models for discrete longitudinal data, in particular, random effects models for longitudinal binary data and longitudinal count data.

All types of variables are assessed in longitudinal studies. Longitudinal 0-1 or binary data are particularly common. In the Morbidity data set, researchers recorded morbidity status of subjects over time. Presence (1) or absence (0) of a number of diseases and symptoms at a number of time points were recorded. This gives us a longitudinal binary response to relate to any of a number of covariates. In the Bolus Count data, researchers recorded a count of the number of doses of pain killer that subjects self-administer in twelve consecutive four-hour periods. The counts are non-negative integers.

The logistic regression model is a generalization of the normal linear model. It is used to relate cross-sectional binary data y_i to a vector of covariates x_i. Similarly, the Poisson regression model allows us to relate a single count y_i to a vector of covariates x_i. The logistic and Poisson regression models can be generalized to handle longitudinal binary and count data by introducing random effects into the model. These models are called the binomial or Bernoulli or logistic random effects model and the Poisson random effects model.

In the next section, we review the logistic regression model and then introduce the logistic random intercept model for longitudinal binary data. The second section does the same for the Poisson model and longitudinal count data.

11.1   The Logistic Random Effects Model

We briefly discuss notation and language for the logistic regression model for cross-sectional data before generalizing it to the logistic random effects model.

Let y_i be a single binary observation on subject i and let x_i be a vector of K covariates. When modeling binary y_i, the linear regression model $y_i = x_i'\alpha + \delta_i$ is mathematically inconvenient to use. We must restrict $x_i'\alpha + \delta_i$ to be only 0 or 1. Because it is potentially possible for $x_i'\alpha$ to take on any value in minus infinity to plus infinity, the possible values of δ_i depend on x_i and the unknown α.

The logistic regression model is simpler and has better statistical properties. We model y_i as Bernoulli

$$y_i \sim \text{Bernoulli}(\pi_i) \qquad (11.1)$$

where $\pi_i = P(y_i = 1|x_i)$ is the probability that y_i is 1, and $1 - \pi_i$ is the probability that y_i is 0. We wish to relate π_i to the covariates x_i. We assume that the probability π_i is a function of a linear combination $x_i'\alpha$ of the covariates x_i, called the *linear predictor*. The linear predictor can vary over the entire real line. The probability π_i is restricted to be between 0 and 1. We transform π_i so that the transformed probability takes values on minus infinity to plus infinity. There are many transformations, called *link functions*, that can do this. Mathematically, a convenient link function is the logit function, defined as

$$\text{logit}(\pi_i) = \log \frac{\pi_i}{1 - \pi_i}.$$

The ratio $\pi_i/(1-\pi_i)$ is called the *odds* and it can be anywhere between 0 and infinity. Taking the log base e of the odds allows the logit of π_i, $\text{logit}(\pi_i)$, to range from minus infinity to plus infinity. The $\text{logit}(\pi_i)$ is called the *log odds* and it is also referred to as the *logit scale*.

We set the log odds equal to the linear predictor $x_i'\alpha$

$$\text{logit}(\pi_i) = x_i'\alpha. \tag{11.2}$$

This makes the connection or *link* between the mean of y_i and the *linear predictor* $x_i'\alpha$. We can invert the formula for log odds to translate linear predictors back to probabilities

$$\pi_i = \frac{\exp(x_i'\alpha)}{1 + \exp(x_i'\alpha)}.$$

This is sometimes called the *expit* function

$$\text{expit}(x_i'\alpha) = \frac{\exp(x_i'\alpha)}{1 + \exp(x_i'\alpha)}.$$

Let subject i have linear predictor $x_i'\alpha$ and subject l have linear predictor $x_l\alpha$. If $x_i'\alpha < x_l'\alpha$ then $\pi_i < \pi_l$; subject l will have a higher linear predictor and will have a higher probability of responding $y_l = 1$.

The logistic model is fit using maximum likelihood (ML) to give ML estimates $\hat{\alpha}$ with estimated variance $\hat{V}(\alpha)$ and standard errors $\text{SE}(\alpha_k)$ that are square roots of the jth diagonal element of the estimated variance matrix $\hat{V}(\alpha)$. We may construct approximate 95% confidence intervals for α_k as $\hat{\alpha}_k \pm 2\text{SE}(\alpha_k)$. To get other confidence intervals, the 2 is replaced by other percentiles of the standard normal distribution.

In the Morbidity data, presence or absence of a number of morbidities were assessed at up to 15 visits over the two-year period. We develop a model to analyze responses over time regarding a single morbidity. The jth observation on subject i is Y_{ij} and is recorded at time t_{ij}. The response Y_{ij} is zero if the subject did not have the particular disease or symptom characterizing that morbidity and $Y_{ij} = 1$ if the subject does have the particular disease or symptom. The vector of covariates at time t_{ij} is x_{ij}. Covariates are specified as we learned in chapter 7.

Like the logistic regression model, the random effects logistic regression model begins with a Bernoulli model for Y_{ij} given the probability π_{ij}

$$Y_{ij} \sim \text{Bernoulli}(\pi_{ij}). \tag{11.3}$$

We set up a model for π_{ij} that relates it to the fixed effects $x_{ij}'\alpha$. We assume that the first element of x_{ij} is a 1 for all i and j and so α_1 will be the intercept.

As with continuous longitudinal responses, observations Y_{ij} and Y_{il} within subject i should be more similar to each other than observations from different subjects, that is, Y_{ij} and Y_{il} should be correlated with each other. Positive correlation for binary data means that given that $Y_{ij} = 1$ then Y_{il} is more likely to be 1 than it would otherwise have been. Similarly, if $Y_{ij} = 0$ then Y_{il} is more likely to be 0 than we would have otherwise expected.

We will model this correlation by introducing a random effect β_{i1} into our linear predictor. This subject-specific level translates into each subject having their own intercept $\alpha_1 + \beta_{i1}$ that deviates from the population intercept α_1. As with the continuous response random effects model, we assume the β_{i1}'s are normally distributed

$$\beta_{i1} \sim N(0, D). \tag{11.4}$$

The random effects term β_{i1} is added to the fixed effect term $x'_{ij}\alpha$ to give the logistic random effects model's linear predictor $x'_{ij}\alpha + \beta_{i1}$. The logit of π_{ij} is set equal to the linear predictor

$$\text{logit}(\pi_{ij}) = x'_{ij}\alpha + \beta_{i1}. \tag{11.5}$$

Equations (11.3)–(11.5) specify our logistic random intercept model.

The output of software to fit the random effects logistic regression model will provide us with estimates and standard errors of the fixed effects. We may construct the usual confidence intervals using quantiles of the normal distribution, and we perform the usual univariate or multivariate tests for coefficients equal to zero.

Data analysis for logistic random effects models is more difficult than for continuous response models. In particular, the profile plot is of little value. Empirical summary plots may be of some value in deciphering the effects of covariates. One may plot for (say) males and for females, the fraction of 1's at each time t_{ij}. We would connect the dots for males and also for females. One would repeat similarly for other covariates in the analysis. For continuous covariates we might cut the distribution of the covariate at one or more points and summarize the fraction of 1's at each time point for subjects in each subset.

11.1.1   Morbidity Data

Malaria sickens roughly 300–500 million people each year and kills roughly one million people each year. Malarial infection is recorded in the Morbidity data set of the Kenya school lunch intervention. We model malarial incidence in our data set with a logistic random intercept model. For covariates we include age_at_baseline in years, gender (girl=0, boy=1), and four intervention-group-by-time (in months) interactions to capture potentially different slopes over time in each intervention group.

Malaria is spread by mosquitos, which might reasonably be expected to be more or less populous in some kind of seasonal pattern. To model seasonal effects, we include a $\sin(2\pi \text{ time }/12)$ and $\cos(2\pi \text{ time }/12)$. A sine or cosine term allows for smooth seasonal variation; they each cycle up and down once every 2π time units but with the sine term lagging behind the cosine term by $\pi/2$. Dividing time in months by 12 to convert to years and multiplying by 2π makes for one up and down cycle each year. Including the two terms in the model allows the annual cycle to start at any time and

Parameter	Est	SE	t	p
Intercept	-.34	.38	-0.9	.37
Age at baseline	-.16	.05	-3.2	.002
Boy	-.16	.12	-1.4	.17
Time×control	-.076	.011	-7.0	<.0001
Time×calorie	-.076	.011	-7.2	<.0001
Time×milk	-.076	.011	-7.1	<.0001
Time×meat	-.076	.011	-6.7	<.0001
Sin	-.15	.07	-2.2	.026
Cos	.20	.06	3.4	.00080
Calorie-control	.00041	.013	0.0	.97
Milk-control	-.00009	.013	0.0	.99
Meat-control	.00001	.014	0.0	1.00
Calorie-milk	.00051	.013	0.0	.97
Calorie-meat	.00040	.013	0.0	.98
Milk-meat	-.00010	.013	0.0	.99
Age/12 - time(ave)	.063	.008	7.5	<.0001
Age/12 - time(cont)	.063	.012	5.4	<.0001
D	.75	.12	6.5	<.0001

Table 11.1. Malaria results. Estimates, standard errors, t-statistics, and p-values for parameters. Older children have less malaria than younger children. There is a strong seasonal trend, a general decreasing trend over time, but no intervention effect. Treatment effect contrasts and two age versus time in study contrasts are given.

allows for a variety of shapes to the cycle. Adding sine and cosine terms can be done with continuous responses as well as discrete responses. The units of time should be years for annual cycles, months for monthly cycles, and days for daily cycles.

Results are given in table 11.1. As with continuous data we have parameter estimates, standard errors, t-statistics, and p-values. Eight different contrasts are given, and the estimated variance D of the random effects is included in the table. There is only one variance term D in the random intercept logistic regression model. Unlike the normal residual error distribution, once we determine π_{ij} for a particular observation Y_{ij}, the Bernoulli error distribution is fully determined.

Age_at_baseline is significant; older children have less malaria than younger subjects. Time is clearly significant; as the study progresses, children are less likely to get malaria in each intervention group. All four slope coefficients are negative and equal to $-.076$. We can easily guess that the four slope coefficients are not significantly different without formally doing a test. The test for differences in the interventions is a chi-square test with 3 degrees of freedom. The test statistic is essentially zero and the

p-value is 1, indicating absolutely no difference in time trends among the four intervention groups.

We have both age_at_baseline and time in study as covariates in this analysis. If the coefficients were equal, then being older at baseline contributes to an equal decrease in malarial incidence as aging within the study. If the coefficients were equal, we would put age in the analysis. We would put terms for age and age by intervention interaction in the model instead of time and time by intervention, and we would omit age_at_baseline from the analysis.

Because age_at_baseline is measured in years, we need to divide its coefficient and standard error by 12 before comparing to the coefficients of time for any intervention group, as time was measured in months. This means that the coefficient of baseline age in months will be a bit above $-.01 \approx -.16/12$ with a very small standard error $.01 = .12/12$ and the coefficient is very different from the $-.076$ coefficient of the age effects. Two different contrasts are given. The first contrast is the coefficient of $age/12 -$ time where the time coefficient is the average of the four intervention time trend coefficients. The second contrast compares $age/12$ to the slope for the control group only, and the standard error is somewhat higher than the previous contrast because there is less data being used to estimate the time slope. For either contrast, the age_at_baseline and time slope are significantly different. We find that the decrease in malaria during the study is greater than the decrease due to increasing baseline age. Because the decline is the same in the control group and in the three intervention groups, it suggests that the decline is not related to the intervention. The cause of the difference could be due to short term weather changes during the study. If the decline is related to being in the study, it could be caused by the regular morbidity surveys about whether the subjects had malaria.

We use an F test to test for an overall seasonal effect. The null hypothesis is that the coefficients of the sine and the cosine terms are both zero. The test statistic is the multivariate test statistic

$$\chi^2 = \hat{\alpha}_I \mathrm{Var}(\hat{\alpha}_I)^{-1} \hat{\alpha}_I$$

where I represents the set of coefficients we wish to test equal to zero, and α_I is that set of coefficient estimates and the variance is the estimated variance matrix of the coefficients. This is the multivariate form of the squared univariate t-test. The test statistic is compared to a chi-square statistic with 2 degrees of freedom. The statistic is $\chi^2_2 = 16.6$ and the p-value is .0007; we have a very significant seasonal effect.

We would like to understand how the slope and the seasonal effect affect the probability of having malaria. We calculate the estimated probability that a particular subject has malaria for a given set of covariate values and a given random effect β_i value. Calculating $\pi_i = \mathrm{expit}(x'_{ij}\hat{\alpha} + \beta_i)$ gives the probability of malaria. We calculate and plot the probabilities over time. A single curve is often rather dull and we typically graph several curves in a single plot. The choices of covariate values are dictated by what information we wish to communicate to the viewer.

Figure 11.1(a) solid line plots the probability of malaria over time in the study for a 6-year-old male in the control intervention whose random effect β_i is zero. The probability starts at around .2 and decreases to around .05 after two years. The decrease is due to the negative slope estimate of $-.076$ for time. The sine and cosine terms induce wiggles that peak around month 10 and 22 and have a trough around months 6 and 18. The dashed lines are for a subject with these same covariates but whose β_is are equal to plus one or minus one times the estimated $\hat{D}^{1/2}$. The dotted lines are for subjects with β_i equal to $\pm 2\hat{D}^{1/2}$ again with the same covariates. The choice of $\beta_i = 0$ sets β_i to the mean and median of the random effects distribution. The five subjects with their different β_i represent a range of possible subjects, from the low end of the random effects distribution with β_i at two standard deviations below its mean to the end high end when β_i is 2 standard deviations above its mean.

The curves are more spread apart at the beginning when $t = 0$ and are closer together at the end when $t = 24$. The curves with $\beta_i = +1 \times \hat{D}^{1/2}$ and $+2 \times \hat{D}^{1/2}$ are farther above the solid line than the $\beta_i = -1 \times \hat{D}^{1/2}$ and $-2 \times \hat{D}^{1/2}$ curves are below the solid line. This is due to the nonlinearity of the logit/expit transformations. If $x'_{ij}\alpha$ is negative and $c > 0$, then $\mathrm{expit}(x'_{ij}\alpha + c) - \mathrm{expit}(x'_{ij}\alpha)$ will be greater than $\mathrm{expit}(x'_{ij}\alpha) - \mathrm{expit}(x'_{ij}\alpha - c)$. Probabilities cannot go below 0 and the curves bunch up as the probabilities approach zero.

In figure 11.1(b), we illustrate another aspect of the nonlinearity of the logit/expit transformations. The two solid lines correspond to girls of ages 5 and 10 in the control group with random effects $\beta_i = 0$. For a normal, the mean is also the median of the distribution. After we transform back from the logit scale back to the probability scale, the median is still the median, but the mean on the probability scale is not the transformation of the mean on the logit scale. These curves are *subject specific* curves. The dotted lines are *population averaged* probabilities for 5 and 10 year old girls in the control group. That is, for any time point and set of covariates, we calculate the mean of $\mathrm{expit}(x'_{ij}\alpha + \beta_i)$ over the distribution of β_i, which is $N(0, \hat{D})$. See problem 6. The probabilities are all less than .5, but the population averaged probabilities are closer to .5 than the individual with $\beta_i = 0$. It will always be the case that the population averaged probabilities will be closer to .5 than the individual with $\beta_i = 0$. One can think of the profile with $\beta_i = 0$ as averaged on the logit scale, while what we call the population averaged probabilities are averaged on the probability scale. If we wish to estimate the fraction of children with malaria in the population, it is the average on the probability scale that we need to calculate. The difference between the subject specific profile and the population average

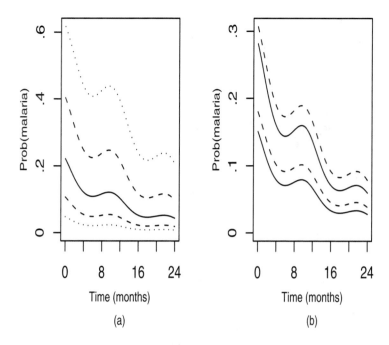

Figure 11.1. Plot of estimated probabilities of having malaria over the course of the study. (a) All curves are probability estimates for boys age 6 at baseline in the control group. Solid curve is the probability for $\beta_i = 0$. Other curves are for $\beta_i = \pm 1$ (dashed) and $\beta_i = \pm 2$ (dotted). (b) Probabilities for girls age 5 (upper solid curve) and age 10 (lower solid curve) at baseline in the control group with $\beta_i = 0$. Dashed lines are population average probabilities for 5-year-old (upper) and 10-year-old (lower) control group girls.

profile is the same issue that was confronted in section 6.6 for making inference about the mean on the original scale after having taking a log transformation of normal data.

11.1.2   Other Models

There are many random effects models that we might consider. The random effects model can be made more complex by introducing more random effects. Let z_{ij} be a vector of r covariates for subject i at time j with z'_{ij} corresponding to the jth row of the Z_i matrix in the normal random effects model. We model r random effects β_i as being distributed $N_r(0, D)$. Initial choices for the columns of Z_i for binary data might or might not be similar to our choices for continuous data.

The RIAS model might seem an obvious second choice for random effects model. Unfortunately it rarely fits longitudinal binary data. The problem is twofold. First, binary data has very little information in it, particularly compared with continuous data, so that we have very little information for estimating parameters of the assumed correlation model. We may not be able to notice or fit a random slope if the variance is small. Second, if a subject-specific time slope is large, then that subject's beginning or ending (or both!) will generally be a sequence entirely of 0's or 1's. The remainder of the sequence may be a mixed area where both 0's and 1's appear. And some subjects will have all 0's and some will have all 1's. This description does not describe many sets of longitudinal binary data, which tend to either be mostly zero's with a few ones or vice versa. The random slopes, if they exist at all must necessarily be close to zero, causing the random slope variance to be close to zero, which in turn means that we have little power or ability to estimate the variance.

A more complex model has

$$\text{logit}(\pi_{ij}) = x'_{ij}\alpha + \gamma_{ij}$$

with the γ_{ij} having a correlation model such as an AR-1 model or other choice from chapter 8

$$\gamma_i = \begin{pmatrix} \gamma_{i1} \\ \vdots \\ \gamma_{in_i} \end{pmatrix} \sim N(0, \Sigma(\theta)).$$

This model is akin to the additive covariance models of section 8.5. One component is the $\Sigma(\theta)$ error, and the second component, an independent additive error, is the Bernoulli error from $Y_{ij} \sim \text{Bernoulli}(\pi_{ij})$.

We used the logit transformation to map probabilities π_{ij} to the linear predictor $x'_{ij}\alpha + \beta_i$. Other transformations are possible. The most popular alternative is the probit transformation, which is the inverse of the standard normal cumulative distribution function. Results are often quite similar to the logit, particularly if most probabilities π_{ij} are between .2 and .8.

Another class of models are *transition models*. These models correspond approximately to the AR(p) models for continuous data. In transition models, Y_{ij} is modeled in a logistic regression where the linear predictor includes terms for previous responses $Y_{i(j-1)}$ and also potentially $Y_{i(j-2)}$ and so on as needed as well as covariates of interest. If K prior responses are used to predict the current response, then typically the first K responses are not used in the estimation. These models can potentially be fit using regular logistic regression software.

11.2   Count Data

Cross-sectional count data y_i is often modeled as a function of covariates x_i using the Poisson regression model. The Poisson distribution has a non-negative mean $\lambda > 0$. A Poisson random variable takes values in the non-negative integers 0, 1, 2, and so on. The probability density of $Y \sim \text{Poisson}(\lambda)$ is

$$P(Y = k) = \frac{\exp(-\lambda)\lambda^k}{k!}.$$

The mean and the variance of this distribution are both λ. We wish to relate the Poisson response to a linear combination $x_i'\alpha$ of covariates. As with the logistic regression model, we take the mean of y_i,

$$\text{E}[y_i|\lambda_i] = \lambda_i$$

and relate it to the linear predictor $x_i'\alpha$. If we take the log of λ_i, we get a quantity that ranges over the whole real line, the same as the natural range of $x_i'\alpha$. Just as the logit function was the *link* function for the Bernoulli mean π_i, the log is the link function for the Poisson mean λ_i and we set $\log \lambda$ equal to the linear predictor

$$\log \lambda = x_i'\alpha.$$

We fit the Poisson regression model using maximum likelihood. This gives an estimate $\hat{\alpha}$ for α and an estimated covariance matrix $\hat{V}$ of $\hat{\alpha}$ whose diagonal elements are the variance estimates of α_j.

The Poisson random effects model arises when we add a random effect into the linear predictor. This is done, for example, when we have longitudinal observations Y_{ij}, $j = 1, \ldots, n_i$ on subject i. The random intercept model adds a single random effect β_i to the linear predictor $x_{ij}'\alpha + \beta_i$ of each Y_{ij}. The full Poisson random intercept model is

$$
\begin{aligned}
Y_{ij} &\sim \text{Poisson}(\lambda_{ij}) \\
\log \lambda_{ij} &= x_{ij}'\alpha + \beta_i & (11.6) \\
\beta_i &\sim N(0, D). & (11.7)
\end{aligned}
$$

The random effect induces a positive correlation across observations within subject. A β_i above zero will induce responses Y_{ij} for subject i that are above $\exp(x_{ij}'\alpha)$ on average, whereas a β_i below zero will induce responses below $\exp(x_{ij}'\alpha)$ on average.

The general Poisson random effects model allows for multiple random effects $z_{ij}'\beta_i$ where z_{ij} is a r-vector of known coefficients corresponding to the transpose of the jth row of the Z_i matrix in the continuous response random effects model. The random effect vector β_i is a $r \times 1$ vector. The general random effects model is

$$
\begin{aligned}
\text{E}[Y_{ij}] &= \lambda_{ij} \\
\log \lambda_{ij} &= x_{ij}'\alpha + z_{ij}'\beta_i & (11.8) \\
\beta_i &\sim N_r(0, D).
\end{aligned}
$$

A Poisson random variable has more information about λ than the Bernoulli distribution does about π, so it is possible to fit complex covariance models in practice, provided the software is able to do the fitting.

The Poisson random intercept model has $z_{ij} = 1$, $r = 1$, and β_i is a scalar. The idea is similar to the random intercept model for continuous or Bernoulli data. Each subject has an intercept β_i, which is how the subject differs from the median $\beta_i = 0$. Like the Bernoulli data model, the population mean and median are different on the λ_{ij} scale, although they are the same on the log scale. Averaged over the random effects distribution, the mean of the linear predictor is $x'_{ij}\alpha$ and this is also the median; $\beta_i = 0$ is both the mean and the median of the $N(0, D)$ random effects distribution. On the λ_{ij} scale, however, the mean of $\exp(x'_{ij}\alpha + \beta_i)$ over the distribution of the random effects will be larger than the median $\exp(x'_{ij}\alpha)$. This means that, like the logit random effects model, the population average of λ_{ij} is not equal to the subject specific mean $\exp(x'_{ij}\alpha)$. Unlike the logistic random effects model, the mean of $\exp(x'_{ij}\alpha + \beta_i)$ is always larger than $\exp(x'_{ij}\alpha)$; in the logistic random intercept model, the mean is closer to .5 than the median. Problem 7 illustrates how to calculate this.

11.2.1   Bolus Count Data

The Bolus Count data is a study of patient controlled analgesia comparing two different dosing regimes. Subjects were allowed to give themselves a dosage of pain medication to control pain. There are two groups, a 1 milligram (mg) per dose group and a 2 milligram per dose group. After each dose, there is a lockout time where the patient may not administer more medication. The lockout time in the 2 milligram dose group is twice as long as in the 1 milligram dose group. The number of doses is recorded for 12 consecutive 4-hour periods. The lockout time allows for a maximum of 30 dosages in the 2 mg group and 60 dosages in the 1 mg group. No response neared the upper limit, so the maximum count in each group for each four hour period is less than the theoretical maximum number of dosages. We use the Poisson distribution to model the counts.

For count data, if the response mean is large enough, it can be useful to draw profile plots. In figure 11.2, we plot the counts in profile plots, separately for each group. We see that the counts are higher for the 1 mg group as might be expected, although the amount does not seem to be twice that in the 2 mg group. There is an appearance that perhaps the counts are decreasing over time, and there may be a bump up in counts at about $t = 5$ and $t = 10$, particularly for the 2 mg group, although neither conclusion is certain.

Figure 11.3 plots an empirical summary plot. We plot means in each group and plus or minus 2 standard errors of the mean, calculated as for normal data, where the sample standard deviation is calculated from the

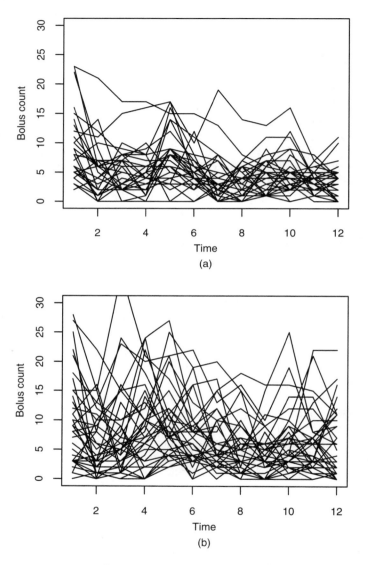

Figure 11.2. Profile plot of Bolus Count data. Upper plot (a) plots data from the 2 mg group, lower plot (b) plots data from the 1 mg group.

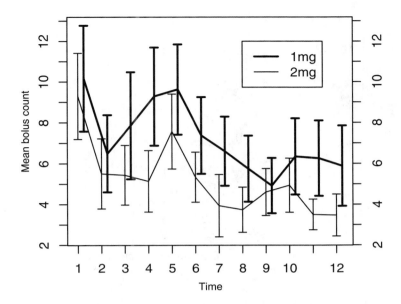

Figure 11.3. Empirical summary plot for Bolus Count data.

data at each time point in each group, and not from assuming that the Poisson distribution holds. This plot confirms that the 2 mg group counts are lower than the 1 mg group counts; the difference seems roughly similar across all twelve periods. Also we see the increase at periods 5 and 10 more clearly and that the increases are in both groups.

I fit a random intercept model to this data. For the fixed effects, I kept a constant difference between groups and an unstructured mean for the time trend, as the increases at times 5 and 10 are otherwise difficult to model. Thus the linear predictor at time j is $\alpha_j + \alpha_{13} \times 1\{1 \text{ mg group}\} + \beta_i$. The parameter α_{13} is the increase from the 2 mg to the 1 mg group, and α_j for $j = 1, \ldots, 12$ is the parameter at time j for the 2 mg group. Estimates and standard errors are given in table 11.2. One standard deviation of the random effects is $.5 = .25^{1/2}$; this is about twice the size of the group differences. If one takes $\exp(\alpha_1 3) = \exp(.27) = 1.31$, that is the ratio of the point estimate $\exp(x'_{ij}\hat{\alpha})$ in the 1 mg group to the ratio of the point estimate in the 2 mg group; we have $31\% = (1.31 - 1)100\%$ greater dosing, substantially less than the 100% greater dosing to have the same total dosage in both groups. In the log linear model, the 31% greater in the 1 mg group applies both to the median $\exp(x'_{ij}\alpha)$ as well as the mean of $\exp(x'_{ij}\alpha + \beta_i)$.

Figure 11.4 presents an inference plot of the fitted values from this model for both groups over time. The point estimates are for an individual whose

Parm	Est	SE
α_1	2.00	.10
α_2	1.51	.11
α_3	1.63	.10
α_4	1.72	.10
α_5	1.88	.10
α_6	1.58	.11
α_7	1.40	.11
α_8	1.29	.11
α_9	1.28	.11
α_{10}	1.46	.11
α_{11}	1.32	.11
α_{12}	1.28	.11
α_{13}	.27	.13
$\text{Var}(\beta_i)$	.25	.05

Table 11.2. Bolus Count data. Estimated parameters using a log linear random intercept model. The group difference α_{13} is significant with $t = 2.1$ and $p = .04$

$\beta_i = 0$, not the population averaged values. The plot gives 95% confidence intervals as well, calculated as $\exp[\text{est} \pm 2\text{SE}(\text{est})]$.

11.3   Issues with Discrete Data Models

There are a number of problems with fitting random effects models to discrete longitudinal data. These models are non-linear, and the models require a complicated mixture of numerical integration and function maximization to estimate the parameters using maximum likelihood. A common method from normal theory models is to integrate out the random effects from the likelihood to produce the marginal model. This can be done in closed form for the normal model as we saw in section 9.3. We then maximize the likelihood of the marginal likelihood. This works well for normal models but not for discrete response models. The reason is that we cannot algebraically calculate the marginal model in discrete response models and must use numerical integration.

The difficulties are such that many different approximations, algorithms and methodologies have been proposed for fitting these models. Software can often be problematic. To compute the likelihood requires an integration; the integration is accomplished through approximations; the approximations can be inaccurate enough to cause the likelihood maximization part of the algorithm to fail. Software will often not converge, leaving the user without usable estimates. Worse, software may converge to, or at least report, a poor if not downright silly estimate. It is incumbent upon the user to confirm that estimates actually make sense.

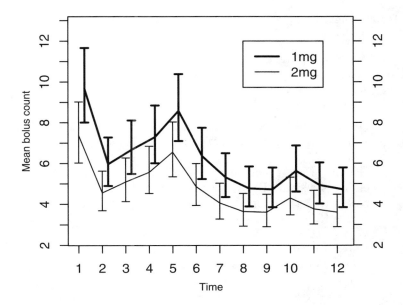

Figure 11.4. Inference plots for the Bolus Count data from the model with an unstructured mean and a fixed difference between the two groups.

There are other approaches besides maximum likelihood for estimating these models that may give better results. The parameter estimates and standard errors that these other methods give may not exactly duplicate maximum likelihood estimates; again one must confirm what is being estimated. Bayesian methods are very competitive with maximum likelihood methods for discrete data random effects model, as Bayesian methods properly adjust for uncertainty in the model, and Bayesian numerical methods encounter little trouble in fitting these models. References for other approaches are given in chapter 14.

11.4 Problems

1. Fit the malaria model to the fever outcome in the Morbidity data set.

   (a) Is there a treatment group difference? Describe it.
   (b) Is there a time trend? Describe it.
   (c) What does the seasonal effect look like? Is it similar to the seasonal effect for malaria?
   (d) Draw plots like 11.1 to illustrate each of your conclusions.

2. Fit the malaria model to the response "sore_in_mouth". Answer the same questions as in problem 1.

3. Fit the malaria model separately to malaria, fever, chills, reduced activity/bedridden, poor appetite, headache, upper respiratory, sore_in_mouth, skin rash, and digestive_system. Several of these outcomes have very similar time patterns.

   (a) Draw the plots like 11.1. Identify which outcomes are similar and describe this pattern. Which are different, and how are they different?
   (b) Explore a more complex time model. In trigonometric models for seasonal effects, we add terms in pairs $\sin(2\pi k t_{ij})$ and $\cos(2\pi k t_{ij})$ to the model. For any model with $sin(2\pi K t_{ij})$ and $cos(2\pi K t_{ij})$ terms in, we always include the terms with smaller $k < K$ also. We start with $k = 1$, then add $k = 2$ and then consider adding $k = 3$. Fit the same malaria model to these outcomes, but include sin and cos terms for $k = 2$ and $k = 3$ as well as $k = 1$.
   (c) Plot the resulting outcomes as in figure 11.1. Describe the patterns.
   (d) Do the extra sin and cos terms significantly improve the fit of the models? For which outcomes?
   (e) Even if the extra terms are not significant, do they change your perception of the outcomes?
   (f) Do you draw different conclusions about the significance of the intervention effect when $k = 1$ as opposed to the model with $k = 3$?
   (g) Do you draw different conclusions about the overall time trend when $k = 1$ versus $k = 3$?

4. The variable Mscore is a score that is one when the subject had any of a number of morbidities, whether mild or severe. The variable S_Mscore is similar, but is 1 only if the subject had a severe morbidity. For each of these questions, as appropriate, report your results in a neat table, report appropriate test statistics or confidence intervals, and draw a plot that illustrates your conclusions.

   (a) Develop a single model that you fit to each of these two variables.
   (b) Is there a time trend?
   (c) Is there a seasonal pattern?
   (d) Is there a difference among intervention groups?
   (e) There is an inequality relationship between these two variables because Mscore is always equal to or greater than S_Mscore. Describe the difference in estimated fraction of morbidities for the two response variables over time. Does the difference appear to change over time?

5. Explore, for your choice of morbidity response(s), whether there are gender differences in the response. Consider intervention by gender interactions, season by gender, and time by gender interactions, among others.

6. We calculate the population averaged prediction in the logistic random effects model for a given set of covariates x_{ij}. Sample a set of (say) 1000 standard normal $N(0,1)$ random variables. Call them b^l for $l = 1, \ldots, 1000$. For each l calculate

$$\text{expit}(x'_{ij}\hat{\alpha} + \hat{D}^{1/2}b^l)$$

and average this over the 1000 draws. Use this technique to reproduce figure 11.1.

7. Calculate the population averaged prediction in the Poisson random effects model for a given set of covariates x_{ij}. Sample a set of (say) 1000 standard normal $N(0,1)$ random variables. Call them b^l for $l = 1, \ldots, 1000$. For each l calculate

$$\exp(x'_{ij}\hat{\alpha} + \hat{D}^{1/2}b^l)$$

and average this over the 1000 draws. Calculate and plot the population averaged mean for the Bolus data for both groups as a function of time.

12
Missing Data

Let not thy mind run on what thou lackest as much as on what thou hast already.

– Marcus Aurelius

Either I've been missing something or nothing has been going on.

– Karen Elizabeth Gordon

Overview

- We illustrate potential effects of missing data,
- define dropout versus in-study non-response, and
- we define missing at random (MAR), missing completely at random (MCAR), and missing not at random (MNAR).
- We discuss modeling missingness.
- We include missingness as a predictor.

In longitudinal data studies, we rarely have complete data on all subjects. Whether dealing with countries or people or rats, all desired data are rarely available. People do not show up for appointments or do not answer or return phone calls. Countries do not pay for data collection. Rats die early or a lab assistant is sick for one day. In some longitudinal studies, we set an overall goal that subjects show up for perhaps 80% to 90% of all desired data collection visits. In some data sets, missing data are actually planned for. Suppose we wish to administer a lengthy questionnaire to subjects every three months over several years, at times $t_j = j$, $j = 1, \ldots, J$. To reduce *respondent burden*, we may ask all subjects to show up for the first visit t_1 and the last visit t_J, but in the middle we only schedule data collection visits at some fraction of the usual three month visits. Different subjects will be scheduled to be missing at different times t_j so that we always have the same fraction of desired interviews at each t_j. The Big Mice data were designed to have missing data; this reduced the data collection burden on the investigator and on the mice. Only two mice had their weights measured for 21 consecutive days, the rest have measurements every three days. We can still estimate the average weight on any given day, by averaging weights of the available mice on that day.

12.1   The Effects of Selective Dropout

Missing data refers to observations that we intended to collect but did not collect or were unable to collect for whatever reason. Intermittent missing data may occur when subjects are temporarily indisposed or otherwise unavailable for data collection. Dropout occurs when a subject dies, refuses to participate in the study any further, or is otherwise unable to supply any additional data to the study. After a subject drops out, we are unable to collect more data from that subject. Intermittent dropout that occurs at the end of the study may well appear to be dropout.

With random times or with missing data some care must be taken in drawing conclusions from summary statistics. We can still take the mean or standard deviation at a time point or in a window about that time. However, care must be taken in interpretation because missing data can be for many reasons and can cause spurious conclusions.

A common circumstance is that very ill, say low Y_{ij}, subjects drop out of our study while healthy, high Y_{ij}, patients continue on. Suppose that subjects follow a pattern like figure 12.1(a), but with the sick low-valued Y_{ij} subjects dropping out of the study. We only get to see data illustrated in figure 12.1(b). By study end only the high responding subjects remain, and apparently our sample mean is increasing over time. However the cause is that ill subjects leave the study and not because subjects are improving over time. Two inferences are potentially possible from inspecting figure 12.1(b). The first is that the population mean of all subjects *had they continued in the study* is constant over time. This is illustrated in figure 12.1(a) where the observations from subjects that left the study in figure 12.1(b) remain in. We would need to look at figure 12.1(b) and mentally fill in the data from the bottom subjects so that the figure appears like figure

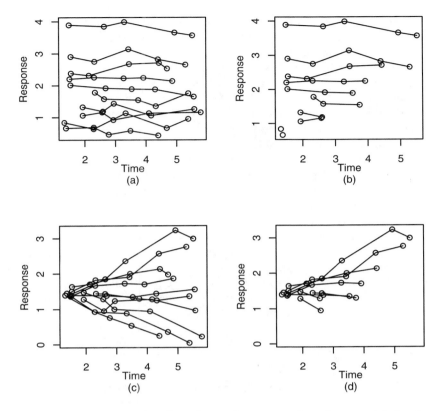

Figure 12.1. Example profile plots illustrating non-random dropout. (a) Random intercept, constant population mean, no dropout. (b) Random intercept, constant population mean; lower valued subjects drop out. (c) Random slope; no dropout. (d) Random slope; lower valued subjects drop out. Observations in (b) and (d) are a subset of the observations in (a) and (c), respectively.

12.1(a). The second possible inference is that the population mean of the *remaining subjects* is increasing. This second inference tells us something about people who remain in the study; it does not tell us something about the people who we originally enrolled, who were the random sample from the population we wished to study. Thus the second inference does not generalize back to the population. Almost exclusively we are interested in the first inference.

Figures 12.1(c) and (d) illustrate a sample of individuals with fixed intercept and random slopes. In figure 12.1(c), everyone stays for the full study duration, while in (d) individuals who are low, drop out. Superficial inspection of (d) suggests that the average response is rising in the study, while in (c), we see that the population variance is increasing but the population mean appears to be roughly constant over time.

Differential dropout can happen in many studies. In studies of the elderly, the oldest patients are more likely to drop out of the study due to illness or death. If the older subjects are lower than younger subjects on the response, then our apparent population mean will increase over time as more and more older subjects drop out. Similar occurrences happen in studies of depressed people. The more depressed someone is, the less likely they are to respond to our study questionnaire. If our main outcome is depression, we may think our subjects are improving when merely the more depressed ones are dropping out.

In randomized behavioral intervention studies, it is usually impossible to hide from subjects which study arm they are in, treatment or control. The control group gets a minimum of attention or training while treatment involves a substantial amount of experimenter-subject contact. Subjects in the control often differentially drop out, and we must ensure that this dropout does not bias results. In other randomized studies, if the treatment works well, the treatment arm subjects may drop out because they no longer need treatment. In end stage cancer studies of subjects with terminal disease, the subjects are enrolling solely for the possibility of getting a new treatment. Those in the control group will more likely drop out of our study in their search for a treatment that works.

12.2   Dropout versus In-Study Non-response

There are several types of missing data. Some subjects refuse or are unable to respond to further surveys; they *drop out* from the study. We are unable to collect further data from subjects that drop out. Death of a subject means no more data collection from that subject. Moving, severe illness, and loss of interest are all reasons a subject may drop out of a longitudinal study. Both a cure to the disease being studied and a lack of improvement in the disease can lead to dropout.

A contrasting type of missing data is called *intermittent* missing or *in-study non-response*. A subject may not drop out of the study but still may not respond to a particular call for data collection. People may be on vacation, too ill to respond one time, or merely busy. In the Pain data, trials alternated between the dominant and non-dominant arm. One subject had a broken arm; data was collected only from the healthy arm. The subject did not drop out of the study, but the broken arm produced two trials with missing data for that subject.

We know that we have in-study non-response when there is a time gap in the data collection record followed by further observations. At the end of the study, a subject may be missing a series of observations. Without further investigation, we cannot determine if the series is in-study non-response or if the subject has dropped out. The longer the series of observations, the more likely that the subject has dropped out of the study.

12.3 Missing at Random and Variants

Another way to classify missing data deals with the mechanism of how the data became missing. The key concept is whether the missingness tells us something about the responses that are missing. Suppose we are studying cancer patients all of whom are dying of their disease. A patient who drops out is likely dead or terminally ill. Our longitudinal response may be a measure of tumor size, a measure of quality of life, or a measure of health care expenditures. A patient who drops out probably has a larger tumor, a lower quality of life, and higher health care expenditures than the average patient. The dropout is likely to be *informative missingness*; the fact that a patient drops out of the study tells us that they are much sicker than the patients who stayed in the study. In a study of depressed seniors, intermittent missingness may be due to the depression or due to a short-term illness, while dropout could be caused by death or severe depression. Either cause is likely to be informative.

Often times subjects may drop out or be unavailable for reasons unrelated to the study. They may go on vacation or move because of a spouse's job. Any data missing resulting from these causes is *uninformative* about the response.

Data are *missing completely at random* (MCAR) if the missingness has nothing to do with any of the data collected in our study. Vacation and moving are possible reasons for MCAR missingness in a study. The missingness in the mice data is MCAR; the investigator decided on what data would be missing in advance of data collection.

Data are *missing at random* (MAR) if the reasons for the missingness are related to data we have observed and are unrelated to the data that is missing. Suppose that in a weight loss study, people who gain weight tend to drop out. If they come in to our clinic and weigh themselves, discover that they are gaining weight, and then decide to drop out, the drop out is related to data that we have observed (i.e., the current observation) and that is MAR missingness.

The third type of missingness is *missing not at random* (MNAR). The dying cancer patients example at the beginning of this subsection illustrated MNAR missing data. In a weight-loss study, some subjects realize that they have gained weight since their last clinic visit and decide to drop out of the study before they visit the clinic again. The weight gain occurs after the last visit. The subjects know about the weight gain while the researchers do not know. Weight losers will stay in the study and have their data recorded, while the gainers drop out without having their data recorded.

This missingness is MNAR as the missingness depends on the missing data. In this situation, if one proceeds to fit a model to the observed data then the estimated trend may well be negative, when in fact the population average is flat, had we observed all of the data.

12.4   Example: the Difference Between MAR and MNAR

Table 4.1 has data constructed to illustrate the differences between MAR and MNAR. The bold observations are weights that are 4 or more pounds higher than the baseline weight. Thereafter, observations are italicized. There are two missing data versions of the data set: (1) italicized data are missing; and (2) bold data and italicized data are both missing. The two data sets with two sets of missing data correspond to MAR and MNAR, respectively. In version (1), subjects come in to the clinic, discover they have gained a lot of weight, and then drop out. The researcher observes the high (boldface) weight. In (2), the subjects realize they gained weight without needing to come in to the clinic. They drop out before the bold observation has been observed by the researcher. In version (1) we have MAR missingness – the missingness depends on the observed (boldface) data but not on the future, unobserved (italicized) weights. In version (2) we have MNAR missingness – the missingness depends on the unobserved (boldface) data.

We fit a compound symmetry (CS) covariance model and a fixed effects model that says that the population weight loss is linear in time, $\alpha_1 + \alpha_2 t_{ij}$, to the complete data version and to the two missing data versions of this data. Time is measured in weeks, and the results are given in table 12.1. The estimates and standard errors for the intercept are nearly identical for the three models.

The estimate of the slope changes dramatically from the complete data model, where the slope is positive but not significant ($\hat{\alpha}_2 = .141$, $t = 1.7$, $p = .1$) to negative and not significant for the MAR data set ($\hat{\alpha}_2 = -.056$, $t = -.6$, $p = .6$) to significantly less than zero for the MNAR data set ($\hat{\alpha}_2 = -.201$, $t = -2.3$, $p = .028$).

In constructing this data set, the true population slope was zero.

The complete data are trivially an example of a data set that is MCAR: nothing is missing. How would we construct a data set that was MCAR that had missing data? One way to do it would be to toss a coin for each subject. Suppose a complete data set with 6 observations per subject and the first observation is a baseline observation. Heads, keep all observations for the subject. Tails, roll a die; if a 1 appears ignore it, as we wish to keep all baseline observations in the data set. If the die shows 2 or higher, then have all observations from the die roll number and on be missing for that

Model	Covariate	Est	SE	t	p
Complete	Intercept	196	6	32.76	<.0001
data	Week	.141	.085	1.7	.10
MAR	Intercept	197	6	33.15	<.0001
	Week	-.056	.097	-.6	.56
MNAR	Intercept	196	6	32.9	<.0001
	Week	-.201	.089	-2.3	.028

Table 12.1. Estimates, standard errors, t-statistics, and p-values from fitting a CS covariance model and a fixed intercept and slope to three versions of the data from table 4.1.

subject. The particular set of probabilities mentioned here, a coin toss, and a die roll, are not special. One could use a biased coin and a biased die for the probabilities. Many probability models to determine the missingness are potentially fine and could have been used to generate which observations are lost to our data set. The key feature of the probabilities determining the missingness is that the probabilities cannot have anything to do with the observed or the missing data values.

12.5 Modeling Missingness

Define some notation for missingness. Let Y_i^* be the complete set of observations for subject i. By complete we mean that if the study is designed to have J observations on each subject, then Y_i^* has J observations. The actually observed observations Y_i may be of length J or it may have fewer observations with $n_i < J$. Thus data Y_i is a subset of the theoretical Y_i^*.

Now define a *missingness* or *missing data indicator* vector R_i of length J with elements R_{ij} associated with observation Y_{ij}^* where

$$R_{ij} = \begin{cases} 0 & Y_{ij}^* \text{ is observed} \\ 1 & Y_{ij}^* \text{ is missing} \end{cases}$$

Consider setting up a statistical model for the R_is, which we write as $f(R_i|\phi)$. This *missingness model* will have unknown parameters ϕ predicting the occurrence of missingness $R_{ij} = 1$ or not $R_{ij} = 0$. The model may have R_i depending on Y_i, as in $f(R_i|\phi, Y_i)$ or it may not as in $f(R_i|\phi)$. We do assume the parameters ϕ are completely distinct from the parameters α and θ already in our model for Y_i.

MCAR missingness says that the model for R_i does not depend on Y_i^* or Y_i, and the sampling density of R_i can be written $f(R_i|\phi)$. MAR missingness says that the model for R_i may depend on the *observed* elements Y_i of Y_i^* but does not depend on the *unobserved* or missing elements of Y_i*, so the density $f(R_i|\phi, Y_i)$ depends on Y_i. MNAR missingness says that the model is $f(R_i|\phi, Y_i*)$ and R_i depends on the missing elements of Y_i* as well as the observed elements.

If we use maximum likelihood or restricted maximum likelihood to fit model (5.6), then if the missing data are MCAR or MAR, we do not need to develop a model $f(R_i|\phi, Y_i)$ for the missingness data. We can *ignore* the missingness and the missingness model $f(R_i|\phi, Y_i)$. Our estimates from model (5.6) are *unbiased*, and standard errors are appropriate if the model is correct.

If the data are MNAR, then fitting model (5.6) will produce *biased* results; the estimates $\hat{\alpha}$ and $\hat{\theta}$ do not correctly estimate α and θ. If the missing data are MNAR, then we must develop the model $f(R_i|\phi, Y_i^*)$ and we must jointly fit this missingness model and model (5.6) and jointly estimate the parameters α, θ, and ϕ. This is problematic; currently statisticians are still in the early stages of developing these models and coming to agreements about how to use them.

The problem with MNAR models is that the model $f(R_i|\phi, Y_i^*)$ cannot be validated; because the missing observations in Y_i^* are (surprise!) missing, we do not have them in our data set; we cannot know whether the model $f(R_i|\phi, Y_i^*)$ is right or not, particularly the part of the model that depends on the missing data. Making a set of plausible assumptions about the missingness and running a *sensitivity analysis* to see if inferences vary from those gotten from only fitting (5.6) is virtually the only solution. If inferences do not vary in an important fashion, then we conclude that the inferences from (5.6) are acceptable. If inferences do vary substantively from fitting (5.6) alone, we know that our inferences are dependent on the MAR or MNAR assumption.

The MNAR model $f(R_i|\phi, Y_i^*)$ includes a model for the missingness conditional on unobserved data, data that we can never collect. It is possible to reverse this dependence, and consider a model of Y_i^* that includes R_i or some function of R_i as a predictor. These models are called *pattern mixture models*. We discuss a simple pattern mixture model in the next section. In the extreme case, one could fit separate parameters for every subset of the data with a distinct R_i pattern. Pattern mixture models, where the model for Y_i depends on R_i is still a MAR model, but one that does take into account something about the missingness for a subject.

One way to improve the usual multivariate linear regression model for Y_i to handle missingness is to determine which predictors predict R_i. If a given covariate predicts missingness, that covariate should be used as a covariate in the analysis of Y_i. Rather than assuming that a particular covariate does or does not predict missingness, an analyst could explore which covariates predict missingness in a statistical model. One way to do this is to treat missingness R_{ij} as a binary outcome variable, and to analyze R_{ij} using the logistic random intercept model of chapter 11.

12.6   Using Missingness as a Covariate

When a study is designed to have specific nominal times t_j for observations, we have missingness when subjects do not respond at time t_j. Subjects that show up for every visit are often different from subjects who skip visits intermittently or who drop out. We can model this difference by including missingness as a covariate in the analysis. Missingness can be parameterized in different ways. In the extreme situation, every missing data pattern R_i could get its own group, but this requires large amounts of data. More simply, we could set up a single indicator variable M_i that is one if any response is missing on subject i or zero if no observations are missing. Often intermittent missingness is thought to be ignorable or to not be of major import to the analysis, while subjects who drop out are thought to possibly be different. Then we might include dropout D_i as a covariate parameterized as $D_i = 0$ for subjects who have the last observation and dropout $D_i = 1$ for subjects missing the last observation.

12.6.1   Schizophrenia Severity Data

The Schizophrenia data has a response severity of illness measured on a 1 to 7 scale where 1 represents normal up to 7, the most severe illness. Subjects were observed at weeks 0 through 6, although most subjects and observations are observed at weeks 0, 1, 3, and 6. There were three drug groups and one placebo group; because the drug groups were similar in response, the three groups were combined into a single drug group, drug=1 or drug=0 for placebo. Several different missingness variables are possible here; we define dropout as an indicator that is zero if the subject is observed at the last time point $t = 6$ and one if the subject is not observed at $t = 6$. Subjects who dropped out were thought to be different from those who stay in the study. Because there is slight curvature in the response over time, we use the square root of time sweek=$\sqrt{(t_{ij})}$ as the time covariate; this straightens out the time trend. An alternative approach is to include a quadratic term in the analysis. The random intercept and slope covariance model are assumed for all models fit here.

All analyses here have separate intercepts and slopes for the drug and placebo groups. The first analysis presented in table 12.2 does not use dropout as a covariate. It pools all subjects together to estimate the drug effect. We have a highly significant effect; the drug group slope is $-.64$ units per square root week lower than the placebo group. Because there is suspicion that dropouts may be different, the next two analyses fit the same model, but separately to subjects who complete the study and subjects who drop out. This means that even the covariance parameters are allowed to be different in the two groups. We see that the sweek×drug interactions are quite different in the two groups. These separate analyses by dropout are one way to fit the three-way interaction between sweek, drug, and dropout,

Analysis	Parameter	Est	SE	t	p
Pooled	Intercept	5.35	.09	60.8	<.0001
	Sweek	-.34	.07	-5.0	<.0001
	Drug	.05	.10	.5	.65
	Sweek×drug	-.64	.08	-8.3	<.0001
Dropout= 0	Intercept	5.22	.11	47.8	<.0001
	Sweek	-.39	.07	-5.4	<.0001
	Drug	.20	.12	1.6	.10
	Sweek×drug	-.54	.08	-6.5	<.0001
Dropout= 1	Intercept	5.53	.14	39.3	<.0001
	Sweek	-.12	.18	-.7	.49
	Drug	-.19	.18	-1.1	.28
	Sweek×drug	-1.17	.22	-5.2	<.0001
Covariate	Intercept	5.22	.11	48.6	<.0001
	Sweek	-.39	.08	-5.2	<.0001
	Drug	.20	.12	1.7	.10
	Sweek×drug	-.54	.09	-6.3	<.0001
	Dropout	.32	.19	1.7	.09
	Sweek×dropout	.25	.16	1.6	.11
	Drug×dropout	-.40	.23	-1.8	.08
	Sweek×drug×dropout	-.63	.20	-3.2	.00
Contrasts	Int placebo nodrop	5.22	.11	48.6	<.0001
	Int placebo drop	5.54	.15	36.4	<.0001
	Int drug nodrop	5.42	.06	98.4	<.0001
	Int drug drop	5.74	.19	29.5	<.0001
	Slope placebo nodrop	-.39	.08	-5.2	<.0001
	Slope placebo drop	-.14	.14	-1.0	.31
	Slope drug nodrop	-.93	.04	-23.8	<.0001
	Slope drug drop	-1.32	.11	-12.3	<.0001

Table 12.2. Analysis of Schizophrenia severity data with various treatments of missingness as a covariate. All analyses use a random intercept and slope covariance model and have a separate intercept and slope for the non-drug and drug groups. The pooled analysis does not account for the missingness. The next two analyses include subjects observed at the last time point, dropout=0, or only subjects missing at the last time point, dropout=1. The last analysis includes missing as a covariate in a full interaction with time (sweek) and drug. The contrasts are from this last analysis and give the intercepts and slopes for each combination of the drug/non-drug and dropout/non-dropout groups.

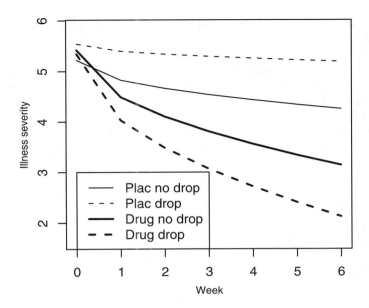

Figure 12.2. Schizophrenia severity data estimated means over time for subjects on drug (bold lines) or on placebo (regular lines) and who drop out (dashed lines) or do not drop out (solid lines).

and they include a separate interaction between dropout and the covariance parameters as well.

The last analysis in table 12.2 includes dropout in a three-way interaction with time and drug variables but with only 1 set of covariance parameters. We also estimate the intercepts and slopes in the four drug×dropout groups. The intercepts are not too different among the four groups, but the slopes are remarkably different and there is a three-way interaction between sweek, drug, and dropout. We see that in the placebo group, the dropout slope is higher than the stayers group; but in the drug group, the dropout slope is even lower than the stayers group. It appears that subjects who do not get well quickly drop out of the study in the placebo group, while those who get well quickly drop out of the drug group.

12.7 Missing Data in the Covariates

Sometimes a covariate will have a modest fraction of subjects who are lacking that covariate. For this subsection only, and for a given analysis, call a subject who has values for all the covariates a *complete covariate subject*. Subjects without complete covariates have *missing data*. Analyzing only the complete covariate subjects can seriously degrade an analysis for either of two reasons. As we decrease our sample size, we lose power in the analysis, as the complete subjects sample size can be a modest fraction of the original sample size. Second, the complete subjects can be substantially different from subjects with some missing data.

In the absence of additional tools to handle missing data, let us consider a data set with 6 covariates that we wish to include in our current model, where each covariate is missing in approximately 10% of the subjects. If we fit a model to the complete subjects, we will lose around 50% to 60% of our subjects, assuming it is not always the same subjects missing all of the variables. Any treatment effects may disappear not because of the adjustment by the covariates, but because of the loss of sample size. It will not be possible to analyze the effects of all the covariates together. Instead, we will have to settle for adding each covariate into our model one at a time, and checking the effects of the 6 covariates individually in 6 additional analyses and not all at once in a single analysis. This way we will only lose approximately 10% of the sample in each additional analysis, rather than 60%.

The proper and more principled approach to analyzing missing covariate data is to develop a series of statistical models involving first the response as a function of the covariates and then taking each covariate with missing data in turn and modeling it as a response variable with predictors that are the remaining covariates in the analysis. Once we model a covariate, it is not included in subsequent models as a predictor. Any covariates with complete data need not be modeled. The series of models are put together and fit as a single model. We end up with a *multivariate regression model*, where the response is our longitudinal response and all the covariates with missing data. Software for doing this is becoming available. An approach called *multiple imputation* is rapidly gaining popularity, and there are already implementations of multiple imputation in commercially and freely available software. See the references in chapter 14 for further description of multiple imputation.

12.8   Problems

1. Suppose we collect annual data on corporations including information such as gross revenue, gross margin, and net income. What causes of missing data could there be? Large companies have reporting requirements imposed by US government regulations. Missing data of any amount may indicate that the company is no longer in business. Are the reasons for missing data in this data set likely to be informative or uninformative? Discuss briefly.

2. Suppose we are collecting annual data such as gross national product, employment rate, etc., on countries. A country stops having data available. What reasons could there be for this? Are the reasons informative about the underlying data values?

3. In the Cognitive data, count how many subjects have each possible missing data pattern. There are 5 rounds, and each subject is supposed to have 5 observations, but some are missing. Because each observation may be observed or missing, there are $2^5 = 32$ potential missing data patterns. How many subjects have each missing data pattern? Are the patterns different depending on whether the response is Raven's, arithmetic, verbal meaning, or total digit span?

4. For the Cognitive Raven's data, do subjects who drop out appear to have a different slope from those who stay in the study? Define an indicator of missing data at the last round (lastmissing=1 if no round 5 observation else lastmissing=0). Use this covariate in an interaction with time and see if it is significant. Report your results in a nicely formatted table.

5. In the Schizophrenia severity data, are there any differences between the genders in drug effect or dropout? Is there a difference in dropout rates? Is there an interaction between dropout, gender, and time?

6. Analyze the Schizophrenia severity data using a quadratic time trend rather than using square root of time. Your analysis should include covariates for drug and dropout and any needed interaction. How should you parameterize the time by group interactions? Report on your results both in table form, graphically and in written form.

7. For the Schizophrenia data, fit the models from table 12.2 using the random intercept and slope model. Report on how the parameter estimates differ in the several models. Draw a plot to show how much the slopes differ in range between subjects who drop out and subjects who stay in the study. Should you pool the data to estimate the covariance parameters when estimating the three-way interaction of sweek×drug×dropout? Explain.

8. For the Schizophrenia data, what is the best fitting covariance model?

9. For the BSI data, consider a response of whether the GSI value is missing or not. Develop a logistic random intercept model using the available covariates for whether the value is missing or not. You will need to make some decisions about what the data is. Clearly state the choices you make in producing a response variable to analyze and in building your data set.

13
Analyzing Two Longitudinal Variables

It takes two to tango.

– Al Hoffman and Dick Manning

Better to have estimated a correlation and found it to be zero than never to have estimated a correlation at all.

Overview

1. Bivariate longitudinal data occur commonly in practice.

2. Using one response as a time-varying covariate for the other response is a flawed approach to analyzing the joint distribution of the two longitudinal variables.

3. Rather, one should model the joint covariance of the two longitudinal variables. We discuss a number of covariance models for the two longitudinal variables:

   (a) Bivariate random intercept and special cases
   (b) Bivariate random intercept and slope
   (c) Unstructured covariance model
   (d) Product correlation models

4. We also discuss models for when the two longitudinal variables are not measured at the same times.

13.1   Thinking About Bivariate Longitudinal Data

Our primary interest has been to understand how time and covariates affect a univariate longitudinal response. Most longitudinal studies assess many variables at first visit, and many variables are typically reassessed over time at follow-up visits. This produces multivariate longitudinal data. Often researchers are interested in how two or more longitudinal variables are interrelated. Are the overall levels correlated? Within a subject, do individual ups and downs track each other?

The correlations between the two responses are called the *cross-correlations*. We may test whether the cross-correlations are zero or not.

We might consider including the second response as a time-varying covariate in the fixed effects when analyzing the first longitudinal response. We test the coefficient of the longitudinal covariate to see if it is significantly different from zero. If the coefficient is positive, the two variables go up together and down together; if the coefficient is negative, the two variables move in opposite directions. We discuss this and some variations in the next section. Unfortunately, this procedure has a number of drawbacks, which we then detail.

In the Pediatric Pain data, we have a second longitudinal variable called pain rating. Each subject is asked to rate the pain of each trial on a 1 to 10 scale. We might anticipate an inverse relationship between pain rating and pain tolerance; the less the pain rating, the greater the pain tolerance and vice versa. This would show up as a negative coefficient in a model that included pain rating as a predictor for log pain tolerance.

A more complex approach is needed for some pairs of longitudinal variables. We might suspect that a subject's average pain rating has no particular meaning, and that average pain rating is uncorrelated with average pain tolerance. On the other hand, we might expect the variation up or down in pain rating from the subject's average rating might have some relationship with pain tolerance. We need a model that allows us to distinguish between over all subject average pain rating level and random variation from trial to trial.

A contrasting example might be when we measure subjects' height and weight repeatedly over time. Subject's average height and weight are likely well correlated, but the random variation in weight might not be particularly correlated with random variation in height. Weight variation might have more to do with daily eating patterns, etc., whereas height variation depends more on slouching and measurement error. The observation to observation variation of one measure probably does not have much to do with the other.

In a more complex example, we might have subject's nutrition level predicting the slope in an anthropometric response such as height or weight. Furthermore, along with the Morbidity data set, we also have Nutrition and

Anthropometric responses measured longitudinally over time, although not at the same times as morbidity.

One way of thinking about the correlation of bivariate longitudinal responses is illustrated in figure 13.1. Four different situations are illustrated. In each figure, a bivariate observation (Y_{ij}, W_{ij}) is measured at time $t_{ij} = j$ for $j = 1, \ldots, J = 4$ times on each of 8 subjects. The first response, Y_{ij}, is plotted on the horizontal axis, and the second response, W_{ij}, is plotted on the vertical axis. Observations from the same subject are plotted using the same symbol; the time sequence of the observations is not indicated. In creating the four examples, the covariation between Y_{ij} and W_{ij} is assumed to have at most two parts. One part is the covariation between the subject means of Y_i and W_i. Then, given the means, observations within a subject (Y_{ij}, W_{ij}) may be correlated. In figures 13.1(a) and 13.1(c), observations within a subject are positively correlated, in contrast, in figures 13.1(b) and (d), observations within a subject are independent. In figures 13.1(a) and (b), the subject means of Y_i and W_i are positively correlated, and in figures 13.1(c) and (d), the subject means are independent. Figure 13.1(c) corresponds to the Pain example, and figure 13.1(b) corresponds to our height and weight example.

The language we are using corresponds to a random intercept model for Y_i and another random intercept model for W_i. Each variable, Y or W, has a random subject intercept, and each observation Y_{ij} or W_{ij} has a residual error. Another way of saying that the subject means are correlated is to say that the random intercept for Y_i is correlated with the random intercept of W_i. Similarly, when we say that the observations within a subject are correlated, we mean that the residual errors at a given time j are correlated.

13.2 Continuous Time-varying Covariates

The simplest approach to bivariate longitudinal data would seem to be to include the second longitudinal response as a time-varying covariate (confer subsection 1.8.3) in the model for the first response. In the log pain tolerance analysis with coping style (CS) by treatment (TMT) interaction, we can include the pain rating as a covariate in the analysis. Doing this gives an estimated pain rating coefficient of $-.055$, negative as hypothesized, indicating that as pain rating goes up, log pain tolerance goes down. The standard error is .025 and the t-statistic is -2.3 with a two-sided p-value of .028. The result is significant and we conclude that pain tolerance is inversely related to pain rating. These results are reported as model 1 in table 13.1 using REML and the unstructured covariance model.

In the previous section, we hypothesized that average log pain tolerance was not correlated particularly with log pain rating, but that the up and down vagaries of pain rating might be correlated with log pain tolerance.

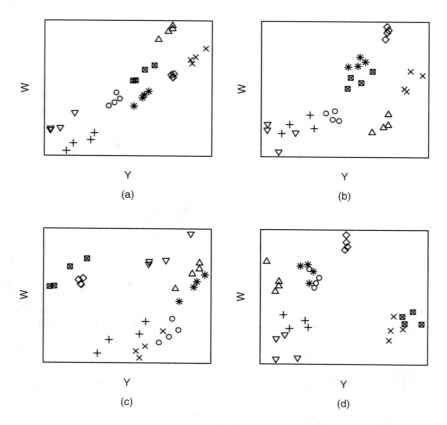

Figure 13.1. Scatterplots of 4 hypothetical bivariate observations from 8 subjects. Observations within a subject are plotted using the same plotting symbol. Time is not represented in the plot. Response Y is on the horizontal axis, response W is on the vertical axis. (a) Figure illustrating bivariate longitudinal data with observations within subject positively correlated and subject averages also positively correlated. (b) Positive correlation between subject averages, observations within subject uncorrelated. (c) Observations within subject correlated, subject averages uncorrelated. (d) Observations within subject uncorrelated and subject averages uncorrelated.

To check on this hypothesis, we split pain rating into two parts, the average pain rating and deviations about the pain rating. Define average pain rating to be the average of the four pain rating observations for each subject. We define the pain rating deviations from the average as

$$\text{rating deviation}_{ij} = \text{rating}_{ij} - \text{average rating}_i.$$

Subjects without complete pain rating and pain tolerance data will be dropped from models that involve average pain rating and pain rating deviations. That way we do not have to worry about a mean that is usually the mean of 4 observations but sometimes is the mean of 3 or even 2 observations.

We hypothesize that average pain rating is not predictive of log pain tolerance, and if we entered it as a time-fixed predictor we would not expect it to be significant. Model 2 in table 13.1 gives the results of using average pain rating as a predictor. The coefficient of average rating is actually farther from zero than that of pain rating, but the standard error is much higher and the coefficient fails to achieve statistical significance.

Even if average rating should not be predictive of log tolerance, rating deviation should be. Model 3 in table 13.1 reports the result. The coefficient of rating deviation is actually identical to that from model 1 and has the same standard error.

Finally, we can include both average rating and rating deviation in the model, and this is reported as model 4 in table 13.1. Results match those of the models with each variable in separately. The test that both coefficients are equal is given in the last line of the table and tests the null hypothesis that both effects are the same. Under H_0, we should therefore have included only the original pain rating variable as a predictor in the first place! This test is not significant, with a p-value of .8.

The results for rating in model 1 and rating deviation in models 3 and 4 are consistent. The higher the subject's pain rating for that trial, the shorter the time of immersion.

13.3 Problems with Using a Response as a Predictor

The models of the previous section enter the second longitudinal response in the model as a predictor for the first response. The first model used all of the observations because if we had pain tolerance, we also had pain rating. For the average pain rating and pain rating deviation analyses, we dropped subjects with missing observations. The response average and response deviations do not have quite the same definitions with differing numbers of observations, and if there is but a single observation, then the deviation must be zero. We could ignore the issue and include all subjects in the

Model	Parameter	Est	SE	t	p
1	Rating	-.055	.025	-2.3	.028
2	Rating average	-.070	.072	-.97	.34
3	Rating deviation	-.055	.027	-2.1	.043
4	Rating deviation	-.055	.027	-2.1	.045
	Rating average	-.069	.072	-1.0	.34
	Deviation - average	.013	.076	.18	.8

Table 13.1. REML estimates using the unstructured covariance model for four models with pain rating as covariate. All models have 8 parameters for the CS×TMT interaction effects. Estimates, standard errors, and inferences are given only for the pain-rating variables. Model 1 has pain rating, model 2 has average pain rating only, model 3 has rating deviation only, and model 4 includes both average rating and rating deviation in the model. The last line presents the hypothesis test that the two coefficients are equal in this last model; the estimate and standard error are for the difference in the two coefficients. Models 2 through 4 use data only from the 58 subjects with complete data.

analysis provided they had at least two observations. Depending on n_i the variable response average has different variance, and the response deviations also have different variance for different n_i. Having varying numbers of observations for different cases is uncomfortable, and a more flexible approach is desirable if available.

Another problem with the time-varying covariate model is that we cannot use observations with one response or the other but not both. Why should these observations not contribute at least partial information to the analysis? As discussed in chapter 12, dropping subjects with some missing data can bias results, and it will reduce the sample size of the analysis, increasing variance. If there is a time trend, then the residuals will show a predictive relationship among the residuals which might be misinterpreted as correlation in the residual variation, when it is actually the time trends of the two variables that are correlated.

Another issue is which response should be treated as the response and which should be treated as the covariate? If we reverse the roles of pain rating and log pain tolerance and use pain rating as the predictor, then do we get the same results? Fitting four models as in table 13.1 with pain rating as the response and with pain tolerance gives slightly different results regarding the association. Log tolerance is significant, but log tolerance deviation is borderline not significant; it comes in with p-values between .05 and .06 for models 2 and 4. If we were compulsive about using a .05 cutoff for significance we would conclude that there was not a significant relationship between pain rating and log pain tolerance deviations.

Finally, another important issue is that whichever response is being used as a predictor may be affected by the other covariates in the analysis. Could it be that treatment predicts pain rating as well as log pain tolerance and that is why the two variables appear related?

All of these problems are caused by trying to treat a response as a predictor. Instead, it is preferable that one should model the two responses jointly. Research questions involving two or more continuous longitudinal variables are often best answered in a multivariate longitudinal framework. Treat the two variables as a bivariate response measured repeatedly over time. One advantage of multivariate models is that we can incorporate subjects with different missing observation patterns on each response. For the pain data, we would use CS, TMT and CS×TMT as predictors for both variables. Any remaining covariation is not the result of CS or TMT being predictive of the two variables.

13.4   The Bivariate Random Intercept Model

Random effects models are particularly useful models for bivariate longitudinal data: they are simple to understand, and current software can often be coerced into fitting them. The random intercept model is the simplest of random effects models for univariate longitudinal data. We generalize the univariate random intercept model to the bivariate random intercept model to fit bivariate longitudinal data.

Let Y_{ij} be the first response measured at time t_{ij} for $j = 1, \ldots n_i$ and let W_{ij} be the second response measured at the same times. Each response is modeled by its own random intercept model. The random effects and error terms will be correlated, inducing correlation between the two responses.

For covariates, we will assume the same K-vector x_{ij} of covariates at time t_{ij} for both responses and the same matrix X_i, n_i by K of covariates with rows x'_{ij}. This is to keep notation and interpretation relatively straightforward but is not a requirement for fitting these models. For convenience, the Z_i matrices for both responses will also be the same. Because we are assuming a random intercept model, Z_i is an n_i-vector of ones for both responses. There will be two fixed effects vectors $\alpha_Y = (\alpha_{Y1}, \ldots, \alpha_{Yp})'$ and $\alpha_W = (\alpha_{W1}, \ldots, \alpha_{Wp})'$ and two random intercepts β_{iY} and β_{iW}, each a scalar. Finally, there are two sets of residual errors $\delta_{iY} = (\delta_{iY1}, \ldots, \delta_{iYn_i})'$ and $\delta_{iW} = (\delta_{iW1}, \ldots, \delta_{iWn_i})'$ both n_i vectors. The model is

$$
\begin{aligned}
Y_{ij} &= x'_{ij}\alpha_Y + \beta_{iY} + \delta_{iYj} \\
W_{ij} &= x'_{ij}\alpha_W + \beta_{iW} + \delta_{iWj}.
\end{aligned}
\tag{13.1}
$$

We make our usual assumption on the random effects that they are distributed as a bivariate normal

$$
\beta_i = \begin{pmatrix} \beta_{iY} \\ \beta_{iW} \end{pmatrix} \sim N_2 \left(\begin{pmatrix} 0 \\ 0 \end{pmatrix}, \begin{pmatrix} D_{YY} & D_{YW} \\ D_{WY} & D_{WW} \end{pmatrix} \right)
\tag{13.2}
$$

with D_{YY} the variance of the β_{iY} and D_{WW} the variance of the W random intercepts β_{iW}. The covariance $D_{YW} = D_{WY}$ of the two random effects allows for the possibility that the average levels of the two effects are correlated. If they are not correlated, then D_{YW} will be zero, and hopefully the REML estimate will be near zero as compared with its standard error. If the covariance D_{YW} is positive, the average levels of Y_i and W_i go up and down together, and if D_{YW} is negative, then when Y_i is high, W_i on average will be low. If D_{YW} is positive, that indicates that either case (a) or (b) applies in figure 13.1, while if D_{YW} is zero, then case (c) or (d) applies.

To complete the specification of the bivariate random intercept model, we must specify the distribution of the residual errors δ_{iYj} and δ_{iWj}. We allow a correlation between residual errors for responses assessed at the same time t_{ij}. Define δ_{ij}, now a 2-vector of residuals to be bivariate normal

$$\delta_{ij} = \begin{pmatrix} \delta_{iYj} \\ \delta_{iWj} \end{pmatrix} \sim N_2 \left(\begin{pmatrix} 0 \\ 0 \end{pmatrix}, \begin{pmatrix} \sigma_{YY} & \sigma_{YW} \\ \sigma_{WY} & \sigma_{WW} \end{pmatrix} \right) \tag{13.3}$$

The residual errors δ_{ij} from different times are independent and are independent of the random effects β_i from the same subject. Residuals and random effects from different subjects are of course independent.

If σ_{YW} is positive, then the residual errors δ_{iYj} and δ_{iWj} are positively correlated, as illustrated in figure 13.1(a) and (c). If σ_{YW} is zero, then there is no within-subject observation correlation as illustrated in figures 13.1(b) and (d). In the pain rating example, if σ_{YW} is negative, when log pain tolerance is unusually high for a given subject, then we expect pain rating to be unusually low for that subject.

The bivariate random intercept model allows a decomposition of the covariation of Y and W into two parts: the correlation among the random intercepts which is a subject level correlation and the residual error correlation which occurs at the observation level within a subject.

13.4.1   Pain Rating and Log Pain Tolerance

For the log pain tolerance and pain rating example, we hypothesized that the subject average log pain tolerance and pain rating were not correlated. For the observation level covariance around those means, we hypothesized that pain rating and log pain tolerance are correlated. We fit the bivariate random intercept model to the Pain data. Parameter estimates for the covariance parameters are presented in table 13.2. Neither covariance parameter is significant when we include both covariance parameters D_{YW} and σ_{YW} in the model. The estimates have K-values of .18 and .06, respectively, although the correlation between the δ's is closer to significance.

The estimated correlation between the random intercepts is

$$-.21 = \frac{-.2740}{(2.1317 \times .7936)^{1/2}}$$

Parameter	Est	SE	t	p
D_{YY}	2.13	.51	4.2	<.0001
D_{YW}	-.27	.20	-1.3	.18
D_{WW}	.79	.16	5.0	<.0001
σ_{YY}	.34	.04	9.4	<.0001
σ_{YW}	-.13	.07	-1.9	.065
σ_{WW}	2.39	.26	9.4	<.0001

Table 13.2. REML estimates for the covariance parameters for the bivariate random intercept model fit to the Y=log pain tolerance and W=pain rating data.

and the estimated correlation between the residuals is

$$-.14 = \frac{-.1279}{(.3432 \times 2.3931)^{1/2}}.$$

Both correlation calculations use more digits of accuracy than were presented in table 13.2. The estimated correlation between the random intercepts is actually higher than that between the residual errors, although it is less significant. This may be because there is less data for estimating the correlation at higher levels of random effects models. We have 64 subjects, and so at most we have 64 pairs of random intercepts contributing to the estimation of the random intercepts correlation. In contrast, we have 245 observations contributing to the calculation of the $-.14$ correlation between the residual errors.

13.4.1.1 Sub-models of the Bivariate Random Intercept Model

Data from the bivariate random intercept model was illustrated in figure 13.1(a). The other three figures illustrate data from three sub-models where one or the other or both of D_{YW} and σ_{YW} are zero. We can fit these three additional models to the Pain data. All models include the CS and TMT main effects and interaction for both responses and REML fitting. Table 13.3 gives summary results from fitting all four models. Listed is a model number; the non-zero covariance parameters, either σ_{YW} or D_{YW} or both; and the total number of covariance parameters, either 4, 5, or 6 in each model. Columns 4, 5, and 6 give the -2 log REML likelihood, AIC, and BIC. The last column is a chi-square test comparing the models to the base model with no correlation between log pain tolerance and pain rating. The degrees of freedom of the test is the difference in the number of parameters between models. Both models that include σ_{YW} are significantly better than model 4 with no correlation at all. In the full model, neither parameter was quite significant, but the hypothesis test that both parameters are zero is rejected with a p-value of .03!

If we prefer a more parsimonious model, the full model 1 is not significantly better than model 3 with $D_{YW} = 0$, and model 3 is the model we originally hypothesized a priori as being most plausible.

M #	Non-zero parameters	df	-2 log REML	AIC	BIC	χ^2 test vs. 4
1	D_{YW}, σ_{YW}	6	1568.8	1580.8	1593.8	$\chi^2 = 6.0$, df= 2, $p = .05$
2	D_{YW}	5	1572.3	1582.3	1593.1	$\chi^2 = 2.5$, df= 1, $p = .11$
3	σ_{YW}	5	1570.7	1580.8	1591.5	$\chi^2 = 4.1$, df= 1, $p = .043$
4		4	1574.8	1582.8	1591.4	$-$

Table 13.3. Model fitting summaries for the bivariate random intercept model fit to the $Y = $ log pain tolerance and $W = $ pain rating data and three sub-models. The first column is the model number, the second column gives the non-zero covariance parameters, and column 3 is the number of covariance parameters. Columns 4 – 6 are -2 log REML, AIC, and BIC. The chi-square test in column 7 compares the models to the null model with no correlation.

Parameter	Est	SE	t
D_{YY}	1.50	.11	14
D_{YW}	1.16	.12	9
D_{WW}	2.46	.23	11
σ_{YY}	1.27	.04	32
σ_{YW}	.34	.06	6
σ_{WW}	5.84	.18	32

Table 13.4. REML estimates for the covariance parameters for the bivariate random intercept model fit to the Cognitive Y=arithmetic and W=Raven's responses. P-values are all significant and less than .0001 and are not reported.

Another special case of the bivariate random intercept model is introduced in subsection 13.7.2.

13.4.2 Cognitive Data

We jointly analyze arithmetic and Raven's. Predictors for both responses are gender, age_at_baseline, time in years, and the time by treatment interaction. We start with the full bivariate random intercept model. Both covariance parameters D_{YW} and σ_{YW} are highly significant, and we do not explore sub-models where one or both covariance parameters are set equal to zero. Table 13.4 reports results for the variance parameters. Both covariance parameters are positive and significant, indicating that average arithmetic and Raven's are positively associated, and residual errors are positively correlated. The random intercepts have correlation .60, while the residual errors have correlation .12.

13.5   Bivariate Random Intercept and Slope

The Raven's data should include a random slope as well as a random intercept, and we presume this is true for arithmetic as well. We can fit a bivariate random intercept and slope model to the arithmetic and Raven's responses. There will be four random effects for each subject, a random intercept and slope for Raven's and a random intercept and slope for arithmetic. The full model allows for all random effects to be correlated, giving a 4 by 4 covariance matrix D. Again the residual errors for arithmetic and Raven's are allowed to be correlated when they are observed at the same t_{ij}.

The full model is

$$
\begin{aligned}
Y_{ij} &= x'_{ij}\alpha_Y + \beta_{iY1} + \beta_{iY2}t_{ij} + \delta_{iYj} \\
W_{ij} &= x'_{ij}\alpha_W + \beta_{iW1} + \beta_{iW2}t_{ij} + \delta_{iWj}.
\end{aligned}
\tag{13.4}
$$

As compare with 13.1, there is an extra term $\beta_{iY1}t_{ij}$ for the random slope for response Y and similarly for response W. Let $\beta_i = (\beta_{iY1}, \beta_{iW1}, \beta_{iY2}, \beta_{iW2})'$ be the 4-vector of random effects. We assume

$$
\beta_i \sim N_4(\mathbf{0}, D).
$$

The residual errors are still correlated as in equation (13.3) and $\delta_{ij'}$ is independent of $\delta_{ij'}$ for $j \neq j'$ and residual errors are independent of the random effects.

13.5.1   Cognitive Data

We fit this model to the Cognitive arithmetic and Raven's data. Fixed effects are the same as in the previous section. Table 13.5 gives the results for the covariance parameters. The covariance for the δ_{ij} is highly significant with a t-statistic of 4 and a corresponding estimated correlation of .11. If at one time the children do unusually well on Raven's, then they tend to do better on arithmetic at that time, too. Among the random intercepts and slope, only one of the covariances is significant, the covariance between the random intercepts for the two measures. This is the covariance we identified as very significant in the bivariate random intercept analysis of the previous section.

Covariances between any random intercept and any random slope are not significant, nor is the covariance between the random slopes. We might consider dropping some or all of these covariance parameters from our model. The $-2 \log$ REML likelihood for this reduced model is 21370.9. We can also fit a model that omits the covariances between the intercepts and slopes, leaving the covariance between the random intercepts and also the covariance between the random slopes. This model has a $-2 \log$ REML likelihood of 21374.5 and has 4 fewer parameters than the full model. The

Parm	Name	Est	SE	t	p
D_{11}	AI AI	1.57	.13	12	
D_{21}	RI AI	1.15	.14	8	
D_{22}	RI RI	2.29	.28	8	
D_{31}	AS AI	-.087	.055	-1.6	.11
D_{32}	AS RI	-.078	.076	-1.0	.30
D_{33}	AS AS	.15	.04	3.5	.0003
D_{41}	RS AI	.05	.11	.4	.68
D_{42}	RS RI	-.04	.18	-.2	.84
D_{43}	RS AS	.11	.07	1.7	.10
D_{44}	RS RS	.77	.20	3.8	
σ_{AA}		1.18	.04	28	
σ_{AR}		.27	.06	4	
σ_{RR}		5.36	.19	28	

Table 13.5. REML estimates for the covariance parameters for the bivariate random intercept and slope model fit to the Y=arithmetic and W=Raven's. Subscripts on D indicate 1 is the arithmetic random intercept (abbreviated AI), 2 is the Raven's intercept (RI), 3 is the arithmetic random slope (AR), and 4 is the Raven's random slope (RS). The σ parameters are the residual variances and covariances. Name gives the abbreviated name of the variance or correlation. P-values less than .0001 are not listed.

chi-square statistic is 3.6, which is not significant on 4 degrees of freedom. In this reduced model, the covariance between the random slopes is still not quite significant with a p-value of .06; the other covariances are very significant. In this reduced model, the estimated correlation between the residual errors is still .11; the estimated correlation between the intercepts is .62 and the estimated correlation between the slopes is .36. We usually do not like to fit a model where the intercept and slope for the same response are uncorrelated, but this reduced model allows a hypothesis test of no correlation between intercepts and slopes and can be fit using current software.

13.6   Non-simultaneously-Measured Observations

Sometimes we have multiple longitudinal sequences measured on the same subjects but not recorded at the same time. Now it does not make sense to ask about the correlation of the residual errors δ_{iYj} and δ_{iWj} at the same time, as the pair of residual errors are never measured simultaneously. We can still ask whether random intercepts and/or random slopes are correlated. For example, in the Nutrition data we have multiple measures over time on nutritional intake for the children. We might well expect that nutrition is correlated with anthropometric measures such as weight or height or cognitive measures such as Raven's. This hypothesis can be

Parm	Est	SE	t	p
D_{WW}	8.09	.51	16	
D_{WI}	.085	.043	1.9	.054
D_{II}	.092	.008	12	
Σ_{WW}	.57	.006	97	
Σ_{WI}	–	–	–	
Σ_{II}	.73	.096	8	

Table 13.6. Bivariate longitudinal analysis of available iron and weight fit using REML estimation. The model has correlated random intercepts and no correlation among residuals. Observations were taken at different times for the two responses.

tested in a random effects framework where the nutritional variable has a random intercept and weight or Raven's has a random intercept and slope. The question becomes whether the nutrition random intercept is correlated with the weight random intercept or with the weight random slope.

13.6.1 Anthropometry and Nutrition

Investigators made a list of all the food that was eaten by subject on a sample of days. The food list is translated into nutrient intake using subject and day specific recipes and a database of food nutrient values that take into account the season and amount of cooking. The nutritional intake data were observed monthly during the first year and bimonthly during the second year. Subjects have up to 17 nutrition intake values. The nutrition, cognitive and anthropometry measures are measured at different times. We ask whether the nutritional random intercept is correlated with the random intercept for the anthropometry data.

To illustrate the model, we chose available iron and weight, and fit the bivariate random intercept model without correlation between the residual errors. Covariance parameter results are given in table 13.6. The subscript W is for the weight data and I for the available iron. The covariance between the intercepts is borderline not significant with $p = .054$.

Perhaps of greater scientific interest would be to fit the model with a random intercept and slope for the weight data, and to see if available iron or other nutrient random intercept is correlated with the random slopes.

13.7 Unstructured Covariance Models for Bivariate Responses

There are several other models that can be fit with commonly existing software to bivariate longitudinal data. Here we discuss three versions of the unstructured model: (i) the fully unstructured covariance model,

		Pain rating				Log pain tolerance			
		1	2	3	4	1	2	3	4
Pain	1	**4.73**	3.02	2.20	1.40	-.62	-.34	-.78	-.44
rating	2	.64	**4.75**	2.08	1.49	-.31	-.31	-.48	-.38
	3	.48	.45	**4.49**	2.27	-.21	.22	-.41	-.04
	4	.32	.34	.54	**4.00**	-.12	.02	-.23	-.21
Log	1	**-.29**	-.14	-.10	-.06	**1.00**	.75	.95	.58
pain	2	-.15	**-.13**	.10	.01	.70	**1.14**	.88	.76
toler-	3	-.31	-.19	**-.17**	-.10	.82	.71	**1.35**	.90
ance	4	-.19	-.17	-.02	**-.10**	.55	.68	.74	**1.11**

Table 13.7. Estimated covariance matrix (upper half) and correlation matrix (lower half) for the Pain data using the unstructured covariance model for bivariate longitudinal data. The long diagonal (in bold) gives the estimated variances for pain rating at each trial and then for log pain tolerance. The upper left and lower right quadrants give the covariance/correlation matrix of the pain ratings and log pain tolerances, respectively. The boldface correlations in the lower left quadrant identify the correlations between simultaneously assessed pain rating and log pain tolerance.

(ii) the unstructured product correlation model, and (iii) the independent unstructured model.

The unstructured model has a separate variance for each response at each time and a separate covariance or correlation parameter between each combination of response and time. This model can be fit to data with non-simultaneously observed responses, although we only discuss it in the context of simultaneously observed responses. As with univariate longitudinal data, fitting the unstructured model can require a large amount of data. For simultaneously observed bivariate data with J times of observation, we have $J(2J + 1)$ variance parameters.

Table 13.7 gives the estimated covariance and correlation matrix for the Pain data. The long diagonal gives the estimated variances for the pain ratings at the four trials and then for the log pain tolerance at the four trials. The pain rating variances are similar across time and the log pain tolerance variances are also similar. The covariances are above the long diagonal and correlations are below the diagonal.

Most of the *cross-correlations* between rating and log tolerance are negative. These are the correlations in the 4 by 4 lower left-hand quadrant. None of the correlations are very large, but that might be expected because one of our measures is pain rating, and measurements such as this are notoriously noisy. There is a suggestion in the table that a bivariate random intercept model might fit, as the correlations within a measure are similar, but there is enough variation among the correlations to not make this a certainty. If the bivariate random intercept model fit, and D_{YW} and σ_{YW} were both negative, then the correlations between Y_{ij} and W_{ij} at the same trial should be larger, i.e., farther from zero, compared to the other

#	Model	df	-2 log REML	AIC	BIC	Chi-square
1	Bivar RI	6	1568.8	1580.8	1593.8	$\chi^2 = 44.4$, df= 30, $p = .044$
2	Ind RI	4	1574.8	1582.8	1591.4	$\chi^2 = 50.4$, df= 32, $p = .02$
3	PCRI	4	1576.4	1584.4	1593.0	$\chi^2 = 52.0$, df= 32, $p = .014$
4	PCUN	12	1553.4	1577.4	1603.3	$\chi^2 = 29.0$, df= 24, $p = .22$
5	INDUN	20	1540.5	1580.5	1623.7	$\chi^2 = 16.1$, df= 16, $p = .45$
6	UN	36	1524.4	1596.4	1674.2	–

Table 13.8. REML log likelihood, AIC, and BIC for several models fit to the bi-variate Pain data. The chi-square test compares each model to the unstructured covariance model. Model 1 is the bivariate random intercept model, model 2 is the independent random intercepts model, and model 3 is the product correla-tion random intercept model. Model 4 is the product correlation unstructured model (PCUN), 5 is the independent unstructured model (INDUN), and 6 is the unstructured model.

correlations between Y_{ij} and Y_{il}, $j \neq l$. Still, we do not have evidence of higher correlations between ratings and log tolerance from the same trial versus between different trials. The bold face correlations in the lower left quadrant are the correlations between the pain rating and log pain toler-ance at the same trial. These correlations are a little larger than average out of the 16 cross-correlations, but among the cross-correlations they are basically not the largest correlations in their row or column. Considering the uncertainty in the correlations, which have a median standard error of .12, the correlations are not hugely different.

Inspecting the cross-correlations, the correlations in the lower left half of the 4 by 4 cross-correlation matrix, i.e., $-.15$, $-.31$, $-.19$, $-.19$, $-.17$, $-.02$, seem to be larger generally than the correlations in the upper right half, i.e., $-.14$, $-.10$, $-.06$, $.10$, $.01$, $-.10$. This suggests that earlier pain rating scores are more predictive of later log pain tolerance times, whereas earlier log pain tolerance times are less predictive of later pain ratings.

Table 13.8 displays fit summary statistics regarding the fit of various models fit to the bivariate Pain data. Model 1 is the bivariate random intercept model and model 2 is the independent random intercept model where $D_{YW} = \sigma_{YW} = 0$. Model 6 is the unstructured covariance model. The chi-square test in the right-hand column compares each model to the unstructured covariance model.

The other three models are discussed in the following subsections.

13.7.1 Unstructured Within Response, Independent Responses

From the fully unstructured model, we can test independence by comparing it to the model where there is a separate unstructured covariance model for each response and no correlation between responses. This bivariate inde-pendent unstructured (INDUN) model has $J(J+1)$ covariance parameters.

We fit this model to provide a global test of no correlation between Y_i and W_i. If the independent model is preferred, we would fit Y_i and W_i in separate analyses. The test has J^2 degrees of freedom. For the Pediatric Pain data, the test statistic is $\chi^2 = 16.1$ on 16 degrees of freedom, and we do not reject the null hypothesis of no correlation. This test is rather underpowered, in that J^2 cross-correlation parameters are probably not needed to describe the correlation between pain rating and pain tolerance. Many covariance models are adequately parameterized by only a few parameters, and these covariance models may offer better power for detecting the cross-correlation.

13.7.2 Product Correlation Models

The *product correlation* unstructured (PCUN) model allows for the same correlation model for each response and essentially the same variance model for the two responses. Let Σ_Y be the covariance matrix of the first response, and let Σ_W be the covariance matrix of the second response. The product correlation model sets

$$\Sigma_W = v_1 \Sigma_Y$$

where v_1 is an unknown scale parameter to be estimated. The correlation between W_{ij} and W_{il} is equal to the correlation between Y_{ij} and Y_{il}

$$\mathrm{Corr}(W_{ij}, W_{il}) = \mathrm{Corr}(Y_{ij}, Y_{il}).$$

The ratio of two variances at different times is also the same

$$\frac{\mathrm{Var}(W_{ij})}{\mathrm{Var}(W_{il})} = \frac{\mathrm{Var}(Y_{ij})}{\mathrm{Var}(Y_{il})}.$$

The variances $\mathrm{Var}(W_{ij})$ and $\mathrm{Var}(Y_{ij})$ are different, because of the v_1 parameter.

The cross-covariance between observations Y_{ij} and W_{il} taken at different times t_{ij} and t_{il} is

$$\mathrm{Cov}(Y_{ij}, W_{il}) = v_2 \Sigma_{jl}$$

where v_2 is a further unknown parameter to be estimated. Because $\Sigma_{jl} = \Sigma_{lj}$,

$$\mathrm{Cov}(Y_{ij}, W_{il}) = \mathrm{Cov}(Y_{il}, W_{ij}),$$

where the j and l subscripts are switched on the right-hand side compared to the left-hand side of the equation. Similarly, it is not too hard to show that the cross-correlations follow the same property

$$\mathrm{Corr}(Y_{ij}, W_{il}) = \mathrm{Corr}(Y_{il}, W_{ij}) = \frac{v_2}{v_1^{1/2}} \frac{\Sigma_{jl}}{(\Sigma_{jj}\Sigma_{ll})^{1/2}}. \tag{13.5}$$

		Pain rating				Log pain tolerance			
	1	**5.25**	**3.68**	**3.83**	**2.32**	-.34	-.24	-.25	-.15
Pain	2	**.67**	**5.70**	**3.65**	**2.96**	-.24	-.36	-.23	-.19
rating	3	**.67**	**.61**	**6.22**	**3.78**	-.25	-.23	-.40	-.24
	4	**.44**	**.54**	**.66**	**5.27**	-.15	-.19	-.24	-.34
Log	5	-.16	-.10	-.10	-.07	**.89**	**.62**	**.65**	**.39**
pain	6	-.10	-.16	-.10	-.08	**.67**	**.96**	**.62**	**.50**
tolerance	7	-.10	-.10	-.16	-.10	**.67**	**.61**	**1.05**	**.64**
	8	-.07	-.08	-.10	-.16	**.44**	**.54**	**.66**	**.89**

Table 13.9. REML covariance parameter estimates from the product correlation unstructured (PCUN) model. Correlations are below, variances are on, and covariances are above the long diagonal. The covariance/correlation matrix for pain rating and separately for log pain tolerance are in bold.

The cross-correlation between Y_{ij} and W_{il} is the same as between Y_{il} and W_{ij}. Unlike what we can potentially see with the unstructured covariance, there cannot be a temporal lead dog. In the product correlation model, we cannot have the situation where Y_{i1} is highly correlated with (hence highly predictive of) W_{i2}, but that W_{i1} does not predict Y_{i2}. This is possible with the unstructured covariance matrix.

When $j = l$, (13.5) simplifies to

$$\text{Corr}(Y_{ij}, W_{ij}) = \frac{v_2 \Sigma_{jj}}{(v_1 \Sigma_{jj} \Sigma_{jj})^{1/2}} = \frac{v_2}{v_1^{1/2}}. \tag{13.6}$$

In the product correlation model, the two observations at the same time Y_{ij} and W_{ij} have the highest correlation in absolute value among all pairs Y_{ij} and W_{il}. This can be seen by comparing the last two equations and realizing that the second term in equation (13.5) is a correlation and so is less than one in absolute value, and this term is missing in the contemporaneous correlation (13.6).

There are $J(J+1)/2$ parameters in Σ_Y plus 2 additional parameters v_1 and v_2, respectively, the variance factor v_1 for W over Y and the cross-covariance factor v_2. The unstructured covariance model has $J(2J+1)$ parameters. We can test the PCUN versus the UN model with a chi-square statistic with $J(2J+1) - [J(J+1)/2 + 2] = 1.5J^2 + J/2 - 2$ degrees of freedom.

For the Pain data and PCUN model, from table 13.8, the -2 log REML likelihood is 1553.4 with 12 parameters. Comparing to the unstructured covariance model, the chi-square statistic is 29 on $36 - 12 = 24$ degrees of freedom, with a p-value of .22, so the unstructured model does not fit significantly better than the PCUN model.

Table 13.9 gives the estimated correlation matrix for the Pain data and the PCUN model. The cross-correlations from this model are all estimated in the range of $-.07$ to $-.16$. The hypothesis test that the cross-covariance v_2 parameter is zero is a test of correlation between the two responses.

Parameter v_2 is estimated to be $-.064$, with a standard error of .027, a
t-statistic of -2.4 and a significant p-value of .017, and we conclude there
is negative correlation between ratings and log tolerances.

There are many versions of the product correlation model. The covariance
matrix Σ_Y can be the unstructured model with $J(J+1)/2$ parameters or
it can be a parameterized covariance model with fewer parameters, for
example, the random intercept model or AR(1) model. Another possible
product correlation model for the Pain data is the product correlation
random intercept (PCRI) model. The covariance matrix Σ_Y is now a RI
covariance model with two parameters, and $\Sigma_W = v_1\Sigma_Y$, and the J by
J covariance matrix $\text{Cov}(Y_i, W_i) = v_2\Sigma_Y$ for a total of 4 parameters. The
correlations within Y_i or within W_i are assumed the same, but the variances
are different between Y_i and W_i. The correlation between Y_{ij} and W_{ij} at the
same time is larger than the correlation between Y_{ij} and W_{il} at different
time points. Fit to the Pain data, this model has the same number of
parameters as the independent RI model, and it fits slightly worse. The
test of $H_0 : v_2 = 0$ again gives a test of the hypothesis that the cross-
correlations are zero. The test is significant, with a test statistic of $t = -2.3$
and a p-value of .02. We can also compare to the unstructured covariance
model. This test has a chi-square statistic $\chi^2 = 52$ on 32 df and a $p = .02$,
so we also prefer the unstructured covariance model to the PCRI model.

Table 13.8 gives fit statistics for the two product correlation models and
the independent unstructured model. The product correlation unstructured
model and the independent unstructured models are the only two for which
the unstructured model is not significantly better than. The correlation
between the two responses is borderline significant. In terms of AIC, the
two best models have correlation between rating and log tolerance, and
the next two have no correlation. In terms of BIC, the best model has no
correlation, but the next two best have correlation. In the models with a
one or two degree of freedom significance test for correlation, the correlated
model is significantly better than the independent model. We are left with
the conclusion that the data support for the hypothesis that pain rating
and log pain tolerance is not overwhelming.

13.8 Discussion

There are many ways that two longitudinal variables can be correlated.
Statisticians have only begun to develop models to describe the interrela-
tionships among two or more longitudinal variables. We have provided an
introduction to a few of the models and ideas in this chapter.

Software can require what at first glance would seem to be an unusual
data structure to analyze bivariate longitudinal. For example, for the Pain
data, we must set up a single data set in long form with 8 observations

for each subject. We would have a single variable "response" that is the pain rating for the first four observations and is the log pain tolerance for the second four observations. A separate variable "type" tells which type of response each observation is. For the Pain data this was "r" or "t" for rating or tolerance. In the Cognitive analyses, we had 10 observations for each subject. The type variable was coded "a" or "r" for arithmetic or Raven's. Whatever fixed effects are desired, they always will have an interaction with type.

These models can be used to identify the type and nature of the correlation across responses. When the responses are measured at the same times, but one response has missing data when the other is observed, the bivariate model may be used to help estimate the missing responses and may improve the precision of the effect estimates.

Each covariance model for bivariate longitudinal data implies a particular model for the univariate longitudinal analysis with the second response as a predictor. For most models for balanced data, the coefficients of the second response are non-linear functions of the covariance parameters. The only way to begin to handle the model and to be sure that it is correct is to include all elements of the second response as predictors for all times of the first response and with separate coefficients at each time. Unfortunately, this is an extremely unwieldy model, and with so many collinear covariates, estimation can be problematic. Missing data makes it quite difficult to handle, as some predictors are not available for some observations. There is one class of covariance models where the univariate longitudinal model with response two as a single time-varying covariate is correctly specified: the product correlation models. Unfortunately, in any of these univariate longitudinal models with covariate the second longitudinal variable, the fixed effects α are now modified to be a linear combination of the fixed effects for the two responses. If a particular predictor is significant in the univariate analysis for response two, it may be significant when response 2 is introduced into the analysis for response 1, whether it was before or not. Also, a previously significant predictor may no longer be significant. These issues go away or are properly handled when the two responses are fit as a bivariate longitudinal response.

13.9   Problems

1. Fit models comparable to those in table 13.1 but with log pain tolerance as a predictor and pain rating as a response. Include a coping style by treatment interaction in the model. What conclusions do you draw? Are they the same as the conclusions drawn from table 13.1?

2. (advanced) Suppose that you have two balanced, equally spaced longitudinal measures W_{ij} and Y_{ij}, for all subjects, $i = 1, \ldots, n$ measured

at times $t_j = j, j = 1, \ldots, J$ where J is moderately large. Discuss how you might develop a model that allows earlier measurements $W_{i(j-1)}$ to predict the current Y_{ij}.

3. Extend the bivariate random intercept model of subsection 13.4 with correlated errors to the situation where there are three longitudinal measures.

   (a) Write out the model.
   (b) How many covariance parameters are there?
   (c) Fit the Cognitive data to this model. Draw conclusions.
   (d) Is there any advantage to this model over the bivariate random intercept model?

4. Fit the bivariate random intercept and slope model to the Cognitive data. Use arithmetic and verbal meaning as responses. Which model is better, the bivariate RI model or the bivariate RIAS model?

5. For bivariate longitudinal response with three time points, write out the 6 by 6 marginal covariance matrix of the observations for the bivariate random intercept model. There are two ways to order the Y and W observations: my preference is to put all the Y observations first then the W observations. The alternative is to alternate Y and W and go sequentially in time. Do it both ways, and decide which way you prefer, and why.

6. (Continued) For each of the 4 special cases of the bivariate random intercept model, write out the corresponding covariance model unconditional on the random effects.

7. Fit all of the models in this chapter to verbal meaning and another Cognitive variable. Construct a table of the fit statistics. What covariance model seems to be best? Print the estimated unstructured covariance and correlation matrix neatly and interpret it. Even though the bivariate RI and RIAS models may not be the best fit, the information that we learn from them may still be helpful in interpreting the covariance from the unstructured covariance model. Interpret the unstructured covariance model.

8. Cognitive data. Digit span is considered to be a different measure than the other three. Digit span assesses memory; the other three measure intellectual ability and academic achievement. See if digit span and Raven's are correlated in the same way that Raven's and arithmetic are correlated. You will need to decide on a model for the fixed effects, and you may need to fit several covariance models to understand the covariance.

14
Further Reading

The end crowns all;
And that old common arbitrator, Time,
Will one day end it.

— William Shakespeare

Not every end is a goal. The end of a melody is not its goal:
but nonetheless, had the melody not reached its end it would
not have reached its goal either.

— Friedrich Nietzsche

Early in the study of longitudinal data, *growth curve models* were popular.
These models allowed for an arbitrary polynomial time trend in each of
several groups. Computation was essentially restricted to balanced data
sets without missing data and a limited number of covariance structures.
See for example Rao (1965) and references therein. Lee and Geisser (1972)
studied prediction in growth curve models. The multivariate analysis of
variance (MANOVA) model is another approach as in the classic paper by
Potthoff and Roy (1964) who introduced the dental data. The multivariate
linear regression formulation (5.6) that we use has overtaken the growth
curve and MANOVA formulations because it is much more flexible. Model
(5.6) allows for parameterized covariance models, unbalanced data, missing
data or even random times, and arbitrary combinations of time-fixed and
time-varying covariates and can be extended to multivariate longitudinal
data. Ware (1985) outlines modern longitudinal data analysis, and Rao
(1987) and discussants provide an interesting discussion of prediction in
longitudinal data analysis.

Harville (1976) extended the Gauss-Markov theorem to random effects models and the estimation of random effects. Lange and Laird (1989) study the effects on inferences of assuming various random effects models. Diggle and Donnelly (1989) give a selected bibliography on the analysis of repeated measurements most of which involve applications to longitudinal data.

Covariance Model Specification

Parameterized covariance model specification as presented here owes much to Jennrich and Schluchter (1986). Zimmerman (2000) and Núñez-Antón and Zimmermann (2000) discuss antedependence covariance models. Núñez-Antón and Woodworth (1994) propose a power transformation of time in an AR(1) model, leading to a particular form of antedependence. Geary (1989) gives an example of a generalization of AR and equicovariance models and Chi and Reinsel (1989) give a test for needing AR(1) correlation in the residuals of the random intercept model. Diggle (1988) has a nice discussion of the autoregressive plus random intercept plus measurement error model. Jones (1993) and Jones and Boadi-Boateng (1991) discusses AR and ARMA correlation structures. Muñoz, Carey, Schouten, Segal, and Rosner (1991) proposed the power AR model of subsection 8.6.3. Izenman and Williams (1989) is an early paper on factor analytic models for longitudinal data and the original source for the Big Mice data.

SAS Proc Mixed® online documentation at, for example, `http://v8doc.sas.com/sashtml/stat/chap41/sect20.htm#mixedrepeat` has a compact description of a wide number of covariance structures. See in particular tables 41.3, 41.4, and 41.5. Pinheiro and Bates (2000) discuss several covariance models in their book on computational algorithms.

Papers by Daniels and Pourahmadi (2002), Daniels and Zhao (2003), and Pourahmadi and Daniels (2002) develop a number of generalizations of models for the covariance matrix; in particular they allow for covariates to affect various aspects of the covariance matrix of longitudinal data and/or the covariance matrix of the random effects. Stram and Lee (1994) explain how to test a more general random effects model against a nested model with fewer random effects. Boscardin and Zhang (2004) and Boscardin, Weiss, and Zhang (2005) give models that allow fitting of the unstructured covariance matrix to sparse longitudinal data.

Computation

Early approaches to longitudinal data analysis stressed non-iterative parameter estimates. Szatrowski (1980) and Szatrowski and Miller (1980) identified the availability of algebraic maximum likelihood estimates for multivariate normal models. Harville (1977) discusses maximum likelihood estimation for random effects models. Laird and Ware (1982), and Laird, Lange, and Stram (1987) explain use of the EM algorithm for random effects models. See Jennrich and Schluchter (1986) and Lindstrom and Bates

(1988) for Newton-Raphson algorithms and Lindstrom and Bates (1990) for algorithms for nonlinear random effects models. Pinheiro and Bates (2000) discuss computation in great detail but require a level of statistical sophistication.

I drew all figures in R® (R Development Core Team 2004). Most figures were prototyped in xlispstat (Tierney 1990). All models were fit in SAS Proc Mixed or Proc Nlmixed (SAS Institute Inc. 1999). Most models in the text can be fit in SAS Proc Mixed (Littell, Milliken, Stroup, and Wolfinger 1996) or in R's NLME package (Pinheiro, Bates, DebRoy, and Sarkar 2004).

REML and Bayesian Methods

A formative paper in longitudinal data analysis is Patterson and Thompson (1971) who introduced restricted maximum likelihood (REML) estimation and Harville (1974) who showed that REML estimation has a partial Bayesian interpretation.

For fully Bayesian methods related to random effects models and hierarchical models, see Gelfand, Hills, Racine-Poon, and Smith (1990), Gilks, Wang, Yvonnet, and Coursaget (1993), and Weiss, Wang, and Ibrahim (1997). Posterior predictive (Rubin 1984; Meng 1994; Gelman, Meng, and Stern 1996) methodology for hypothesis testing and model checking is very flexible. For general information on Bayesian modeling see Gelman, Carlin, Stern, and Rubin (2003) and for information on Bayesian computing see the texts Gilks, Richardson, and Spiegelhalter (1995), Robert and Casella (2004) or Gamerman (1997).

Complexities: Hierarchical, Multivariate, Nonlinear

The *hierarchical* and *multi-level model* literature discusses the analysis of data with one or often more levels of nesting. Usually they limit discussion to random effects models; unfortunately this is often inappropriate when one of the levels is time within subject. See Raudenbush and Bryk (2001).

Research in *multivariate longitudinal data* is just beginning. Most of what there is involves random effects (Shah, Laird, and Schoenfeld 1997; Reinsel 1982; Reinsel 1984). Two exceptions are Sy, Taylor, and Cumberland (1997) and Wang and Belin (2002). Multivariate longitudinal is likely to be a prolific research area in the future.

Nonlinear normal longitudinal models are discussed in Davidian and Giltinan (1995) and information on *cross-over trials* can be found in Senn (2002) and Jones and Kenward (1989).

Discrete Data

There are a number of difficulties in the computational aspect of fitting discrete longitudinal data, hence there are many approaches to the computations. Most approaches discuss only random effects models. One key computational problem is that maximum likelihood approaches need to use integration methods for integrating the random effects out of the likelihood to produce a marginal likelihood of the parameters α and D. This marginal likelihood must then be maximized. Unfortunately, the integration can only be done numerically, and the maximization algorithm is rather fragile because common approximate integration methods create a likelihood that is not continuous in α and D and the maximization algorithms often fail to converge. One popular package can produce a large number of different parameter estimates depending on the options specified for the integration algorithm; the parameter estimates can vary quite dramatically.

Some of the more popular approaches for fitting longitudinal data models to discrete data are outlined in Breslow and Clayton (1993) and Wolfinger and O'Connell (1993) implemented in SAS Macro Glimmix® and Proc Glimmix®. Moulton and Zeger (1989) proposed a bootstrap approach.

One class of solutions proposed has been what are called *generalized estimating equations* (GEE) methods (Liang and Zeger 1986; Zeger, Liang, and Albert 1988; Zeger, Liang, and Self 1985; Hardin and Hilbe 2002). GEE methods are quite popular because the software runs easily and quickly. However, GEE methods tend not to correspond to a completely specified sampling model for the data, and inference is entirely dependent on asymptotics. Because the sampling model is not always completely specified, what assumptions are being made when using GEE methods can be unclear. On occasion the methods can correspond to no possible sampling distributions for the data. GEE uses a "working correlation matrix" for the data. This working correlation matrix is not necessarily assumed to actually describe the correlation matrix of the data. Instead, a fix is applied to the estimated correlation matrix of the fixed effects to correct the standard errors. This fix is the *sandwich estimator*. I do not like statistical methods that do not completely specify the sampling distribution of the data, or worse, that assume that you have made a mistake in specifying the distribution of the data and then try to fix it. My preference is, if the model for the data is not correct, then I attempt to correct it and to develop as accurate a model as possible.

From a Bayesian perspective, the problem in analyzing discrete data is caused mainly by the twentieth-century aversion to integration and consequent over-reliance on function maximization. If one relaxes the antipathy to Bayesian methods dating to Fisher and Pearson, one can make rapid progress in the analysis of multivariate discrete data. Bayesian methods are becoming easy to use with current software such as WinBUGS (Gilks, Thomas, and Spiegelhalter 1994). Random effects models for discrete data (Zeger and Karim 1991; Chen, Ibrahim, Shao, and Weiss 2003) are examples where Bayesian methods particularly shine because they do not depend on asymptotics and they can handle both complex models and small data sets, neither of which work particularly well with classical or likelihood based methods. Additionally, Bayesian methods allow you to make inference about complex functions of the parameters and are easily extensible, while classical methods tend to be difficult to extend and inferences are usually restricted to simple functions of the parameters. Chen, Ibrahim, Shao, and Weiss (2003) illustrates a Bayesian approach to model selection in longitudinal discrete data including a covariance model other than a random effects model.

Splines, Semi-parametric Analysis, and Smooth Time Trends

There has been quite a bit of work in the area of modeling the time trend of longitudinal data by a smooth curve. Early work concentrated on the population mean function (Zeger and Diggle 1994; Wang and Taylor 1995a; Wang and Taylor 1995b). Later work also smoothed the covariance model, using high dimensional random effects to approximate the covariance model as well (Anderson and Jones 1995; Shi, Weiss, and Taylor 1996; James, Hastie, and Sugar 2000; Rice and Wu 2001; Grambsch, Randall, Bostick, Potter, and Louis 1995). A precursor paper to this literature was Izenman and Williams (1989). Methods from the functional data analysis literature (Ramsay, Wang, and Flanagan 1995; Ramsay and Silverman 1997) can be applied to the analysis of longitudinal data, particularly when there are a large number of repeated measures. Perhaps the most ambitious model to date is DiMatteo, Genovese, and Kass (2001) who allow their model to estimate both the order of the spline approximating the mean and the location of the knots as well.

Missing Data

Rubin (1976) introduced the language of missing at random (MAR), missing completely at random (MCAR), and missing not at random (MNAR). Little (1993; 1994) introduced *pattern mixture models* as one way of accommodating the problem of handling missing data in an analysis. Hedeker and Gibbons (1997) analyze the schizophrenia data and we borrowed their analysis in section 12.6. Little and Rubin (2002) discuss general strategies for handling missing data. Schafer (1997) generally discusses statistical modeling with missing data. Rubin (1987) introduced *multiple imputation* as a strategy for dealing with missing data in general data structures. The general approach is to create several complete data sets that are analyzed using the same complete data analysis; inferences from the multiple analyses are then combined in relatively easy fashion (Rubin and Schenker 1986; Schafer 1997). For purposes of longitudinal data analysis, we need multiple imputation when we have missing covariates, not missing responses. Because we so frequently are missing covariates in longitudinal data analysis, multiple imputation is frequently applied in longitudinal data analysis. Diggle and Donnelly (1989) and Little (1988b) propose tests for MCAR in longitudinal data.

Data Analysis: Graphics, Residuals, Outliers, Accommodations, Influence, and Transformations

Segal (1994) and Jones and Rice (1992) give methodologies for picking a few representative subjects rather than plotting all subjects in a profile plot. Dempster and Ryan (1985) and Lange and Ryan (1989) propose methods for checking normality in random effects models. Early longitudinal data analysis examples include various models for the mean and covariance, often some graphics, and usually some discussion of computing (Louis and Spiro III 1986; Waternaux, Laird, and Ware 1989; Verbyla and Cullis 1990; Cullis and McGilchrist 1990; Diggle and Donnelly 1989; Gregoire, Schabenberger, and Barrett 1995).

For information on outliers and residuals particularly in random effects models, see Weiss and Lazaro (1992) and Zhang and Weiss (2000) and also Louis (1988). Little (1988a) and Lange, Little, and Taylor (1989) discuss tests for outlying observations and subjects. Modeling approaches for accommodating outliers use long-tailed error distributions, for example t-errors (Seltzer, Wong, and Bryk 1996; Seltzer 1993; Lange, Little, and Taylor 1989) or mixtures of small and large variance normals (Sharples 1990; Weiss, Wang, and Ibrahim 1997).

Influence assessment methodology mostly has been developed for random effects models. Confer Beckman, Nachtsheim, and Cook (1987), Christensen, Pearson, and Johnson (1992), Lesaffre and Verbeke (1998), and Lindsey and Lindsey (2000).

The literature on transformation in regression models is quite rich, for example (Box and Cox 1964; Bickel and Doksum 1981; Box and Cox 1982; Duan 1983; Taylor 1986). Hamilton and Taylor (1993), Taylor, Cumberland, and Meng (1996), and Lipsitz, Ibrahim, and Molenberghs (2000) discuss transformation in the context of longitudinal data.

Bibliography

Anderson, S. J. and R. H. Jones (1995). Smoothing splines for longitudinal data. *Statistics in Medicine 14*, 1235–1248.

Beckman, R. J., C. J. Nachtsheim, and R. D. Cook (1987). Diagnostics for mixed-model analysis of variance (Corr: 1990 v32 p241). *Technometrics 29*, 413–426.

Bickel, P. J. and K. A. Doksum (1981). An analysis of transformations revisited. *Journal of the American Statistical Association 76*, 296–311.

Boscardin, W., R. Weiss, and X. Zhang (2005). Fitting unstructured covariance matrices to sparse longitudinal data. Technical report, UCLA School of Public Health Department of Biostatistics.

Boscardin, W. J. and X. Zhang (2004). Modelling the covariance and correlation matrix of repeated measures. In A. Gelman and X.-L. Meng (Eds.), *Applied Bayesian Modeling and Causal Inference from Incomplete-Data Perspectives*, pp. 215–226. John Wiley & Sons.

Box, G. E. P. and D. R. Cox (1964). An analysis of transformations. *Journal of the Royal Statistical Society, Series B 26*, 211–252.

Box, G. E. P. and D. R. Cox (1982). An analysis of transformations revisited, rebutted. *Journal of the American Statistical Association 77*, 209–210.

Breslow, N. E. and D. G. Clayton (1993). Approximate inference in generalized linear mixed models. *Journal of the American Statistical Association 88*, 9–25.

Chen, M., J. Ibrahim, Q. Shao, and R. Weiss (2003). Prior elicitation for model selection and estimation in generalized linear mixed models. *Journal of Statistical Planning and Inference 111*, 57–76.

Chi, E. M. and G. C. Reinsel (1989). Models for longitudinal data with random effects and AR(1) errors. *Journal of the American Statistical Association 84*, 452–459.

Christensen, R., L. M. Pearson, and W. Johnson (1992). Case-deletion diagnostics for mixed models. *Technometrics 34*, 38–45.

Cook, R. D. and S. Weisberg (1999). *Applied Regression Including Computing and Graphics*. New York, NY: Wiley.

Cullis, B. R. and C. A. McGilchrist (1990). A model for the analysis of growth data from designed experiments. *Biometrics 46*, 131–142.

Daniels, M. and M. Pourahmadi (2002). Bayesian analysis of covariance matrices and dynamic models for longitudinal data. *Biometrika 89*, 553–566.

Daniels, M. and Y. Zhao (2003). Modelling the random effects covariance matrix in longitudinal data. *Statistics in Medicine 22*, 1631–1647.

Davidian, M. and D. M. Giltinan (1995). *Nonlinear Models for Repeated Measurement Data*. Boca Raton, FL: Chapman & Hall/CRC.

Dempster, A. P. and L. M. Ryan (1985). Weighted normal plots. *Journal of the American Statistical Association 80*, 845–850.

Diggle, P. J. (1988). An approach to the analysis of repeated measurements. *Biometrics 44*, 959–971.

Diggle, P. J. and J. B. Donnelly (1989). A selected bibliography on the analysis of repeated measurements and related areas. *The Australian Journal of Statistics 31*, 183–193.

DiMatteo, I., C. Genovese, and R. Kass (2001). Bayesian curve fitting with free-knot splines. *Biometrika 88*, 1055–1073.

Duan, N. (1983). Smearing estimate: A nonparametric retransformation method. *Journal of the American Statistical Association 78*, 605–610.

Fanurik, D., L. Zeltzer, M. Roberts, and R. Blount (1993). The relationship between childrens coping styles and psychological interventions for cold pressor pain. *Pain 53*, 213–222.

Fox, J. (1997). *Applied Regression Analysis, Linear Models and Related Methods*. Thousand Oaks, CA: Sage Publications.

Gamerman, D. (1997). *Markov Chain Monte Carlo: Stochastic Simulation for Bayesian Inference*. Boca Raton, FL: Chapman & Hall/CRC.

Geary, D. N. (1989). Modelling the covariance structure of repeated measurements. *Biometrics 45*, 1183–1195.

Gelfand, A. E., S. E. Hills, A. Racine-Poon, and A. F. M. Smith (1990). Illustration of Bayesian inference in normal data models using Gibbs sampling. *Journal of the American Statistical Association 85*, 972–985.

Gelman, A., J. B. Carlin, H. S. Stern, and D. B. Rubin (2003). *Bayesian Data Analysis* (2nd ed.). Boca Raton, FL: Chapman & Hall/CRC.

Gelman, A., X.-L. Meng, and H. Stern (1996). Posterior predictive assessment of model fitness via realized discrepancies (disc: p760-807). *Statistica Sinica 6*, 733–760.

Gilks, W., S. Richardson, and D. Spiegelhalter (1995). *Markov Chain Monte Carlo in Practice*. Boca Raton, FL: Chapman & Hall/CRC.

Gilks, W., A. Thomas, and D. Spiegelhalter (1994). A language and program for complex Bayesian modelling. *The Statistician 43*, 169–178.

Gilks, W. R., C. C. Wang, B. Yvonnet, and P. Coursaget (1993). Random-effects models for longitudinal data using Gibbs sampling. *Biometrics 49*, 441–453.

Grambsch, P. M., B. L. Randall, R. M. Bostick, J. D. Potter, and T. A. Louis (1995). Modeling the labeling index distribution: An application of functional data analysis. *Journal of the American Statistical Association 90*, 813–821.

Gregoire, T. G., O. Schabenberger, and J. P. Barrett (1995). Linear modelling of irregularly spaced, unbalanced, longitudinal data from permanent-plot measurements. *Canadian Journal of Forestry Research 25*, 137–156.

Hamilton, S. A. and J. M. G. Taylor (1993). A comparison of the Box-Cox transformation method and nonparametric methods for estimating quantiles in clinical data with repeated measures. *Journal of Statistical Computation and Simulation 45*, 185–201.

Hardin, J. and J. Hilbe (2002). *Generalized Estimating Equations*. Boca Raton, FL: Chapman and Hall/CRC.

Harville, D. A. (1974). Bayesian inference for variance components using only error contrasts. *Biometrika 61*, 383–385.

Harville, D. A. (1976). Estension of the Gauss-Markov theorem to include the estimation of random effects. *The Annals of Statistics 4*, 384–395.

Harville, D. A. (1977). Maximum likelihood approaches to variance component estimation and to related problems (c/r: p338-340). *Journal of the American Statistical Association 72*, 320–338.

Hedeker, D. and R. Gibbons (1997). Application of random-effects pattern-mixture models for missing data in longitudinal studies. *Psychological Methods 2*, 64–78.

Henderson, R. and S. Shimakura (2003). A serially correlated gamma frailty model for longitudinal count data. *Biometrika 90*, 355–366.

Izenman, A. and J. Williams (1989). A class of linear spectral models and analyses for the study of longitudinal data. *Biometrics 45*, 831–849.

James, G. M., T. J. Hastie, and C. A. Sugar (2000). Principal component models for sparse functional data. *Biometrika 87*, 587–602.

Jennrich, R. I. and M. D. Schluchter (1986). Unbalanced repeated-measures models with structured covariance matrices. *Biometrics 42*, 805–820.

Jones, B. and M. G. Kenward (1989). *Design and Analysis of Cross-Over Trials*. London: Chapman & Hall/CRC.

Jones, M. C. and J. A. Rice (1992). Displaying the important features of large collections of similar curves. *The American Statistician 46*, 140–145.

Jones, R. H. (1993). *Longitudinal Data with Serial Correlation: A State-Space Approach*. Boca Raton, FL: Chapman & Hall/CRC.

Jones, R. H. and F. Boadi-Boateng (1991). Unequally spaced longitudinal data with AR(1) serial correlation. *Biometrics 47*, 161–175.

Kutner, M. H., C. J. Nachtsheim, and J. Neter (2004). *Applied Linear Regression Models* (4th ed.). New York, NY: McGraw-Hill/Irwin.

Laird, N., N. Lange, and D. Stram (1987). Maximum likelihood computations with repeated measures: Application of the EM algorithm. *Journal of the American Statistical Association 82*, 97–105.

Laird, N. M. and J. H. Ware (1982). Random-effects models for longitudinal data. *Biometrics 38*, 963–974.

Lange, K. L., R. J. A. Little, and J. M. G. Taylor (1989). Robust statistical modeling using the *t*-distribution. *Journal of the American Statistical Association 84*, 881–896.

Lange, N. and N. M. Laird (1989). The effect of covariance structure on variance estimation in balanced growth-curve models with random parameters. *Journal of the American Statistical Association 84*, 241–247.

Lange, N. and L. Ryan (1989). Assessing normality in random effects models. *The Annals of Statistics 17*, 624–642.

Lee, J. C. and S. Geisser (1972). Growth curve prediction. *Sankhyā, Series A, Indian Journal of Statistics 34*, 393–412.

Lesaffre, E. and G. Verbeke (1998). Local influence in linear mixed models. *Biometrics 54*, 570–582.

Liang, K.-Y. and S. L. Zeger (1986). Longitudinal data analysis using generalized linear models. *Biometrika 73*, 13–22.

Lindsey, P. J. and J. K. Lindsey (2000). Diagnostic tools for random effects in the repeated measures growth curve model. *Computational Statistics & Data Analysis 33*, 79–100.

Lindstrom, M. J. and D. M. Bates (1988). Newton-Raphson and EM algorithms for linear mixed-effects models for repeated-measures data (Corr: 1994 v89 p1572). *Journal of the American Statistical Association 83*, 1014–1022.

Lindstrom, M. J. and D. M. Bates (1990). Nonlinear mixed effects models for repeated measures data. *Biometrics 46*, 673–687.

Lipsitz, S. R., J. Ibrahim, and G. Molenberghs (2000). Using a Box-Cox transformation in the analysis of longitudinal data with incomplete responses. *Applied Statistics 49*, 287–296.

Littell, R. C., G. A. Milliken, W. W. Stroup, and R. D. Wolfinger (1996). *SAS System for Mixed Models*. Cary, NC: SAS Institute, Inc.

Little, R. J. A. (1988a). Commentary. *Statistics in Medicine 7*, 347–355.

Little, R. J. A. (1988b). A test of missing completely at random for multivariate data with missing values. *Journal of the American Statistical Association 83*, 1198–1202.

Little, R. J. A. (1993). Pattern-mixture models for multivariate incomplete data. *Journal of the American Statistical Association 88*, 125–134.

Little, R. J. A. (1994). A class of pattern-mixture models for normal incomplete data. *Biometrika 81*, 471–483.

Little, R. J. A. and D. B. Rubin (2002). *Statistical Analysis with Missing Data* (2nd ed.). New York, NY: John Wiley & Sons.

Louis, T. A. (1988). General methods for analysing repeated measures. *Statistics in Medicine 7*, 29–45.

Louis, T. A. and A. Spiro III (1986). Fitting and assessing first-order auto-regressive models with covariates. Technical report, Harvard School of Public Health, Department of Biostatistics.

Mallon, G. (1994). Mixed linear models and applications. Honours thesis, University of Queensland, Department of Mathematics.

Meng, X.-L. (1994). Posterior predictive p-values. *The Annals of Statistics 22*, 1142–1160.

Moulton, L. H. and S. L. Zeger (1989). Analyzing repeated measures on generalized linear models via the bootstrap. *Biometrics 45*, 381–394.

Muñoz, A., V. Carey, J. P. Schouten, M. Segal, and B. Rosner (1991). A parametric family of correlation structures for the analysis of longitudinal data. *Biometrics 48*, 733–742.

Neumann, C., N. Bwibo, S. Murphy, M. Sigman, S. Whaley, L. Allen, D. Guthrie, R. Weiss, and M. Demment (2003). Animal source foods improve dietary quality, micronutrient status, growth and cognitive function in Kenyan school children: Background, study design and baseline findings. *Journal of Nutrition 11*, 3941S–3949S.

Núñez-Antón, V. and G. G. Woodworth (1994). Analysis of longitudinal data with unequally spaced observations and time-dependent correlated errors. *Biometrics 50*, 445–456.

Núñez-Antón, V. and D. L. Zimmermann (2000). Modeling nonstationary longitudinal data. *Biometrics 56*, 699–705.

Patterson, H. D. and R. Thompson (1971). Recovery of inter-block information when block sizes are unequal. *Biometrika 58*, 545–554.

Pinheiro, J., D. Bates, S. DebRoy, and D. Sarkar (2004). *nlme: Linear and Nonlinear Mixed Effects Models*. R package version 3.1-52.

Pinheiro, J. C. and D. M. Bates (2000). *Mixed-Effects Models in S and S-PLUS*. New York, NY: Springer-Verlag.

Potthoff, R. and S. Roy (1964). A generalized multivariate analysis of variance model useful especially for growth curve problems. *Biometrika 51*, 313–326.

Pourahmadi, M. and M. Daniels (2002). Dynamic conditionally linear mixed models. *Biometrics 58*, 225–231.

R Development Core Team (2004). *R: A Language and Environment for Statistical Computing*. Vienna, Austria: R Foundation for Statistical Computing. 3-900051-07-0.

Ramsay, J. O. and B. W. Silverman (1997). *Functional Data Analysis*. New York, NY: Springer-Verlag.

Ramsay, J. O., X. Wang, and R. Flanagan (1995). A functional data analysis of the pinch force of human fingers. *Applied Statistics 44*, 17–30.

Rao, C. R. (1965). The theory of least squares when the parameters are stochastic and its application to the analysis of growth curves. *Biometrika 52*, 447–468.

Rao, C. R. (1987). Prediction of future observations in growth curve models (c/r: p448-471). *Statistical Science 2*, 434–447.

Raudenbush, S. W. and A. S. Bryk (2001). *Hierarchical Linear Models: Applications and Data Analysis Methods* (2nd ed.). Thousand Oaks, CA: SAGE Publications.

Reinsel, G. (1982). Multivariate repeated-measurement or growth curve models with multivariate random-effects covariance structure. *Journal of the American Statistical Association 77*, 190–195.

Reinsel, G. (1984). Estimation and prediction in a multivariate random effects generalized linear model. *Journal of the American Statistical Association 79*, 406–414.

Rice, J. and C. Wu (2001). Nonparametric mixed effects models for unequally sampled noisy curves. *Biometrics 57*, 253–259.

Robert, C. and G. Casella (2004). *Monte Carlo Statistical Methods* (2nd ed.). New York: Springer-Verlag.

Rotheram-Borus, M. J., M. Lee, Y. Y. Lin, and P. Lester (2004). Six-year intervention outcomes for adolescent children of parents with the human immunodeficiency virus. *Archives of Pediatrics & Adolescent Medicine 158*, 742–748.

Rubin, D. B. (1976). Inference and missing data. *Biometrika 63*, 581–590.

Rubin, D. B. (1984). Bayesianly justifiable and relevant frequency calculations for the applied statistician. *The Annals of Statistics 12*, 1151–1172.

Rubin, D. B. (1987). *Multiple Imputation for Nonresponse in Surveys*. New York, NY: John Wiley & Sons.

Rubin, D. B. and N. Schenker (1986). Multiple imputation for interval estimation from simple random samples with ignorable nonresponse. *Journal of the American Statistical Association 81*, 366–374.

SAS Institute Inc. (1999). *SAS/STAT® Users Guide, Version 8, Chapter 41*. Cary, NC. Proc Mixed is Chapter 41.

Schafer, J. L. (1997). *Analysis of Incomplete Multivariate Data*. Boca Raton, FL: Chapman & Hall/CRC.

Segal, M. R. (1994). Representative curves for longitudinal data via regression trees. *Journal of Computational and Graphical Statistics 3*, 214–233.

Seltzer, M. H. (1993). Sensitivity analysis for fixed effects in the hierarchical model: A Gibbs sampling approach. *Journal of Educational Statistics 18*, 207–235.

Seltzer, M. H., W. H. Wong, and A. S. Bryk (1996). Bayesian analysis in applications of hierarchical models: Issues and methods. *Journal of Educational and Behavioral Statistics 21*, 131–167.

Senn, S. (2002). *Cross-over Trials in Clinical Research* (2nd ed.). New York, NY: John Wiley & Sons.

Shah, A., N. Laird, and D. Schoenfeld (1997). A random-effects model for multiple characteristics with possibly missing data. *Journal of the American Statistical Association 92*, 775–779.

Sharples, L. D. (1990). Identification and accommodation of outliers in general hierarchical models. *Biometrika 77*, 445–453.

Shi, M., R. E. Weiss, and J. M. G. Taylor (1996). An analysis of paediatric CD4 counts for acquired immune deficiency syndrome using flexible random curves. *Applied Statistics 45*, 151–163.

Stram, D. O. and J. W. Lee (1994). Variance components testing in the longitudinal mixed effects model (Corr: 1995 v51 p1196). *Biometrics 50*, 1171–1177.

Sy, J. P., J. M. G. Taylor, and W. G. Cumberland (1997). A stochastic model for the analysis of bivariate longitudinal AIDS data. *Biometrics 53*, 542–555.

Szatrowski, T. H. (1980). Necessary and sufficient conditions for explicit solutions in the multivariate normal estimation problem for patterned means and covariances. *The Annals of Statistics 8*, 802–810.

Szatrowski, T. H. and J. J. Miller (1980). Explicit maximum likelihood estimates from balanced data in the mixed model of the analysis of variance. *The Annals of Statistics 8*, 811–819.

Taylor, J. M. G. (1986). The retransformed mean after a fitted power transformation. *Journal of the American Statistical Association 81*, 114–118.

Taylor, J. M. G., W. G. Cumberland, and X. Meng (1996). Components of variance models with transformations. *The Australian Journal of Statistics 38*, 183–191.

Tierney, L. (1990). *LISP-STAT: An Object-Oriented Environment for Statistical Computing and Dynamic Graphics*. New York, NY: John Wiley & Sons.

Verbyla, A. P. and B. R. Cullis (1990). Modelling in repeated measures experiments. *Applied Statistics 39*, 341–356.

Wang, J. and T. Belin (2002). Handling incomplete high dimensional multivariate longitudinal data by multiple imputation using a longitudinal factor analysis model. *Proceedings of the American Statistical Association Section on Statistical Computing*, 3615–3620.

Wang, Y. and J. M. G. Taylor (1995a). Flexible methods for analysing longitudinal data using piecewise cubic polynomials. *Journal of Statistical Computation and Simulation 52*, 133–150.

Wang, Y. and J. M. G. Taylor (1995b). Inference for smooth curves in longitudinal data with application to an AIDS clinical trial. *Statistics in Medicine 14*, 1205–1218.

Ware, J. H. (1985). Linear models for the analysis of longitudinal studies. *The American Statistician 39*, 95–101.

Waternaux, C., N. M. Laird, and J. H. Ware (1989). Methods for analysis of longitudinal data: Blood-lead concentrations and cognitive development. *Journal of the American Statistical Association 84*, 33–41.

Weisberg, S. (2004). *Applied Linear Regression* (3rd ed.). New York, NY: John Wiley & Sons.

Weiss, R. (1994). Pediatric pain, predictive inference and sensitivity analysis. *Evaluation Review 18*, 651–678.

Weiss, R. (1996). An approach to Bayesian sensitivity analysis. *Journal of the Royal Statistical Society, Series B 58*, 739–750.

Weiss, R. E. and M. Cho (1998). Bayesian marginal influence assessment. *Journal of Statistical Planning and Inference 71*, 163–177.

Weiss, R. E., M. Cho, and M. Yanuzzi (1999). On Bayesian calculations for mixture likelihoods and priors. *Statistics in Medicine 18*, 1555–1570.

Weiss, R. E. and C. G. Lazaro (1992). Residual plots for repeated measures. *Statistics in Medicine 11*, 115–124.

Weiss, R. E., Y. Wang, and J. G. Ibrahim (1997). Predictive model selection for repeated measures random effects models using Bayes factors. *Biometrics 53*, 592–602.

Wolfinger, R. and M. O'Connell (1993). Generalized linear mixed models: A pseudo-likelihood approach. *Journal of Statistical Computation and Simulation 48*, 233–243.

Zeger, S. L. and P. J. Diggle (1994). Semiparametric models for longitudinal data with application to CD4 cell numbers in HIV seroconverters. *Biometrics 50*, 689–699.

Zeger, S. L. and M. R. Karim (1991). Generalized linear models with random effects: A Gibbs sampling approach. *Journal of the American Statistical Association 86*, 79–86.

Zeger, S. L., K.-Y. Liang, and P. S. Albert (1988). Models for longitudinal data: A generalized estimating equation approach (Corr: 1989 v45 p347). *Biometrics 44*, 1049–1060.

Zeger, S. L., K.-Y. Liang, and S. G. Self (1985). The analysis of binary longitudinal data with time-independent covariates. *Biometrika 72*, 31–38.

Zhang, F. and R. E. Weiss (2000). Diagnosing explainable heterogeneity of variance in random-effects models. *The Canadian Journal of Statistics 28*, 3–18.

Zimmerman, D. L. (2000). Viewing the correlation structure of longitudinal data through a PRISM. *The American Statistician 54*, 310–318.

Appendix
Data Sets

Knowledge is invariably a matter of degree: you cannot put your
finger upon even the simplest datum and say "this we know."
— T. S. Eliot

Data is what distinguishes the dilettante from the artist.
— George V. Higgins

Data sets are discussed where they are introduced, but usually only for as
much as needed. This section provides more comprehensive and centralized
information about each data set. Data sets are available via the book Web
site (http://www.biostat.ucla.edu/books/mld).

A.1. Basic Symptoms Inventory (BSI)

The basic symptoms inventory (BSI) is a psychological inventory composed
of 53 items rated on a 0 to 4 integer scale. Each item represents a psychiatric
symptom or a negative state of mind, and subjects indicate how much that
particular symptom has troubled them during the past weeks. A rating of
0 means that symptom has not troubled them, a 4 indicates that they have
had lots of trouble with that symptom. The BSI has 9 sub-scales as well as
the over-all BSI scale called the global severity index (GSI). The score is
the average rating on the items making up the scale or sub-scale. Because
the resulting sample histogram is skewed and zero is possible, we add a
small constant 1/53 to GSI and then take logs base 2 when analyzing GSI.

Variable name	Description
Age	Adolescent age in years
Age_at_baseline	Adolescent age at baseline
Bsi_anx	BSI Anxiety
Bsi_dep	BSI Depression
Bsi_gsi	BSI Global Severity Index
Bsi_host	BSI Hostility
Bsi_ocd	BSI Obsessive-Compulsive
Bsi_par	BSI Paranoid Ideation
Bsi_phob	BSI Phobic Anxiety
Bsi_psych	BSI Psychoticism
Bsi_sens	BSI Interpersonal Sensitivity
Bsi_soma	BSI Somatization
Diagnosis	Baseline parent diagnosis
Drug_status	Parent drug use last 3 months
Gender	Adolescent gender
Hispanic	Adolescent Hispanic or not
Parent	Parent ID
Parent_alcohol	Parent alcohol use past 3 months 1=yes
Parent_base_age	Parent baseline age
Parent_died	Parent died during study
Parent_gender	
Parent_hispanic	
Parent_marijuana	Parent marijuana use past 3 months
Pid	Adolescent ID
Rounded3_true_month	Rounded time to nearest 3 months
Season	Winter/spring/summer
Study_month	Nominal time
Treatment	1 =yes, 0 =no
True_month	Time in months

Table A.1. BSI variables.

For sub-scales made up of K items, we add $1/K$ to the scale before taking a log. The sample studied in this data set consists of longitudinal measures on adolescent children of HIV+ parents (Rotheram-Borus, Lee, Lin, and Lester 2004).

In addition to the 10 BSI scales and sub-scales, there are adolescent covariates and parent covariates. Variables age, age_at_baseline, gender, Hispanic (yes/no), and treatment are adolescent variables. Treatment is 1 if the family was in the treatment group. There is also parent age at baseline, gender, and Hispanic (yes/no) variables as well. Three time variables are given. True_month is the time-since-baseline in months for the adolescent's observation. Study_month is the nominal time of the observation, and rounded3_true_month is the true time rounded to the nearest multiple of three months. Pid is the adolescent id, and parent is the parent id. Adolescents are nested inside parent. A number of time-varying parent variables are given. These are available only if there is a parent observation within the previous 3 months of the adolescent observation or 1 month forward of the adolescent observation. Drug_status has three categories. Parents who never reported using hard drugs (i.e., not marijuana, alcohol, or cigarettes) during the study are nonusers. Parents who reported hard drug use sometime during the study can be either using-users or non-using-users, depending on whether they reported hard drug use during the previous three months or not. Parent_marijuana and parent_alcohol are binary indicators of whether the parent reported marijuana or alcohol use respectively during the previous three months. Parent_died indicates that the parent died sometime during the study.

A.2 Big Mice and Small Mice

The Big Mice data consist of weights in milligrams of new-born male mice one each from 35 litters of mothers from a single strain (Izenman and Williams 1989). Thirty-three mice were weighed every three days up to day 20. Eleven mice were weighed starting on day 0, 10 mice starting on day 1, and 12 mice starting on day 2 for a total of seven observations each. These 3 sets of mice define groups 1, 2 and 3. The last two mice are in group 4 and were weighed daily from day 0 to day 20. All weighings were performed by a single person using a single scale. There is no expectation of any differences among the four groups.

A subset of the Big Mice data forms the *Small Mice* data consisting of the group three mice plus the group four observations on the same days. The Small Mice form a balanced subset of 14 mice measured 7 times each every third day starting at 2 through day 20.

A.3 Bolus Counts

The Bolus Count data is a study of patient controlled analgesia (PCA) comparing two different dosing regimes (Henderson and Shimakura 2003). Subjects are allowed to self-medicate with a pain medication. The number of doses in a 4-hour period is recorded for 12 consecutive periods. After each dose, there is a lockout period where the patient may not administer more medication. There are two groups, a 1 milligram (mg) per dose group and a 2 milligram per dose group. The lockout time in the 2 milligram dose group is twice as long as in the 1 milligram dose group. The lockout period allows for a maximum of 30 dosages in the 2 mg group and 60 dosages in the 1 mg group. No patient came near these upper limits.

A.4 Dental

The Dental data set comes from the classic paper by Potthoff and Roy (1964). It reports the distance in millimeters from the center of the pituitary gland to the pteryomaxillary fissure. The subjects are 16 boys and 11 girls. Data was taken every two years starting at age 8 and ending at age 14.

A.5 Kenya School Lunch Intervention

In the Kenya school lunch intervention (Neumann, Bwibo, Murphy, Sigman, Whaley, Allen, Guthrie, Weiss, and Demment 2003), first-form children were given one of four school lunch interventions: meat, milk, calorie, or control. The first three groups were fed a school lunch of a stew called githeri supplemented with either meat, milk, or oil to create a lunch with a given caloric level. The control group did not receive a lunch. Instead, control group families received a goat at the end of the study. Three schools were randomized to each group. The baseline cognitive measurement was taken prior to the lunch program onset. At time zero the school lunch programs are started. For subjects in the meat, milk, or calorie groups, a school lunch is delivered to students in attendance provided school is in session. The lunch program is the same for all children within a school.

The Kenya data is broken up into sub data sets from four domains: Anthropometry, Cognitive, Morbidity, and Nutrition. Each sub data set is described here in its own subsection. It was intended that all data be collected on all subjects, and for the most part this occurred. The subject id numbers are the same across the data sets, allowing for the possibility of multivariate longitudinal data analyses. Covariates are described in the covariates subsection.

There are 547 subjects.

A.5.1 Anthropometry

Anthropometry assessed height, weight, and several measures of skin fold, fat, and muscle area. Table A.2 gives details.

A.5.2 Cognitive

Cognitive measures such as Raven's were assessed at up to five times, called rounds, for each subject. Raven's refers to the subjects score on the Raven's colored progressive matrices, a measure of cognitive ability. Responses range from 0 to 31 with an overall mean of 18.2 and a standard deviation of the 2555 observations of 3.0. There are three additional response variables: arithmetic score (arithmetic), verbal meaning (vmean-

Variable name	Description
Height	Height [cm]
Mfa	Mid-upper-arm fat area [mm^2]
Mma	Mid-upper-arm muscle area [mm^2]
Muac	Mid-upper-arm circumference [cm]
Ssf	Subscapular skin-fold thickness [mm]
Tsf	Triceps skin-fold thickness [mm]
Weight	Weight (adjusted for clothes) [kg]

Table A.2. Anthropometry response variables.

Variable name	Description
Arithmetic	Arithmetic score
Dstotal	Digit span, total
Raven's	Raven's colored matrices
Vmeaning	Verbal meaning

Table A.3. Cognitive data variables.

ing), and total digit span score (dstotal). Digit span is a test of memory while others are considered measures of intelligence or education. Table A.3 summarizes the variables. The round 1 average assessment time was -1.6 months; the average times of assessment for round 2 through 5 were at 1.1, 5.6, 13.4, and 21.6 months. Times for individual subjects have a standard deviation of approximately .8 months around the average round times.

A.5.3 Covariates

Covariates include gender and a measure of socio-economic status (SES). SES ranges from a low of 28 to a high of 211 with a median of 79. It is constructed from an extensive baseline survey of social and economic variables. Assessments of both father and mother educational status were made, but there was substantial missing data in the father data, and even the mother covariates have a fair amount of missing data. Anthropometry baseline variables height, weight, and head circumference as well as age at baseline may be used as covariates in analyses of cognitive and other variables to adjust for differences among subjects. Table A.4 lists some available covariates.

A.5.4 Morbidity

Morbidities in the subjects were assessed using an extensive home interview with the parent. Presence or absence of a number of different symptoms and diseases were recorded, plus information on their existence on the current day and over the previous week. Morbidity visits were monthly during the first year and bimonthly for the second year. Table A.5 lists some of the

Variable name	Description
Age_at_time0	Age at baseline
Head_circ	Head circumference at baseline
Id	Subject id number
Readabil	Mother's reading ability
Readtest	Mother's reading test
Gender	Female, male
Treatment	Calorie, meat, milk, zcontrol
Writeabil	Mother's writing ability
Writetest	Mother's writing test
Yrsofsch	Mother's years of educations
Ht_base	Height at baseline
Relyear	Time in years from baseline
School	School id 1-12
Ses	Socio-Economic Status score
Wt_base	Weight at baseline

Table A.4. Kenya data covariates.

variables available. A number of morbidity items were classified as mild or severe morbidities. If any severe morbidity was present then the variable morbscore was coded as "severe." If any mild morbidity was present, but no severe morbidity was present, then morbscore was coded as "mild." Otherwise morbscore was coded as "none." Mscore is a 0-1 recoding of morbscore with severe and mild as 1 versus none as 0. The variable S_Mscore is 0-1 as well, but with severe=1 versus mild and none is 0. The average morbscore, coding severe=2, mild=1 and none=0 across all measurements for the subject is used as a covariate in some analyses, see for example section 7.5. In about 40% of all visits, subjects reported no morbidity. In about 45% of visits, subjects had mild morbidity. In the remaining visits, subjects had a report of some severe morbidity.

Malaria facts in the text came from the Centers for Disease Control Web site on malaria.

A.5.5 Nutrition

An enormous number of nutrition variables have been constructed from food intake surveys. Twenty-four-hour food recall surveys were taken regularly. Subject's food intake was assessed and converted to nutritional content using detailed household specific recipes and seasonally adjusted food content micronutrient data bases. Zinc and iron are particularly important variables, as well as information on the amount of animal source nutrients. Interest lies in whether the general level of a nutrient differs by treatment and whether the nutrient levels affect other response variables.

Variable Name	Description
Q1	Fever
Q2	Chills
Q3	Reduced activity/bed ridden
Q4	Poor appetite
Q5	Headache
Q6	Upper respiratory
Q11	Sore in mouth
Q12	Skin rash
Q13	Digestive
Q15	Malaria
Q23	Typhoid
Vn	Morbidity visit number
Mscore	Severe and mild versus no morbidity
S_Mscore	Severe versus mild or no morbidity
Morbscore	Morbidity score: none/mild/severe

Table A.5. Morbidity response variables.

A.6 Ozone

The Ozone data set records ozone over a three-day period during late July 1987 in and around Los Angeles, California, USA, at 20 sites. Twelve hourly recordings are recorded starting from 0700 hours to 1800 hours giving us $20 \times 12 \times 3$ ozone readings. Measurement units are in parts per hundred million. Table 2.1 gives four letter abbreviation for the sites, the full names, and the longitude, latitude, and altitude of each site. Also given is a valley indicator to indicate whether the site is in the Simi or San Fernando Valleys (SF) or San Gabriel Valleys (SG). The remaining sites are adjacent to the ocean or otherwise do not have mountain ranges between them and the ocean. Figure 2.11 shows a map of the site locations. Each night ozone returns to baseline values, and we treat the data as having $60 = 20 \times 3$ subjects with 12 longitudinal measures each.

A.7 Pediatric Pain

The Pediatric Pain data has up to four observations on 64 elementary school children aged eight to ten (Fanurik, Zeltzer, Roberts, and Blount 1993). The response is the length of time in seconds that a child can tolerate keeping his or her arm in very cold water (the *cold pressor* task), a proxy measure of pain tolerance. After the cold becomes intolerable, the child removes his or her arm, the arm is toweled off, and no harm is caused. There is some missing data due to kids having casts on an arm or being absent, but no one dropped out for reasons related to the experiment. Subjects underwent two trials during a first visit followed by two more trials during a second visit after a two-week gap. The first trial uses the dominant arm, the right arm for right-handed children, and the left arm for left-handers. The second trial is with the non-dominant arm.

Subjects were asked what they were thinking about during the first two trials. Those who were thinking about the experiment, the experimental apparatus, the feelings from their arms, and so on, were classified as having an attender (A) coping style (CS). Those who thought about other things: the wall, homework from school, going to the amusement park, or things unrelated to the experiment were classified as having a distracter (D) coping style.

A randomized treatment (TMT) was administered prior to the fourth trial. The treatment consisted of a ten-minute counseling intervention where coping advice was given either to attend (A), distract (D) or no advice (N). The N TMT consisted of a ten-minute discussion with the child without any advice regarding a coping strategy. Interest lies in the main effects of TMT and CS and interactions between TMT and CS. CS-TMT combinations are indicated by the two letter sequence AA, AD, AN, DA, DD, and DN. Interactions between TMT and CS were anticipated. In particular, matched CS-TMT combinations AA and DD were expected to better than mis-matched AD and DA combinations.

At each trial, the children were asked to rate the pain on a 1 to 10 pain rating scale. This pain rating is a second response. Additional covariates, gender, SES, and age are available. I have used the Pain data to motivate and illustrate statistical methodology in a number of papers (Weiss 1994; Weiss 1996; Weiss, Wang, and Ibrahim 1997; Weiss and Cho 1998; Weiss, Cho, and Yanuzzi 1999; Zhang and Weiss 2000).

A.8 Schizophrenia

The Schizophrenia data and the analysis presented in section 12.6 comes from Hedeker and Gibbons (1997). The Schizophrenia data has a response severity of illness measured on a 1 to 7 scale where 1 represents normal up to 7, the most severe illness. Subjects were observed at weeks 0 through 6, although most observations were taken at weeks 0, 1, 3, and 6. There were three drug groups (chlorpromazine, fluphenazine, thioridazine) and one placebo group. Because the drug groups were similar in response, Hedeker and Gibbons (1997) combine the three groups into a single drug group coded drug=1 and coded drug=0 for placebo. Subjects who dropped out were thought to be different from those who stay in the study. The data set includes subject id, illness severity on a seven-point scale, time in weeks from 0 to 6, treatment, and gender (0=female, 1=male).

Anim	Animal id number
Sex	1=male, 2=female
Loca	Location of animal
Leng	Length of animal (tenths of a millimeter)
Head	Head length
Ear	Ear length
Arm	Arm length
Leg	Leg length
Pres	Pes (foot) length
Tail	Tail length
Weight	Weight (tenths of a gram)
Age	Age in days from birth

Table A.6. Variables available in the Wallaby data set.

A.9 Small Mice

Section A.2 talks about both the Big Mice and the Small Mice data sets.

A.10 Vagal Tone

Vagal tone is supposed to be high. In response to stress it gets lower. The subjects in this study were a group of 21 very ill babies who were undergoing cardiac catheterization, an invasive, painful procedure. All available data can be found in table 2.9 with subject id number, gender, age in months, length of time in minutes of cardiac catheterization procedure, five vagal tone measures, and a medical severity measure (higher is worse). The first vagal tone measure was taken the night before, the second measure the morning before the catheterization. The third measure was taken right after the catheterization, the fourth was taken the evening after, and the last measure was taken the next day. There is a substantial amount of missing data; blanks in the table indicate missing data; subject 12 is missing all variables.

A.11 Wallaby

The Wallaby data set comes from the Australian Data Set and Story library (OZDASL) at http://www.statsci.org/data/oz/wallaby.html. It contains measurements over time of the lengths of various body parts of Tammar wallabies usually in tenths of millimeters and also the weight of the wallabies (Mallon 1994). The full data set contains the information given in table A.6.

A.12 Weight Loss

The Weight Loss data consist of weekly weights in pounds from women enrolled in a weight-loss trial. Patients were interviewed and weighed the first week and enrolled in the study at the second week. There are from 4 to 8 measurements per subject. Weights range from roughly 140 pounds to 260 pounds.

Study protocol called for the subjects to visit the clinic at weeks 1, 2, 3, and 6 and weigh themselves on the clinic scale. At weeks 4, 5, 7, and 8, study personnel called subjects at home and asked subjects to weigh themselves on their home scales and report the measurement. Week 1 was a screening visit; participation in the actual weight-loss regimen did not start until week 2. Some subjects do not have observations for the last few weeks because the data was acquired prior to the end of the study.

Two versions of the data set are available. Version one has the data with time measured in integer weeks. Version two has the actual day from the initial visit rather than the week of each measurement and also has the clinic/phone visit type.

A hypothetical diet study data set is given in problem 2. It is designed to illustrate concepts surrounding missing data. Although vaguely modeled on the Weight Loss data, it is strictly hypothetical. Another hypothetical diet study is discussed in problem 10.

Index

adjusting for baseline, 228–230
AIC, *see* Akaike information criterion
Akaike information criterion, 147–150
algorithms, 396
analysis of variance, 86, 87, 89, 90, 95
Anthropometry
  description, 414
  problem, 84
  variables, 415
approximately equal to, 39
arithmetic, *see* Cognitive
asymptote, 192

back-transformation, 68, 69, 88–90,
  163–167, 356
backward elimination, 151–152, 232
  problems, 153, 231
balanced data, 19, 23
  with equal spacing, 19
  with missing data, 19–20
banded correlation matrix, 61
basic symptoms inventory, *see* BSI
Bayes
  computation, 163
  empirical, 113, 318, 333
  inference, 134, 150, 151, 153, 210,
    357, 397–399

references, 397
testing, 150
Bayesian information criterion,
  147–151, 153, 275, 278, 282,
  283
  for fixed effects, 153
  Pediatric Pain, 278, 279, 283, 383,
    384, 389, 392
  Small Mice, 275–276
  Weight Loss, 315
bell-curve, 6, 120, 154
bent line, 220–222, 225–226
bias, 104–105
BIC, *see* Bayesian information
  criterion
Big Mice, 30–33, 43, 44, 53, 56–58,
  67, 84, 121, 157, 162, 187, 273
  analysis, 227
  boxplots, 32
  cubic spline, 226–227
  description, 30, 413
  empirical prediction plot, 68
  empirical summary plot, 68
  inference plot, 335
  problem, 26, 73, 77, 84, 123, 241,
    296
  residual plot, 335

Springer Texts in Statistics *(continued from page ii)*

springeronline.com

Models for Discrete Longitudinal Data
G. Molenberghs and G. Verbeke

This book provides a comprehensive treatment on modeling approaches for non-Gaussian repeated measures, possibly subject to incompleteness. The authors begin with models for the full marginal distribution of the outcome vector. This allows model fitting to be based on maximum likelihood principles, immediately implying inferential tools for all parameters in the models. At the same time, they formulate computationally less complex alternatives, including generalized estimating equations and pseudo-likelihood methods. They then briefly introduce conditional models and move on to the random-effects family, encompassing the beta-binomial model, the probit model and, in particular the generalized linear mixed model. Several frequently used procedures for model fitting are discussed and differences between marginal models and random-effects models are given attention.

2005. 744 p. (Springer Series in Statistics) Hardcover ISBN 0-387-25144-8

The Evaluation of Surrogate Endpoints
T. Burzykowski, G. Molenberghs, and M. Buyse (Editors)

Surrogate endpoints are useful when they can be measured earlier, more conveniently, or more frequently than the "true" endpoints of primary interest. Regulatory agencies around the globe are introducing provisions and policies relating to the use of surrogate endpoints in registration studies. But how can one establish the adequacy of a surrogate? What kind of evidence is needed, and what statistical methods portray that evidence most appropriately? This book offers a balanced account on this controversial topic. The text presents major developments of the last couple of decades, together with a unified, meta-analytic framework within which surrogates can be evaluated from several angles. Methodological development is coupled with perspectives on various therapeutic areas.

2005. 416 p. (Statistics for Biology and Health) Hardcover ISBN 0-387-20277-3

Easy Ways to Order ▶ Call: Toll-Free 1-800-SPRINGER • E-mail: orders-ny@springer.sbm.com • Write: Springer, Dept. S8113, PO Box 2485, Secaucus, NJ 07096-2485 • Visit: Your local scientific bookstore or urge your librarian to order.